Novel Nanomaterials for Energy Storage and Catalysis

Novel Nanomaterials for Energy Storage and Catalysis

Editors

Zhenyu Yang
Jinsheng Zhao

Basel • Beijing • Wuhan • Barcelona • Belgrade • Novi Sad • Cluj • Manchester

Editors

Zhenyu Yang
School of Biological and
Chemical Engineering
Zhejiang University of
Science and Technology
Hangzhou
China

Jinsheng Zhao
College of Chemistry and
Chemical Engineering
Liaocheng University
Liaocheng
China

Editorial Office
MDPI
St. Alban-Anlage 66
4052 Basel, Switzerland

This is a reprint of articles from the Special Issue published online in the open access journal *Materials* (ISSN 1996-1944) (available at: www.mdpi.com/journal/materials/special_issues/5Q06JY37R1).

For citation purposes, cite each article independently as indicated on the article page online and as indicated below:

Lastname, A.A.; Lastname, B.B. Article Title. *Journal Name* **Year**, *Volume Number*, Page Range.

ISBN 978-3-7258-0766-6 (Hbk)
ISBN 978-3-7258-0765-9 (PDF)
doi.org/10.3390/books978-3-7258-0765-9

© 2024 by the authors. Articles in this book are Open Access and distributed under the Creative Commons Attribution (CC BY) license. The book as a whole is distributed by MDPI under the terms and conditions of the Creative Commons Attribution-NonCommercial-NoDerivs (CC BY-NC-ND) license.

Contents

About the Editors vii

Preface ix

Zhenyu Yang and Jinsheng Zhao
Advancing Energy Storage and Catalysis with Novel Nanomaterials
Reprinted from: *Materials* **2023**, *16*, 6425, doi:10.3390/ma16196425 1

Zhanjun Chen, Tao Wang, Xianglin Yang, Yangxi Peng, Hongbin Zhong and Chuanyue Hu
TiO_2 Nanorod-Coated Polyethylene Separator with Well-Balanced Performance for Lithium-Ion Batteries
Reprinted from: *Materials* **2023**, *16*, 2049, doi:10.3390/ma16052049 3

Hao Zhang, Youkui Wang, Ruili Zhao, Meimei Kou, Mengyao Guo, Ke Xu, et al.
Fe^{III} Chelated with Humic Acid with Easy Synthesis Conditions and Good Performance as Anode Materials for Lithium-Ion Batteries
Reprinted from: *Materials* **2023**, *16*, 6477, doi:10.3390/ma16196477 17

Yeong A. Lee, Kyu Yeon Jang, Jaeseop Yoo, Kanghoon Yim, Wonzee Jung, Kyu-Nam Jung, et al.
Three-Dimensional Flower-like MoS_2 Nanosheets Grown on Graphite as High-Performance Anode Materials for Fast-Charging Lithium-Ion Batteries
Reprinted from: *Materials* **2023**, *16*, 4016, doi:10.3390/ma16114016 33

Ravindra N. Bulakhe, Anh Phan Nguyen, Changyoung Ryu, Ji Man Kim and Jung Bin In
Facile Synthesis of Mesoporous Nanohybrid Two-Dimensional Layered Ni-Cr-S and Reduced Graphene Oxide for High-Performance Hybrid Supercapacitors
Reprinted from: *Materials* **2023**, *16*, 6598, doi:10.3390/ma16196598 47

Elianny Da Silva, Ginebra Sánchez-García, Alberto Pérez-Calvo, Ramón M. Fernández-Domene, Benjamin Solsona and Rita Sánchez-Tovar
Anodizing Tungsten Foil with Ionic Liquids for Enhanced Photoelectrochemical Applications
Reprinted from: *Materials* **2024**, *17*, 1243, doi:10.3390/ma17061243 61

Meifang Zhang, Xiangfei Liang, Yang Gao and Yi Liu
C_{60}- and CdS-Co-Modified Nano-Titanium Dioxide for Highly Efficient Photocatalysis and Hydrogen Production
Reprinted from: *Materials* **2024**, *17*, 1206, doi:10.3390/ma17051206 78

Xiaoyan Pei, Jiang Liu, Wangyue Song, Dongli Xu, Zhe Wang and Yanping Xie
CO_2-Switchable Hierarchically Porous Zirconium-Based MOF-Stabilized Pickering Emulsions for Recyclable Efficient Interfacial Catalysis
Reprinted from: *Materials* **2023**, *16*, 1675, doi:10.3390/ma16041675 89

Qian Liu, Wen-Yong Deng, Lie-Yuan Zhang, Chang-Xiang Liu, Wei-Wei Jie, Rui-Xuan Su, et al.
Modified Bamboo Charcoal as a Bifunctional Material for Methylene Blue Removal
Reprinted from: *Materials* **2023**, *16*, 1528, doi:10.3390/ma16041528 101

Fen Wei, Jiaxiang Qiu, Yanbin Zeng, Zhimeng Liu, Xiaoxia Wang and Guanqun Xie
A Novel POP-Ni Catalyst Derived from PBTP for Ambient Fixation of CO_2 into Cyclic Carbonates
Reprinted from: *Materials* **2023**, *16*, 2132, doi:10.3390/ma16062132 121

Cecylia Wardak, Klaudia Morawska and Karolina Pietrzak
New Materials Used for the Development of Anion-Selective Electrodes—A Review
Reprinted from: *Materials* **2023**, *16*, 5779, doi:10.3390/ma16175779 132

About the Editors

Zhenyu Yang

Zhenyu Yang is a professor at the School of Biological and Chemical Engineering, Zhejiang University of Science and Technology, and the School of Chemistry and Chemical Engineering, Nanchang University, China. He received his PhD degree at the Technical Institute of Physics and Chemistry, Chinese Academy of Sciences, in 2005. He used to work as a visiting research fellow at Rensselaer Polytechnic Institute in the USA from 2012 to 2013 and at Nanyang Technological University in Singapore in 2019. Currently, his main research focuses on energy storage materials for power sources, including Li-ion batteries, Li-S batteries and supercapacitors.

Jinsheng Zhao

Jinsheng Zhao is a professor at the School of Chemistry and Chemical Engineering, Liaocheng University, China. He received his PhD degree at the Technical Institute of Physics and Chemistry, Chinese Academy of Sciences, in 2005. Currently, his main research focuses on the preparation and application of organic conjugated polymer materials, including electrochromic materials and devices, Li-ion batteries and photocatalysis.

Preface

In recent years, the field of nanomaterials has undergone remarkable advancements, offering unprecedented opportunities to revolutionize energy storage and catalysis technologies. The integration of novel nanomaterials into these domains has not only enhanced performance, but also opened new avenues for addressing pressing global challenges related to energy sustainability and environmental protection.

This reprint delves into the cutting-edge developments and applications of nanomaterials in energy storage and catalysis. Authored by leading experts in the field, the insightful works aim to provide a comprehensive overview of the state-of-the-art research, innovative methodologies and promising technologies driving significant progress in these critical areas.

The journey through this reprint begins with an exploration of advanced nanomaterials designed for energy storage applications, including batteries, supercapacitors and fuel cells. It delves into the fundamental principles governing their design, synthesis, characterization and performance optimization, highlighting key breakthroughs and emerging trends.

Transitioning into catalysis, the reprint showcases the pivotal role of nanomaterials in catalytic processes for energy conversion, environmental remediation and sustainable chemical engineering. Readers gain valuable insights into design strategies, catalytic mechanisms and performance enhancements enabled by nanomaterial-based catalysts, showcasing their potential to revolutionize diverse sectors.

Throughout this reprint, an emphasis is placed on the interdisciplinary nature of nanomaterial research, highlighting the synergistic integration of materials science, chemistry, physics, engineering and environmental science. Moreover, it underscores the importance of collaboration, knowledge exchange and continuous innovation in driving impactful solutions for global energy and environmental challenges.

As editors of this reprint, we are honored to present a compilation of contributions from esteemed researchers and practitioners whose dedication and expertise have significantly contributed to advancing nanomaterials for energy storage and catalysis. We hope this reprint serves as a valuable resource for researchers, students, educators and industry professionals alike, inspiring continued exploration, discovery and innovation in this exciting and rapidly evolving field.

Zhenyu Yang and Jinsheng Zhao
Editors

Editorial
Advancing Energy Storage and Catalysis with Novel Nanomaterials

Zhenyu Yang [1,*] and Jinsheng Zhao [2]

1. School of Chemistry and Chemical Engineering, Nanchang University, Nanchang 330031, China
2. School of Chemistry and Chemical Engineering, Liaocheng University, Liaocheng 252059, China; j.s.zhao@163.com
* Correspondence: zyyang@ncu.edu.cn

In the dynamic realm of materials science, novel nanomaterials possess the transformative potential to reshape various industries, ranging from energy storage to catalysis. The objective of this Special Issue, titled "Innovative Nanomaterials for Energy Storage and Catalysis", is to facilitate the exchange of groundbreaking research and ideas related to the synthesis, characterization, and application of innovative nanomaterials.

The articles featured in this Special Issue encompass a diverse spectrum of topics, thereby showcasing the multifaceted capabilities of nanomaterials in addressing challenges within the domains of energy storage and catalysis. Noteworthy breakthroughs include the utilization of three-dimensional flower-like MoS_2 nanosheets and TiO_2 nanorod-coated polyethylene separators, both of which mark significant advancements in the creation of high-performance materials designed for rapid-charging lithium-ion batteries. Furthermore, a comprehensive review delves into the realm of new materials tailored for anion-selective electrodes, offering insights into a multitude of potential applications. Our Special Issue also highlights the innovative POP-Ni catalyst for CO_2 fixation, derived from PBTP, which offers a groundbreaking approach for the ambient fixation of CO_2 into cyclic carbonates—a notable contribution to the ongoing endeavors related to carbon capture and utilization. Additionally, we delve into the realm of CO_2-switchable hierarchically porous zirconium-based MOF-stabilized Pickering emulsions, elucidating the prospects of recyclable and efficient interfacial catalysis through the use of advanced materials.

These contributions highlight the diverse nature of nanomaterial research, covering various aspects such as material synthesis, hierarchical organization, device fabrication, and characterization. As readers explore this Special Issue, we encourage them to discover the incredible potential inherent in nanomaterials and their pivotal role in shaping the future of energy storage and catalysis. It is our sincere hope that the articles presented here will serve as a source of inspiration, encouraging further exploration and innovation in the field of nanomaterials. Ultimately, these efforts have the potential to bring about transformative advancements in energy storage and catalysis.

Funding: This work was supported by the Natural Science Research Programs of Jiangxi Province (20202ACB202004, 20212BBE53051, 20213BCJ22024).

Conflicts of Interest: The authors declare no conflict of interest.

Short Biography of Authors

Zhenyu Yang is a professor at the School of Chemistry and Chemical Engineering, Nanchang University, China. He received his PhD degree at the Technical Institute of Physics and Chemistry, Chinese Academy of Sciences, in 2005. He used to work as a visiting research fellow at Rensselaer Polytechnic Institute in the USA from 2012 to 2013 and at Nanyang Technological University in Singapore in 2019, respectively. Currently, his main research focuses on energy storage materials for power sources, including Li-ion batteries, Li-S batteries, and supercapacitors.

Jinsheng Zhao is a professor at the School of Chemistry and Chemical Engineering, Liaocheng University, China. He received his PhD degree at the Technical Institute of Physics and Chemistry, Chinese Academy of Sciences, in 2005. Currently, his main research focuses on the preparation and application of organic conjugated polymer materials, including electrochromic materials and devices, Li-ion batteries, and photocatalysis.

Disclaimer/Publisher's Note: The statements, opinions and data contained in all publications are solely those of the individual author(s) and contributor(s) and not of MDPI and/or the editor(s). MDPI and/or the editor(s) disclaim responsibility for any injury to people or property resulting from any ideas, methods, instructions or products referred to in the content.

Article

TiO$_2$ Nanorod-Coated Polyethylene Separator with Well-Balanced Performance for Lithium-Ion Batteries

Zhanjun Chen [1,†], Tao Wang [2,†], Xianglin Yang [1,3,*], Yangxi Peng [1], Hongbin Zhong [1] and Chuanyue Hu [1]

1. Modern Industry School of Advanced Ceramics, Hunan Provincial Key Laboratory of Fine Ceramics and Powder Materials, School of Materials and Environmental Engineering, Hunan University of Humanities, Science and Technology, Loudi 417000, China
2. School of Materials Science and Engineering, Dongguan University of Technology, Dongguan 523808, China
3. Western Australia School of Mines, Curtin University, Kalgoorlie, WA 6430, Australia
* Correspondence: xianglin.yang@curtin.edu.au; Tel.: +86-0738-8325065
† These authors contributed equally to this work.

Abstract: The thermal stability of the polyethylene (PE) separator is of utmost importance for the safety of lithium-ion batteries. Although the surface coating of PE separator with oxide nanoparticles can improve thermal stability, some serious problems still exist, such as micropore blockage, easy detaching, and introduction of excessive inert substances, which negatively affects the power density, energy density, and safety performance of the battery. In this paper, TiO$_2$ nanorods are used to modify the surface of the PE separator, and multiple analytical techniques (e.g., SEM, DSC, EIS, and LSV) are utilized to investigate the effect of coating amount on the physicochemical properties of the PE separator. The results show that the thermal stability, mechanical properties, and electrochemical properties of the PE separator can be effectively improved via surface coating with TiO$_2$ nanorods, but the degree of improvement is not directly proportional to the coating amount due to the fact that the forces inhibiting micropore deformation (mechanical stretching or thermal contraction) are derived from the interaction of TiO$_2$ nanorods directly "bridging" with the microporous skeleton rather than those indirectly "glued" with the microporous skeleton. Conversely, the introduction of excessive inert coating material could reduce the ionic conductivity, increase the interfacial impedance, and lower the energy density of the battery. The experimental results show that the ceramic separator with a coating amount of ~0.6 mg/cm^2 TiO$_2$ nanorods has well-balanced performances: its thermal shrinkage rate is 4.5%, the capacity retention assembled with this separator was 57.1% under 7 C/0.2 C and 82.6% after 100 cycles, respectively. This research may provide a novel approach to overcoming the common disadvantages of current surface-coated separators.

Keywords: TiO$_2$ nanorods; polyethylene; ceramic separator; lithium ion batteries; thermal stability

1. Introduction

Lithium-ion batteries (LIBs) have attracted extensive attention in recent years due to their balanced electrochemical performance and high energy density. However, with the ever-lasting demand for high-power applications, the safety and reliability of LIBs have become critical. In a lithium-ion battery system, the separator, which functions as the ion conductor and electronic insulation between the anode and the cathode, is of paramount importance for the safety of LIBs [1]. Generally, an ideal separator should possess high porosity and excellent electrolyte wettability for rapid lithium-ion migration as well as desired mechanical strength and toughness for facile manufacturing [2,3]. Currently, the conventional separators for LIBs consist mainly of polyethylene (PE), polypropylene (PP), and their blends, which have suitable mechanical strength, chemical stability, and membrane thickness. However, these separators would suffer from severe thermal shrinkage under abnormal conditions such as overheating or overcharging, resulting in catastrophic thermal runaway, which may cause gas emission, rupture, fire, or explosion [4,5].

In order to enhance the thermal stability of commercial separators, intensive efforts have been made in recent years. On the one hand, alternatives to polyolefin separators (e.g., non-woven separators [6] and solid electrolytes) have been developed. However, the large pore size and poor mechanical properties of non-woven separators severely restrict their further application, and there remain many technical challenges associated with the solid electrolytes, such as interface impedance, processability, and electrode/electrolyte interface stability [7–9]. On the other hand, researchers try to improve the thermal stability of polyolefin-based separators by various methods, such as surface grafting, surface coating, blending, and so on. Among them, the surface coating of PE separators with inorganic nanoparticles (e.g., SiO_2 [10–12], Al_2O_3 [13,14], metal hydroxides [15,16], zeolite [17,18], ZrO_2 [19,20] and TiO_2 [21]) have attracted considerable attention, because it's an industrially more competitive method to improve the thermal stability of conventional PE separators. However, surface coating of separators with nanoparticle materials still faces many problems. For example, the nanoparticles often block the micropores and inhibit the free migration of lithium ions during the charging and discharging process, leading to the increase of battery internal resistance [13]. Moreover, some inorganic nanoparticles would detach from the separator surface because of the interfacial stress resulting from the manufacturing process of LIBs, which gives rise to nonuniform impedance distribution, potentially resulting in the thermal runaway of LIBs [22]. Additionally, these electrochemical inert solids (ceramics and binders) are not effective enough in the state-of-the-art research on achieving better electrochemical performance, such as higher energy density, higher power density, and so on [23,24]. Minimizing the usage amount of the electrochemical inert solids can improve the electrochemical performances, but the ultrathin layers are not effective in improving thermal stability.

Within the structure of surface-coated separators, the inorganic nanoparticle is bound onto the base film (i.e., PE membrane) via the action of a binder, which is called the "point bonding" pattern. Such a bonding pattern entails a thicker coating or higher adhesive usage to achieve remarkably reduced thermal shrinkage. Therefore, only by changing the interaction mode between the coating layer and the base film can the thermal stability of the PE membrane be substantially improved while simultaneously decreasing the use of electro-chemically inert solids. Recently, the use of one-dimensional [25] or two-dimensional nanomaterials instead of inorganic nanoparticles as the surface coating of the base film can transform the "point bonding" pattern into the "inter-line or inter-plane bonding" mode, thus greatly enhancing the adhesive force and thermal stability. For example, Zhao [26] prepared nanofibers-coated polypropylene (PP) separator with a three-dimensional network of interlacing $Mg_2B_2O_5$ bundles, which hardly shrank after heating at 160 °C for 30 min. Hu [27] transplanted catechol functional groups on the surface of a PP separator with dopamine to interact with aryl nanofibers, and the results showed that high thermal stability could be achieved even if a small amount of aryl nanofibers were immersed (0.005% concentration emulsion). Zhang [28] decorated the PP membrane using hexagonal $Mg(OH)_2$ nanosheets and found that it could maintain its original shape after heating at 180 °C for 30 min even at a coating thickness of 0.035 µm. Inspired by the aforementioned research findings, in this paper, TiO_2 nanorods were used for the surface modification of the PE membrane. Under the condition of "wire bonding" interaction, the influence of the thickness of the modification layer on the performance of PE base film is discussed, and the balance point between the two is sought to obtain a ceramic separator of high thermal stability with little effect on battery power and energy density.

2. Experimental

2.1. Material Preparation

The TiO_2 nanorods were synthesized based on the previously published method [29]. The detailed preparation process is as follows: 1.0 g TiO_2 powder (Analytical Reagent, Aladdin- Reagent) was dispersed into 200 mL of 10 M sodium hydroxide solution, well stirred, and then transferred to a Teflon-lined autoclave for hydrothermal reaction at 180 °C for 24 h. After the reaction, the resulting white precipitate was washed with dilute HCl and deionized water until the filtrate became neutral. Finally, the product was dried at 80 °C for 10 h.

2.2. Preparation of Ceramic Separator

The coating slurry is prepared as follows: 1.0 g PVDF is dissolved in a certain amount of NMP solvent at 50 °C to form a uniform solution, then 1.5 g TiO_2 nanorods and 0.01 g polyvinylpyrrolidone (PVP) are added and stirring is continued to form a uniform solution. After carefully adjusting the viscosity and solid content of the suspension, the as-prepared slurry was coated on the surface of the pristine PE membrane by the doctor-blade technique to obtain a ceramic separator. After drying, the targeted ceramic separator is prepared. The ceramic separators with a designed coating amount of 0.6, 0.9, 1.2, and 1.5 mg/cm^2 are named C-0.6, C-0.9, C-1.2, and C-1.5, respectively, and the actual coating amount is obtained by the subsequent weighing method, which is 0.6, 0.97, 1.24, and 1.83 mg/cm^2, respectively. For comparison and analysis, the uncoated blank PE base film was named C-0. In addition, the same coating method was used to coat the PE base film with TiO_2 nanoparticles (Aladdin reagent, about 50 nanometers in diameter). The sample is named C-P.

2.3. Characterization Analysis

The crystal structure of the prepared TiO_2 material was examined with X-ray powder diffraction (XRD) using a Bruker D8 Advance diffractometer with Cu-Kα source (λ = 1.54056 Å) from 10° to 90° with a step size of 0.02° s^{-1}. The surface morphology, cross-section morphology, and SEM-EDS linear scanning of the TiO_2 material and the coated layer were investigated using a scanning electron microscope (SEM, Philip-XL30). The thermostability of the samples was measured with differential scanning calorimetry (DSC, TA-Q200) in a temperature range of 60–200 °C at a heating rate of 10 °C min^{-1} under N_2 flow. The thermal shrinkage of the separators (original size: 3 × 3 cm) was determined by measuring their dimensional changes after storage at 140 °C for 30 min. The degree of thermal shrinkage was calculated by using the equation (Thermal shrinkage ratio (%) = ($A_1 - A_2$)/A_1 × 100%), where A_1 is the initial area and A_2 is the final area of the separator after the storage test. The tensile properties were tested using an Instron Universal Testing machine (RGWT-4002, Reger, Shenzhen, China). At least five independent measurements were performed for each sample with a constant rate of 20 mm s^{-1}, a 20 mm length. The Liquid electrolyte uptakes of the separators were measured in 1 mol L^{-1} LiPF$_6$ (EC:DMC = 3:7) solution at room temperature inside the glove box for 6 h. Liquid electrolyte-soaked membranes were weighed immediately after removing the excrescent surface electrolyte by wipes. The liquid electrolyte uptakes were calculated using the equation of ($M_1 - M_0$)/M_0 × 100%, where M_0 and M_1 were the weight of the membrane before and after immersion in the liquid electrolyte, respectively. The air permeability of separators was examined with a Gurley densimeter (UEC, 1012 A) by measuring the time for 100 cc of air to pass through under a given pressure.

The ionic conductivities of the separators were measured by electrochemical impedance spectroscopy (EIS, CHI-660 E). A total of 2025 coin-type test cells were assembled by sandwiching the separator between two stainless steel (SS) electrodes and soaking it into the liquid electrolyte (1 M LiPF$_6$ in 3:7 (volume ratio) mixture of ethylene carbonate (EC) and dimethyl carbonate (DMC)) for AC impedance measurements. Impedance data were obtained in the frequency range of 1 Hz–100 kHz with an amplitude of 10 mV at room temperature. The ionic conductivity (κ) was calculated using the equation ($\kappa = L/(R \times S)$). Here, R is the electrolyte resistance measured by AC impedance, and L and S are the thickness and area of the separators, respectively. The electrochemical stability window of the separators was estimated by a linear sweep voltammetry program of the CHI-660E electrochemical workstation to check oxidation decomposition, where the stainless steel was used as the working electrode and the lithium metal was used as the counter electrode at a scan rate of 10 mV s^{-1} from 2.5 V to 6.0 V versus Li/Li$^+$. The interfacial resistance between lithium electrodes was determined from the AC impedance spectrum recorded for Li | separators | Li cell over storage for up to 2 days. The measurement was carried out over a frequency range of 65,000 Hz to 0.01 Hz, with an amplitude of 10 mV.

The electrochemical performance of the prepared separators was examined using 2025 coin-type cells, comprising of the prepared separators, a cathode [LiCoO$_2$ (active material):polyvinylidene fluoride (PVDF, binder):Super-P (conducting agent) = 80:10:10 wt%], and a lithium foil as an anode. Then, 1 M LiPF$_6$ in EC/DMC 3:7 by volume was employed as an electrolyte. All the test cells were assembled in a dry, argon-filled glove box. The assembled coin cells were charged/discharged in the voltage range of 3.0~4.3 V on the CT2001A cell testing instrument (Land Electronic Co., Ltd.) at currents of 0.2, 0.4, 1.0, 2.0, 3.0 C, 5.0 C, and 7.0 C to test the rate capability. For the cycle stability, the charge/discharge current density was fixed at 0.5 C. All electrochemical tests were conducted at room temperature. Electrochemical impedance spectroscopy (EIS) was performed on an electrochemical workstation, while the impedance spectra were recorded under a 0.02 V amplitude and a frequency range of 50 mHz~10^5 Hz.

3. Results and Discussion

The crystal structure of TiO$_2$ nanorods was analyzed by XRD as shown in Figure 1a. It was found that all the peaks were completely consistent with the standard diffraction peaks (JCPDS: 46–1237), indicating that the prepared TiO$_2$ sample is a pure phase. And the broad diffraction peaks illustrate that its grain size is small. The SEM image of the TiO$_2$ sample (Figure 1b) shows that lots of nanorods with a diameter of about 10–100 nanometers and a length of about tens of microns are uniformly dispersed. In ancient China, straw or bamboo, having similar morphology to the TiO$_2$ sample, was commonly mixed with clay into a slurry and then coated on the wall surface, as shown in Figure 1c. The coating layer not only has good permeability but also binds strongly with the wall because the straw or bamboo interacts with the wall through a "wire bonding". Therefore, if the synthesized TiO$_2$ nanorods are prepared into a viscous slurry and coated on the surface of the PE separator, a protective layer with strong binding force and high permeability can also be obtained, which further improves the thermal stability of the PE separator without affecting the free migration of lithium ions.

The surface morphologies of the pristine PE separator and ceramic separators are shown in Figure 2. The pristine PE separator (Figure 2a) has a typical interconnected submicron pore structure originating from the wet method. This structure can facilitate the storage of electrolytes and allow the free migration of lithium ions inside the separator. Compared with the pristine separators, as depicted in Figure 2b,d–f, all the ceramic separators have similar surface morphologies for the inorganic coating layer, where the TiO$_2$ nanorods are uniformly dispersed and interlaced on the surface of the PE separator, forming a three-dimensional network with porous structure. The surface morphology of the reverse side of the coating layer for the sample of C-0.6 was also examined, as shown in Figure 2c. It was found that the pore structures of the PE separator can be maintained after the coating process, indicating that the TiO$_2$ nanorods, unlike other nanoparticles, did not clog the micropores because the nanorods with a high ratio of length to diameter tend to bridge over the micropores while the nanoparticles with a small size can be embedded into the micropores preventing the Li-ions from migrating through the separator. Besides, the morphologies of the cross-sectional and the thicknesses of the coating layer for the ceramic separators were observed with SEM and liner SEM methods, respectively, as shown in Figure 2g–j. The structures of the coating can be observed clearly, and the thickness for the samples of C-0.6, C-0.9, C-1.2, and C-1.5 are 0.5, 0.9, 1.3, and 2.1 μm, respectively. These results are consistent with the actual coating amount tested by weighing method, where the actual coating amount for sample C-0.6, C-0.9, C-1.2, and C-1.5 are 0.6, 0.97, 1.24, and 1.83 mg/cm^2, respectively.

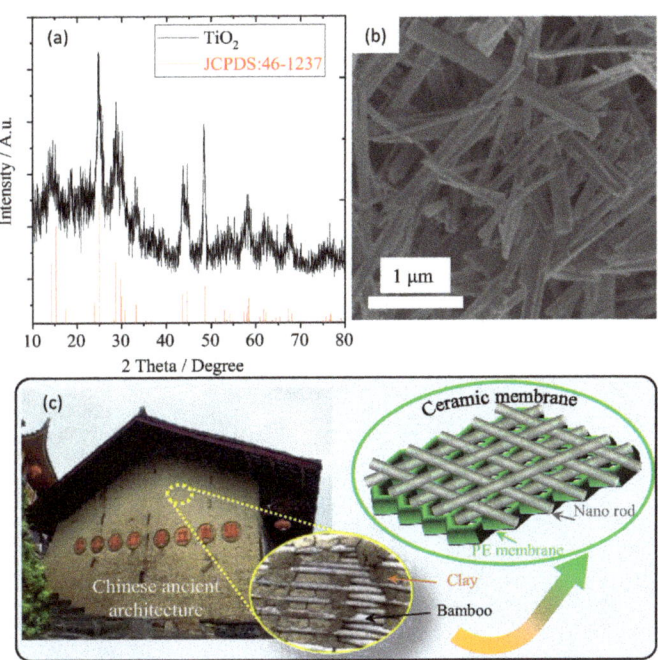

Figure 1. (**a**) XRD pattern and (**b**) SEM image of the as-synthesized TiO$_2$ nanorods, (**c**) is the schematic diagram for ceramic separator obtained from TiO$_2$ nanorods.

In order to investigate the thermal-resistant characteristics of the ceramic separators with different coating amounts, thermal shrinkage behaviors are observed by measuring the dimensional change (area-based) after storing the separators at 140 °C for 0.5 h. The results are shown in Figure 3a–g. It can be seen that the pristine PE separator (in Figure 3a) easily loses dimensional stability due to its low melting point of around 130 °C, which may cause an internal short circuit in the battery, further resulting in thermal runaway. When the synthesized TiO$_2$ nanorods are used to modify the pristine PE separator, their thermal stability at high temperatures is significantly improved even under low coating amounts, as shown in Figure 3b. The thermal shrinkage of C-0.6 is about 4.5%, which is within the acceptable value (5%) for commercial cell [30–32]. When the coating amount increases, as shown in Figure 3c–e, their thermal shrinkage decreases gradually. The shrinkage of C-1.5 is almost negligible, while the coating amount increases to 1.83 mg/cm^2. In addition, the morphology for the reverse side of the coating of sample C-0.6 after the thermal shrinkage test was also investigated by SEM measurement. Compared with the surface characteristics before the thermal shrinkage test (Figure 2a,c), most of the micropores can still maintain the original structure except for some micropores closed by melting in Figure 3f, indicating that the ultrathin coating layer composed of TiO$_2$ nanorods can effectively inhibit the shrinkage of the pristine PE separator at high temperature. It is observed in Figure 3b–f that the optimum coating amount is about 0.6 mg/cm^2 (C-0.6) because it will not introduce too much inert material and reduce the energy density of the battery while can obtain decent thermal shrinkage. In order to study the influence of different coating materials on the thermal stability of the separator, TiO$_2$ nanoparticles with a diameter of about 50 nm were also used to modify the PE surface. The coating amount was tailored based on its thermal shrinkage, which is exactly equivalent to that of sample C-0.6, as depicted in Figure 3g,h. The coating thickness for C-P is about 4 μm which is much thicker than that of C-0.6 (0.5 μm, observed in Figure 2g). It unequivocally demonstrated that the coating material with nanorods compared with nanoparticles could not only effectively reduce the coating thickness and the use of inert substances, improve the energy density of batteries, but also effectively inhibit thermal shrinkage of the separator at high temperatures. Moreover, the differential scanning calorimeter (DSC) analysis was carried out to illustrate the effect of surface coating on the thermal stability of ceramic separators with different coating amounts. As shown in Figure 3i, the melting point of the pristine PE separator is 134.9 °C, which is consistent with previous literatures [33,34]. While the melting point of the ceramic separators was significantly improved after coating with TiO_2 nanorods, a smaller increase was observed after the coating thickness increased to more than 1.3 μm (C-1.2 and C-1.5). Therefore, it can be concluded that the thermal stability for all the ceramic separators was greatly improved, but the optimum coating amount is about 0.6 mg/cm^2 because when the coating amount is greater than this value, the introduction of massive non-electrochemical substances and the resulting reduction of the energy density will offset the slightly improved thermal stability of separators.

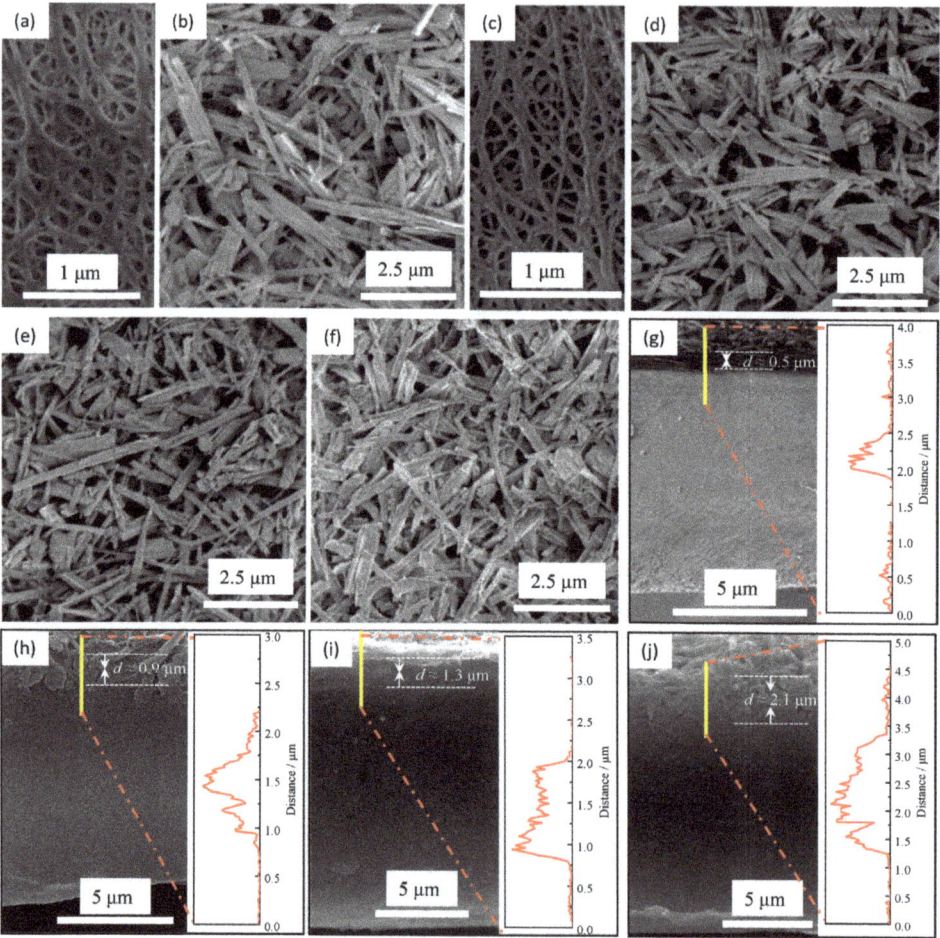

Figure 2. The SEM images for the pristine PE separator (**a**) and the ceramic separators with different coating amounts (**b**–**j**): the surface morphologies for the coating layer (**b**) and the reverse side of the coating layer (**c**) of sample C-0.6. (**d**–**f**) are the surface morphologies for coating of samples C-0.9, C-1.2, and C-1.5, respectively. (**g**–**j**) are the cross-section morphologies and the thicknesses of the coating layer for the ceramic separators C-0.6, C-0.9, C-1.2, and C-1.5, respectively.

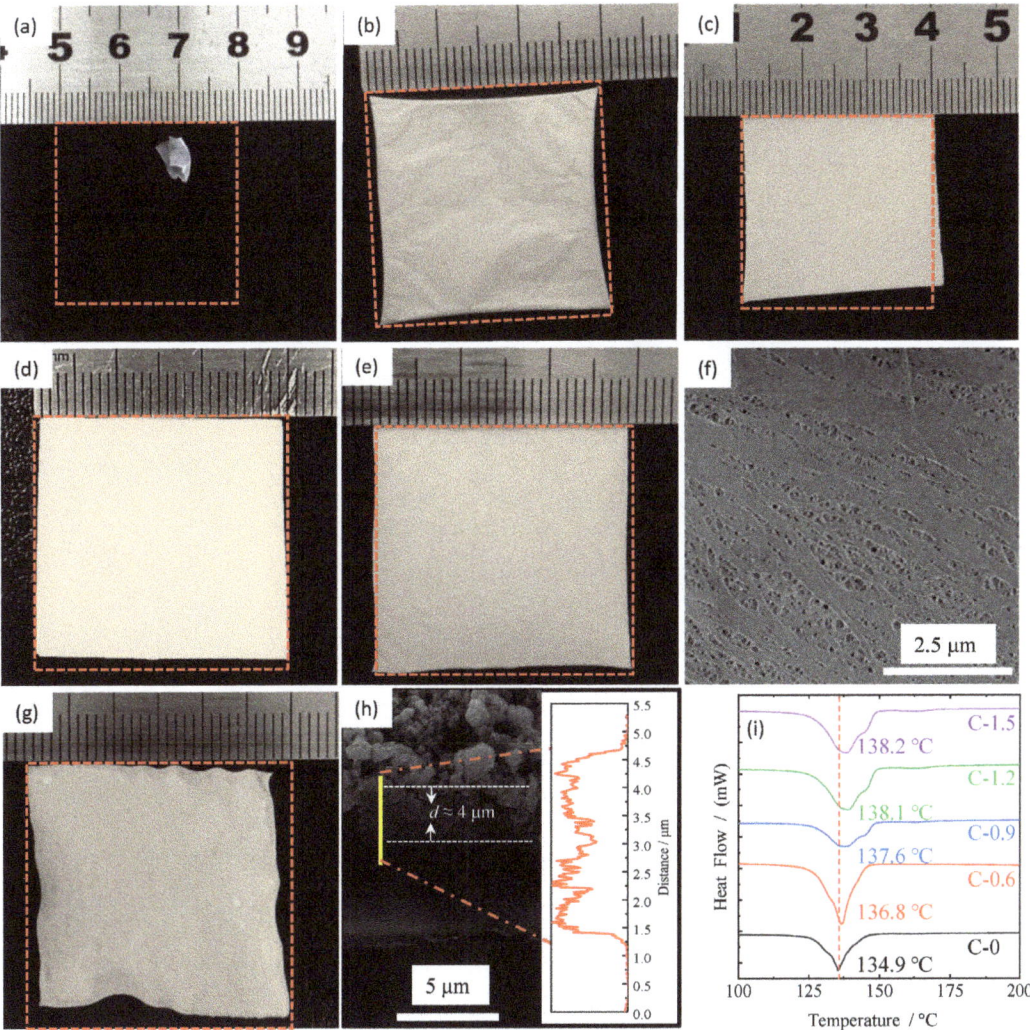

Figure 3. Thermal shrinkage (%) of PE separator (**a**) and ceramic separators at 140 °C for 0.5 h: (**b**) C-0.6, (**c**) C-0.9, (**d**) C-1.2, (**e**) C-1.5 and (**g**) C-P. (**f**) is the SEM image for the reverse side of the coating of sample C-0.6 after thermal shrinkage test. (**h**) is the cross-section morphology and the thicknesses of the coating layer for C-0.6. (**i**) is the DSC curves for PE separators and ceramic separators.

The influence of coating amount on the mechanical properties of separators is analyzed by tension testing. As shown in Figure 4a, the tensile strength of the pristine separator (sample C-0) is 15.5 MPa, and it can be improved after coating with the synthesized TiO_2 nanorods, as evidenced by the tensile strengths of 17.0 (C-0.6), 17.36 (C-0.9), 17.98 (C-1.2) and 18.05 (C-1.5) MPa, respectively, which is consistent with the aforementioned thermal stability analysis. Therefore, based on the analysis results of mechanical properties and thermal stability, a possible mechanism was proposed in Figure 4b. The coating can be divided into a surface layer and a stacking layer. Firstly, in the surface layer, some TiO_2 nanorods "bridged" on the skeletons of the microporous of the pristine membrane through the binder. Then, the other TiO_2 nanorods will stack on the surface layer to form a stacking layer, where these TiO$_2$ nanorods do not interact directly with the pristine membrane. When the separator is subjected to external force (mechanical stretching or thermal shrinkage), the micropores can maintain the original shape and keep themselves from thermal shrinkage or mechanical stretching mainly due to the interaction between the skeletons of the microporous and TiO$_2$ nanorods in the surface layer rather than the interaction between the skeletons of the microporous and TiO$_2$ nanorods in stacking layer. Therefore, smaller improvements in thermal stability and mechanical strength were observed when the coating amount of TiO$_2$ nanorods increased to above 0.6 mg/cm^2. However, if the TiO$_2$ nanoparticles are used to modify the pristine separator, the interaction between TiO$_2$ nanoparticles and the skeleton of the pristine separator can only occur by "point" gluing rather than "bridging" gluing when the separator is subjected to external force (mechanical stretching or thermal shrinkage), this bonding mode is very inefficient in preventing the shrinkage or extension of the separator.

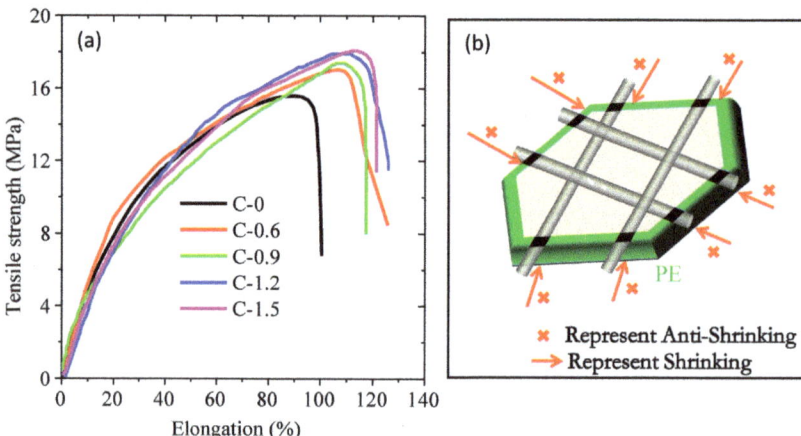

Figure 4. The stress-strain curves of the ceramic separators with different coating amounts (**a**) and anti-shrinkage mechanism diagram (**b**).

The electrochemical performances of ceramic separators with different coating amounts were characterized by the electrochemical workstation. The electrochemical window is a vital parameter to evaluate the electrochemical stability of separators, which were usually investigated by linear sweep voltammetry (LSV) tests. Generally, the onset of the suddenly increasing current was caused by the oxidative reaction of electrolyte decomposition, and the corresponding voltage indicates the maximum electrochemical stable voltage [35,36]. As shown in Figure 5a, there were no obvious current changes during the potential sweeps at 4 V, whereas the current showed a dramatic difference between 4.0 and 5.5 V. The electrochemical stabilities of PE separators are about 4.1 V (vs. Li$^+$/Li). In comparison, the current onsets of C-0.6, C-0.9, C-1.2, and C-1.5 separators are 5.3 V. The TiO$_2$ nanorods-coated separator has a wider electrochemical stability window, which indicates that the TiO$_2$ nanorod-modified separator possesses better electrochemical stability. The stability enhancement means better compatibility with the electrolyte of the lithium-ion battery, which should be attributed to the excellent electrolyte affinity of TiO$_2$-coated PE separator and the stabilization of electrolyte anions by Ti-O units acting as the Lewis acid centers [37–39]. Ionic conductivity is another important indicator to evaluate the electrochemical performance of the separator. Figure 5b shows the Nyquist plots of the stainless steel (SS)/separator-electrolyte/SS molds assembled by sandwiching the pristine PE separator or coated separators soaking in liquid electrolyte between two pieces of SS. The high-frequency intercept on the real axis reflects the bulk resistance (R_b), which can be used to calculate the ionic conductivity in Table 1. According to the results, all

TiO$_2$ nanorod-modified PE separators show higher ionic conductivity than the pristine PE separator, and the highest ionic conductivity was observed for sample C-0.6, which may benefit from the synergistic contributions of the significantly increased electrolyte uptake and the well-preserved porous structure [13,40], as observed in Figure 2c. The compatibility of liquid electrolyte-soaked separators with a lithium electrode is also a very important factor in the C-rate capability of lithium-ion batteries, which can be investigated by evaluating the impedance variation of Li/liquid electrolyte-soaked separator/Li cells. As shown in Figure 5c, a semicircle was observed from the impedance spectra of cells with all separators that represent the Li/electrolyte interfacial resistance (R_{int}), which was related to the charge transport across the passivation layer (solid electrolyte film) and the charge transfer reaction, Li$^+$ + e$^-$ = Li [3,37]. The R_{int} for C-0, C-0.6, C-0.9, C-1.2, and C-1.5 samples are 243 Ω, 118 Ω, 124 Ω, 141 Ω and 161 Ω, respectively. Compared with the uncoated PE separator, it can be seen that all TiO$_2$ nanorod-modified PE separators had lower interfacial resistance, indicating smooth ion transport between the ceramic nanoparticle-coated separators and electrodes. It can be attributed to the TiO$_2$ nanorod-modified PE separators capable of retaining the original porous structure and the layer of TiO$_2$ nanorods able to obtain higher electrolyte uptake (Table 1), which can effectively decrease the interaction between electrolyte components and the lithium electrode, and gradually stabilizes the interface [41]. Moreover, the C-0.6 separator exhibited the smallest interfacial impedance and therefore had the best separator-electrode compatibility. This is consistent with previous research [38,42] that a very thin inorganic oxide layer can negate the interfacial impedance between the electrolyte and lithium metal, owing to the high binding energy between lithium and the oxides layer.

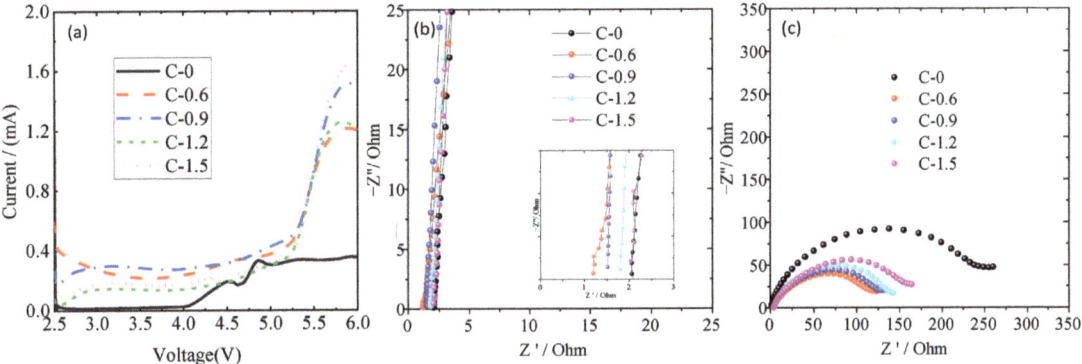

Figure 5. The Linear scan voltammetry (LSV) curve of Li/separator- liquid electrolyte/SS cells (**a**), AC impedance spectra of the SS/separator-liquid electrolyte/SS cell (**b**), the interfacial resistances of the Li/separator-liquid electrolyte/Li cell (**c**).

Table 1. Properties of the surface-modified separators.

Sample	Thickness of Coating (μm)	Gurley Value (s 100 cc^{-1})	Electrolyte Uptake (%)	Thermal Dimensional Shrinkage (%, 140 °C)	Melting Temperature (°C)	Ionic Conductivity (mS cm^{-1})
C-0	0	178	132	~98.0	134.9	0.33
C-0.6	~0.5	191	269	~10.0	136.8	0.57
C-0.9	~0.9	213	283	~5.65	137.6	0.47
C-1.2	~1.3	226	299	~5.69	138.1	0.40
C-1.5	~2.1	241	311	~3.0	138.2	0.38

From the results of thermal stability, mechanical properties, and electrochemical performance tests for TiO$_2$ nanorod-modified separators with different coating amounts, sample

C-0.6 has the best overall performance, such as the best ionic conductivity and interfacial impedance, excellent electrochemical stability window, acceptable thermal stability and mechanical properties, the minimum introduction of inert ingredients. Therefore, the electrochemical performance of the half-cell composed of the sample C-0.6 separator, a negative electrode (lithium metal), and a positive electrode (LiCoO$_2$) was further studied. For better comparative analysis, the electrochemical performance of the half-cell assembled by pristine PE membrane (C-0) was also tested. As shown in Figure 6a, the discharge specific capacity at low rates and its capacity recovering property (0.2 C) after 7 C for C-0.6 and C-0 are very close, but their differences gradually become larger with the increase of discharge current density. When the discharge current density is 7 C, the specific capacities of C-0.6 and C-0 are ~86.4 and ~60.4 mAh g^{-1}, respectively. Moreover, from their discharge curve in Figure 6b,c, it can be further seen that there is little difference in the curve shape and the average discharge voltage platform under low rates, while the voltage platform decrease for the C-0 sample is more than that of C-0.6 at high rates. These results indicate that the polarization resistance of the battery assembled by TiO$_2$ nanorod-modified separator is smaller than that of the battery composed of the pristine separator at high rates, which is consistent with the test results of ionic conductivity (Figure 5b), interfacial impedance (Figure 5c) and electrolyte uptake (Table 1). Furthermore, the cycle performance of these half-cells was studied, as depicted in Figure 6d. At the discharge current density of 0.5 C, the capacity retention of the C-0.6 sample is 82.6% after 100 cycles, while that of the C-0 sample is only 77.8%. Similarly, from their representative discharge curves (Figure 6e,f), it can be seen that the C-0.6 sample can maintain a good discharge capacity after cycling.

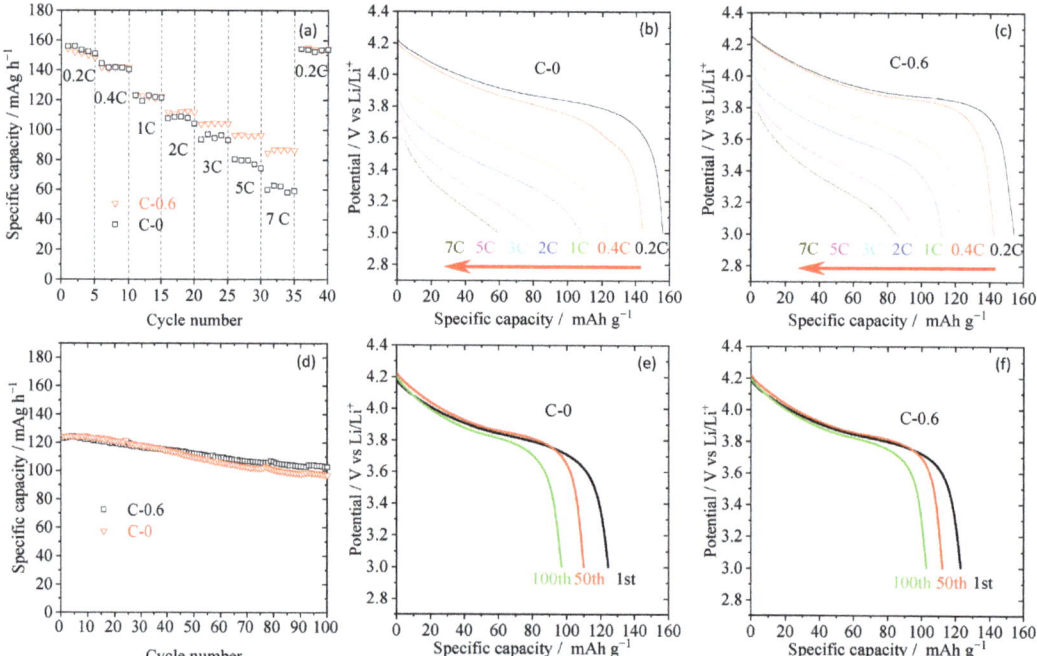

Figure 6. The C-rate capabilities (**a**) and their corresponding discharge profiles (**b,c**) of half-cells assembled with pristine PE separator (C-0 sample) and TiO$_2$ nanorod-modified PE separator (C-0.6 sample), where charge/discharge current densities are varied from 0.2/0.2–7/7 C under a voltage range between 3.0 and 4.2 V. the cyclic performance at 0.5 C (**d**) and their corresponding discharge profiles (**e,f**) of half-cells assembled with pristine PE separator (C-0 sample) and TiO$_2$ nanorod-modified PE separator.

In order to further understand the influence of the TiO$_2$ nanorod-modified layer on the electrochemical performance of the PE separator, AC impedance spectrum analysis was performed on the half-cell assembled by the C-0.6 and C-0 separators. As can be seen from the Nyquist plot in Figure 7, the semicircle in the high-middle frequency region represents the interfacial resistance (R_{int}) modified by the combined effect of solid-electrolyte interface resistance (R_{SEI}) and charge-transfer resistance (R_{ct}). The TiO$_2$ nanorod-modified PE separator (C-0.6) exhibited a smaller semicircle at high-middle frequency, indicating that their interfacial resistance is relatively low compared with that of the pristine PE separator. These results were ascribed to surface modification leading to enhanced hydrophilicity and affinity with electrolyte, resulting in thinner and more compact SEI formation by TiO$_2$ nanorod-modified PE separator, which agreed with the findings of the ionic conductivity (Figure 5b), interfacial resistance (Figure 5c) and electrolyte uptake (Table 1).

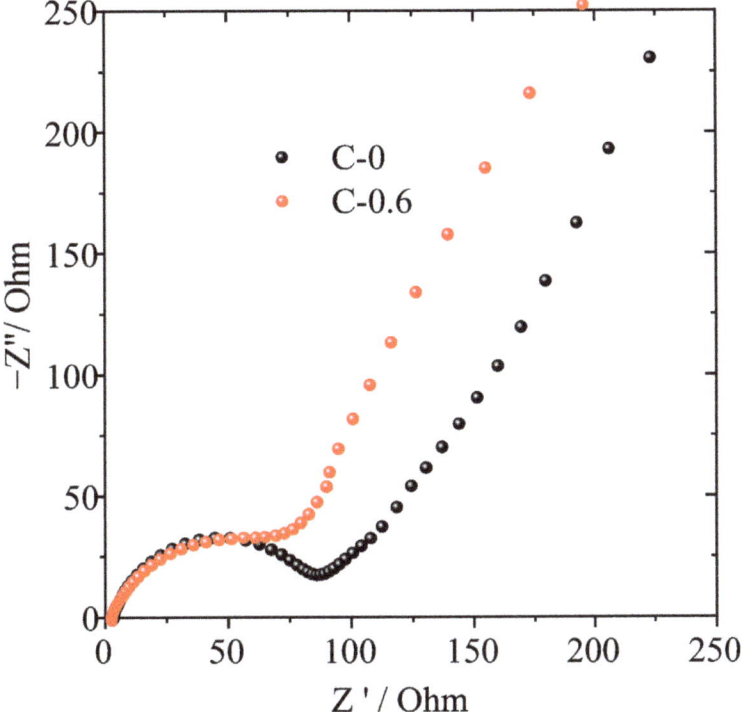

Figure 7. Nyquist plots of half-cells assembled with pristine PE (C-0) and TiO$_2$ nanorod-modified PE (C-0.6) separator.

4. Conclusions

In this paper, the TiO$_2$ nanorods with a diameter of about 10–100 nanometers and a length of about tens of microns are used to modify the PE separator. It can be observed from SEM tests that the TiO$_2$ nanorods are "bridged" on the microporous skeleton of pristine PE separator to form a coating structure with a three-dimensional porous network, while the TiO$_2$ nanoparticles are "embedded" into the micropores, which will cause a series of problems such as micropore blockage, easy detaching, and introduction of excessive inert substances. Then, multiple analytical techniques (e.g., SEM, DSC, EIS, LSV, and so on.) are also utilized to investigate the effect of coating amount on the physicochemical and electrochemical properties of ceramic separator. The results showed that these properties can be effectively improved by coating TiO$_2$ nanorods, but the degree of improvement is not directly proportional to the coating amount. For example, the thermal stability and mechanical properties of the ceramic separator increase with the increase of the coating amount, however, smaller improvements are observed when the coating amount is above 0.6 mg/cm^2. In fact, when the separator is subjected to external force (mechanical stretching or thermal contraction), the forces inhibiting micropore deformation are derived from the interaction of TiO$_2$ nanorods directly "bridging" with the microporous skeleton rather than those indirectly "glued" with the microporous skeleton. In addition, when the loading level is 0.6 mg/cm^2, the ceramic separator can achieve optimal performance in terms of ionic conductivity, electrochemical stability window, and interface compatibility because the introduction of excessive inert coating material can reduce the ionic conductivity, increase the interfacial impedance, and lower the energy density of the battery. Moreover, the capacity retention assembled by the ceramic separator with a loading of 0.6 mg/cm^2 TiO$_2$ nanorods was 57.1% under 7 C/0.2 C and 82.6% after 100 cycles, respectively, indicating that the ceramic separator with a thin coating layer has well-balanced performances. This research may provide a novel approach to overcoming the common disadvantages of current surface-coated separators.

Author Contributions: Conceptualization, Z.C. and X.Y.; Methodology, T.W.; Validation, Y.P.; Investigation, C.H.; Data curation, H.Z.; Writing—original draft, Z.C. All authors have read and agreed to the published version of the manuscript.

Funding: Z. Chen wants to acknowledge the financial support from the Natural Science Foundation of Hunan Province of China (No. 2021JJ30374), Hunan Provincial Education Office Foundation of China (No. 19A261), and Key R & D projects in Hunan Province (No. 2021GK2015). T. Wang wants to acknowledge the financial support from Natural Science Foundation of Guangdong Province of China-Regional joint fund (No. 2021B1515140025) and Natural Science Foundation of Guangdong Province of China-General Program (No. 2022A1515010972).

Data Availability Statement: Not applicable.

Conflicts of Interest: The authors declare no conflict of interest.

References

1. Liu, K.; Liu, Y.; Lin, D.; Pei, A.; Cui, Y. Materials for lithium-ion battery safety. *Sci. Adv.* **2018**, *4*, eaas9820. [CrossRef] [PubMed]
2. Lagadec, M.F.; Zahn, R.; Wood, V. Characterization and performance evaluation of lithium-ion battery separators. *Nat. Energy* **2018**, *4*, 16–25. [CrossRef]
3. Jana, K.K.; Lue, S.J.; Huang, A.; Soesanto, J.F.; Tung, K.-L. Separator Membranes for High Energy-Density Batteries. *ChemBioEng Rev.* **2018**, *5*, 346–371. [CrossRef]
4. Feng, X.; Ouyang, M.; Liu, X.; Lu, L.; Xia, Y.; He, X. Thermal runaway mechanism of lithium ion battery for electric vehicles: A review. *Energy Storage Mater.* **2018**, *10*, 246–267. [CrossRef]
5. Lu, W.; Yuan, Z.; Zhao, Y.; Zhang, H.; Zhang, H.; Li, X. Porous Membranes in Secondary Battery Technologies. *Chem. Soc. Rev.* **2017**, *46*, 2199–2236. [CrossRef]
6. Wang, W.; Liao, C.; Liew, K.M.; Chen, Z.; Song, L.; Kan, Y.; Hu, Y. A 3D flexible and robust HAPs/PVA separator prepared by a freezing-drying method for safe lithium metal batteries. *J. Mater. Chem. A* **2019**, *7*, 6859–6868. [CrossRef]
7. Chen, R.; Qu, W.; Guo, X.; Li, L.; Wu, F. The pursuit of solid-state electrolytes for lithium batteries: From comprehensive insight to emerging horizons. *Mater. Horiz.* **2016**, *3*, 487–516. [CrossRef]
8. Jeong, K.; Park, S.; Lee, S.-Y. Revisiting polymeric single lithium-ion conductors as an organic route for all-solid-state lithium ion and metal batteries. *J. Mater. Chem. A* **2019**, *7*, 1917–1935. [CrossRef]
9. Fan, L.; Wei, S.; Li, S.; Li, Q.; Lu, Y. Recent Progress of the Solid-State Electrolytes for High-Energy Metal-Based Batteries. *Adv. Energy Mater.* **2018**, *8*, 1702657. [CrossRef]
10. Na, W.; Koh, K.H.; Lee, A.S.; Cho, S.; Ok, B.; Hwang, S.-W.; Lee, J.H.; Koo, C.M. Binder-less chemical grafting of SiO$_2$ nanoparticles onto polyethylene separators for lithium-ion batteries. *J. Membr. Sci.* **2019**, *573*, 621–627. [CrossRef]
11. Liao, H.; Zhang, H.; Qin, G.; Hong, H.; Li, Z.; Lin, Y.; Li, L. Novel Core-Shell PS-co-PBA@SiO$_2$ Nanoparticles Coated on PP Separator as "Thermal Shutdown Switch" for High Safety Lithium-Ion Batteries. *Macromol. Mater. Eng.* **2017**, *302*, 1700241. [CrossRef]
12. Liao, C.; Wang, W.; Han, L.; Mu, X.; Wu, N.; Wang, J.; Gui, Z.; Hu, Y.; Kan, Y.; Song, L. A flame retardant sandwiched separator coated with ammonium polyphosphate wrapped by SiO$_2$ on commercial polyolefin for high performance safety lithium metal batteries. *Appl. Mater. Today* **2020**, *21*, 100793. [CrossRef]
13. Wang, Q.; Yang, J.; Wang, Z.; Shi, L.; Zhao, Y.; Yuan, S. Dual-Scale Al$_2$O$_3$ Particles Coating for High-Performance Separator and Lithium Metal Anode. *Energy Technol.* **2020**, *8*, 1901429. [CrossRef]

14. Qiu, Z.; Yuan, S.; Wang, Z.; Shi, L.; Jo, J.H.; Myung, S.-T.; Zhu, J. Construction of silica-oxygen-borate hybrid networks on Al$_2$O$_3$-coated polyethylene separators realizing multifunction for high-performance lithium ion batteries. *J. Power Source* **2020**, *472*, 228445. [CrossRef]
15. Yeon, D.; Lee, Y.; Ryou, M.H.; Lee, Y.M. New flame-retardant composite separators based on metal hydroxides for lithium-ion batteries. *Electrochim. Acta* **2015**, *157*, 282–289. [CrossRef]
16. Cui, J.; Liu, J.; He, C.; Li, J.; Wu, X. Composite of polyvinylidene fluoride-cellulose acetate with Al(OH)$_3$ as a separator for high-performance lithium ion battery. *J. Membr. Sci.* **2017**, *541*, 661–667. [CrossRef]
17. Shekarian, E.; Nasr, M.R.J.; Mohammadi, T.; Bakhtiari, O.; Javanbakht, M. Preparation of 4A zeolite coated polypropylene membrane for lithium-ion batteries separator. *J. Appl. Polym. Sci.* **2019**, *136*, 47841. [CrossRef]
18. Dong, X.; Mi, W.; Yu, L.; Jin, Y.; Lin, Y.S. Zeolite coated polypropylene separators with tunable surface properties for lithium-ion batteries. *Microporous Mesoporous Mater.* **2016**, *226*, 406–414. [CrossRef]
19. Liu, L.; Wang, Y.; Gao, C.; Yang, C.; Wang, K.; Li, H.; Gu, H. Ultrathin ZrO$_2$-coated separators based on surface sol-gel process for advanced lithium ion batteries. *J. Membr. Sci.* **2019**, *592*, 117368. [CrossRef]
20. Kim, K.J.; Kwon, H.K.; Park, M.S.; Yim, T.; Yu, J.S.; Kim, Y.J. Ceramic composite separators coated with moisturized ZrO$_2$ nanoparticles for improving the electrochemical performance and thermal stability of lithium ion batteries. *Phys. Chem. Chem. Phys.* **2014**, *16*, 9337–9343. [CrossRef]
21. Peng, K.; Wang, B.; Li, Y.; Ji, C. Magnetron sputtering deposition of TiO$_2$ particles on polypropylene separators for lithium-ion batteries. *RSC Adv.* **2015**, *5*, 81468–81473. [CrossRef]
22. Kim, P.S.; Le Mong, A.; Kim, D. Thermal, mechanical, and electrochemical stability enhancement of Al$_2$O$_3$ coated polypropylene/polyethylene/polypropylene separator via poly(vinylidene fluoride)-poly(ethoxylated pentaerythritol tetraacrylate) semi-interpenetrating network binder. *J. Membr. Sci.* **2020**, *612*, 118481. [CrossRef]
23. Zhu, X.; Jiang, X.; Ai, X.; Yang, H.; Cao, Y. A Highly Thermostable Ceramic-Grafted Microporous Polyethylene Separator for Safer Lithium-Ion Batteries. *ACS Appl. Mater. Interfaces* **2015**, *7*, 24119–24126. [CrossRef] [PubMed]
24. Kim, J.Y.; Lim, D.Y. Surface-Modified Membrane as A Separator for Lithium-Ion Polymer Battery. *Energies* **2010**, *3*, 866–885. [CrossRef]
25. Wang, W.; Yuen, A.C.Y.; Yuan, Y.; Liao, C.; Li, A.; Kabir, I.I.; Kan, Y.; Hu, Y.; Yeoh, G.H. Nano architectured halloysite nanotubes enable advanced composite separator for safe lithium metal batteries. *Chem. Eng. J.* **2023**, *451*, 138496. [CrossRef]
26. Wang, X.; Peng, L.; Hua, H.; Liu, Y.; Zhang, P.; Zhao, J. Magnesium borate fiber coating separators with high Li-ion transference number for lithium ion batteries. *ChemElectroChem* **2020**, *7*, 1187–1192. [CrossRef]
27. Hu, S.; Lin, S.; Tu, Y.; Hu, J.; Wu, Y.; Liu, G.; Li, F.; Yu, F.; Jiang, T. Novel aramid nanofiber-coated polypropylene separators for lithium ion batteries. *J. Mater. Chem. A* **2016**, *4*, 3513–3526. [CrossRef]
28. Han, D.-H.; Zhang, M.; Lu, P.-X.; Wan, Y.-L.; Chen, Q.-L.; Niu, H.-Y.; Yu, Z.-W. A multifunctional separator with Mg(OH)$_2$ nanoflake coatings for safe lithium-metal batteries. *J. Energy Chem.* **2021**, *52*, 75–83. [CrossRef]
29. Deng, Q.; Wei, M.; Ding, X.; Jiang, L.; Wei, K.; Zhou, H. Large single-crystal anatase TiO$_2$ Bipyramids. *J. Cryst. Growth* **2010**, *312*, 213–219. [CrossRef]
30. Arora, P.; Zhang, Z. Battery Separators. *Chem. Rev.* **2004**, *104*, 4419–4462. [CrossRef]
31. Heimes, H.H.; Kampker, A.; Lienemann, C.; Locke, M.; Offermanns, C. *Lithium-Ion Battery Cell Production Process*, 3rd ed.; PEM of RWTH Aachen and VDMA: Frankfurt am Main, Germany, 2019.
32. Liu, Y.; Zhang, R.; Wang, J.; Wang, Y. Current and future lithium-ion battery manufacturing. *iScience* **2021**, *24*, 102332. [CrossRef] [PubMed]
33. Gao, X.; Sheng, W.; Wang, Y.; Lin, Y.; Luo, Y.; Li, B.-G. Polyethylene battery separator with auto-shutdown ability, thermal stability of 220 °C, and hydrophilic surface via solid-state ultraviolet irradiation. *J. Appl. Polym. Sci.* **2015**, *132*, 42169. [CrossRef]
34. Peng, L.; Kong, X.; Li, H.; Wang, X.; Shi, C.; Hu, T.; Liu, Y.; Zhang, P.; Zhao, J. A Rational Design for a High-Safety Lithium-Ion Battery Assembled with a Heatproof–Fireproof Bifunctional Separator. *Adv. Funct. Mater.* **2020**, *31*, 2008537. [CrossRef]
35. Chen, W.; Shi, L.; Zhou, H.; Zhu, J.; Wang, Z.; Mao, X.; Chi, M.; Sun, L.; Yuan, S. Water-Based Organic-Inorganic Hybrid Coating for a High-Performance Separator. *ACS Sustain. Chem. Eng.* **2016**, *4*, 3794–3802. [CrossRef]
36. Huang, F.; Xu, Y.; Peng, B.; Su, Y.; Jiang, F.; Hsieh, Y.-L.; Wei, Q. Coaxial Electrospun Cellulose-Core Fluoropolymer-Shell Fibrous Membrane from Recycled Cigarette Filter as Separator for High Performance Lithium-Ion Battery. *ACS Sustain. Chem. Eng.* **2015**, *3*, 932–940. [CrossRef]
37. Wang, Z.; Guo, F.; Chen, C.; Shi, L.; Yuan, S.; Sun, L.; Zhu, J. Self-assembly of PEI/SiO$_2$ on polyethylene separators for Li-ion batteries with enhanced rate capability. *ACS Appl. Mater. Interfaces* **2015**, *7*, 3314–3322. [CrossRef]
38. Wu, S.; Ning, J.; Jiang, F.; Shi, J.; Huang, F. Ceramic Nanoparticle-Decorated Melt-Electrospun PVDF Nanofiber Membrane with Enhanced Performance as a Lithium-Ion Battery Separator. *ACS Omega* **2019**, *4*, 16309–16317. [CrossRef]
39. Tan, L.; Li, Z.; Shi, R.; Quan, F.; Wang, B.; Ma, X.; Ji, Q.; Tian, X.; Xia, Y. Preparation and Properties of an Alginate-Based Fiber Separator for Lithium-Ion Batteries. *ACS Appl. Mater. Interfaces* **2020**, *12*, 38175–38182. [CrossRef]
40. Rahman, M.M.; Mateti, S.; Cai, Q.; Sultana, I.; Fan, Y.; Wang, X.; Hou, C.; Chen, Y. High temperature and high rate lithium-ion batteries with boron nitride nanotubes coated polypropylene separators. *Energy Storage Mater.* **2019**, *19*, 352–359. [CrossRef]

41. Wei, Z.; Gu, J.; Zhang, F.; Pan, Z.; Zhao, Y. Core-Shell Structured Nanofibers for Lithium Ion Battery Separator with Wide Shutdown Temperature Window and Stable Electrochemical Performance. *ACS Appl. Polym. Mater.* **2020**, *2*, 1989–1996. [CrossRef]
42. Han, X.; Gong, Y.; Fu, K.K.; He, X.; Hitz, G.T.; Dai, J.; Pearse, A.; Liu, B.; Wang, H.; Rubloff, G.; et al. Negating interfacial impedance in garnet-based solid-state Li metal batteries. *Nat. Mater.* **2017**, *16*, 572–579. [CrossRef] [PubMed]

Disclaimer/Publisher's Note: The statements, opinions and data contained in all publications are solely those of the individual author(s) and contributor(s) and not of MDPI and/or the editor(s). MDPI and/or the editor(s) disclaim responsibility for any injury to people or property resulting from any ideas, methods, instructions or products referred to in the content.

Article

Fe[III] Chelated with Humic Acid with Easy Synthesis Conditions and Good Performance as Anode Materials for Lithium-Ion Batteries

Hao Zhang [1], Youkui Wang [1], Ruili Zhao [1], Meimei Kou [1], Mengyao Guo [1], Ke Xu [1], Gang Tian [2], Xinting Wei [2], Song Jiang [1], Qing Yuan [1,3,*] and Jinsheng Zhao [1,3,*]

[1] School of Chemistry and Chemical Engineering, Liaocheng University, Liaocheng 252059, China; zh1836958@163.com (H.Z.); m19861904240@163.com (Y.W.); zhao1784308@163.com (R.Z.); kmm2018705344@163.com (M.K.); gmy15275641513@163.com (M.G.); x15866598163@163.com (K.X.); jiangsong006@163.com (S.J.)

[2] Shandong Tianyi New Energy Co., Ltd., Liaocheng 252059, China; tg1999@163.com (G.T.); 18863588007@139.com (X.W.)

[3] Shandong Provincial Key Laboratory of Chemical Energy Storage and Novel Cell Technology, Liaocheng University, Liaocheng 252059, China

* Correspondence: yuanqing@lcu.edu.cn (Q.Y.); j.s.zhao@163.com (J.Z.)

Abstract: In this work, we prepared a green, cheap material by chelating humic acid with ferric ions (HA-Fe) and used it as an anode material in LIBs for the first time. From the SEM, TEM, XPS, XRD, and nitrogen adsorption–desorption experimental results, it was found that the ferric ion can chelate with humic acid successfully under mild conditions and can increase the surface area of materials. Taking advantage of the chelation between the ferric ions and HA, the capacity of HA-Fe is 586 mAh·g^{-1} at 0.1 A·g^{-1} after 1000 cycles. Moreover, benefitting from the chelation effect, the activation degree of HA-Fe (about 8 times) is seriously improved compared with pure HA material (about 2 times) during the change–discharge process. The capacity retention ratio of HA-Fe is 55.63% when the current density increased from 0.05 A·g^{-1} to 1 A·g^{-1}, which is higher than that of HA (32.55%) and Fe (24.85%). In the end, the storage mechanism of HA-Fe was investigated with ex-situ XPS measurements, and it was found that the C=O and C=C bonds are the activation sites for storage Li ions but have different redox voltages.

Keywords: lithium-ion battery; anode material; organic anode; metal–organic compound; humic acid; Fe chelate

1. Introduction

Rechargeable LIBs have attracted a great deal of attention due to their high energy density and good cycle stability as a new type of energy storage device [1–6]. LIBs are used in many applications, including portable electronics, electric vehicles, energy storage devices, and many others [7–12]. Anode material is a critical part of high-power LIBs. At present, due to poor rate performance or low theoretical capacity, inorganic materials struggle to meet the demands of high-performance batteries [13]. For example, graphite carbon materials have been used as anode materials for commercial lithium-ion batteries due to their good cycle stability and accessibility. However, further use in high-performance LIBs is hindered by the low theoretical capacity (373 mAh·g^{-1}) and poor rate performance [14,15]. Silicon materials have a serious volume effect in the repeated charge and discharge process, resulting in a rapid capacity fading and a poor cycle stability, which is also a problem to be overcome before the commercial application of LIBs [16,17]. In addition, most inorganic material production processes are based on the redox reaction of inorganic compounds, containing expensive transition metal elements [18–20].

Citation: Zhang, H.; Wang, Y.; Zhao, R.; Kou, M.; Guo, M.; Xu, K.; Tian, G.; Wei, X.; Jiang, S.; Yuan, Q.; et al. Fe[III] Chelated with Humic Acid with Easy Synthesis Conditions and Good Performance as Anode Materials for Lithium-Ion Batteries. *Materials* 2023, 16, 6477. https://doi.org/10.3390/ma16196477

Academic Editor: Satyam Panchal

Received: 24 August 2023
Revised: 24 September 2023
Accepted: 27 September 2023
Published: 29 September 2023

Copyright: © 2023 by the authors. Licensee MDPI, Basel, Switzerland. This article is an open access article distributed under the terms and conditions of the Creative Commons Attribution (CC BY) license (https://creativecommons.org/licenses/by/4.0/).

Organic electrode materials, due to their low cost, light weight, and abundance in nature, are considered as substitute materials to overcome the disadvantages of inorganic materials [21–26]. Until now, electrode materials for LIBs have consisted of many organic materials, including organosulfur materials [27–30], free radical compounds [31–34], conducting polymers [35–37], aromatic amines [38], quinine-type carbonyl compounds [39,40], and other organic materials [41]. Among them, the metal–organic polymer is an important material due to the π-d conjugated coordination. Using tetraaminobenzoquinone (TABQ) as the organic ligand and M^{2+} (M=Co, Ni, and Cu) as the metal ligands, Li et al. developed the metal–organic polymer as a cathode material of lithium-ion batteries with good rate performance (even 237.2 mAh·g^{-1} at 2 A·g^{-1}) [42]. Wu et al. reported two Ni-based conjugated coordination polymers with N and S as co-chelating atoms, and it was found that the co-existence of N and S resulted in high electrical conductivity and high stability, leading to a high capacity, excellent cyclic performance, and a high rate of capability [43]. Yang et al. reported a highly stable Zn, Ni-bimetallic porous nanocomposite via a one-step pyrolysis of a metal–organic framework as an efficient anode material (1105.2 mAh·g^{-1} at 0.5 A·g^{-1} after 400 cycles) [44]. However, the practical application of organic electrodes in lithium-ion batteries has been hampered by the fact that the majority of reported organics are derived from chemical feedstock, which are expensive and environmentally problematic [45–48]. Therefore, it is very urgent to develop new electrode materials for LIBs with higher energy density, environmental friendliness, low cost, good rate performance, and long cycle life [49]. Developing new materials originating from natural sources is an important way to exploit new green, economic anode. Zhang et al. generated a novel anode material by chelating the tannic acid via ferric ions (TA-Fe) and used the anode materials with high reversible capacity (1105.2 mAh·g^{-1} at 0.1 A·g^{-1}) and ultra-long cycling stability (10.0 A·g^{-1} over 16,000 cycles with a capacity retention of 78.8%) [50] Zhang et al. chelated the rhodizonic acid disodium salt (RA) using ferric ions and generated a novel organic anode material (RAFe) for the first time. The strong chelation interaction between ferric ions and rhodizonic acid changes its initial structure and characteristics, enabling the obtained organic RAFe compound with outstanding electrochemical performance as an anode for lithium-ion batteries (LIB) (1283 mAh·g^{-1} at 0.1 A·g^{-1}) [51]. Zhu et al. used the humic acid as anode material for lithium-ion batteries and sodium ion batteries because of its richness in oxygen functional groups (carboxylic, quinonic, phenolic, and ketonic). However, the capability and rate performance of HA at higher currents are not excellent. The reported specific capacity of HA is 420 mAh·g^{-1} at 20 mA·g^{-1}, while it is only 60 mAh·g^{-1} at 400 mA·g^{-1}, whose capacity retention is 14.29% [45]. In addition, the long cycle performance is also worse than that of the traditionally reported organic materials at the current density of 40 mA·g^{-1}. Inspired by the promotion method of TA-Fe anode material and the RA-Fe using ferric ions, a Fe^{3+} chelated humic acid (named as HA-Fe) in this work at easy synthesis conditions is developed as the anode material for Li-ion batteries. Unlike previously reported organic compounds or metal oxides as anodes, which frequently involve a large number of complex synthetic processes, high temperatures, and toxic pollutant emissions [50], the HA-Fe in this work can be obtained under easy conditions without high energy consumption and has no environmental impact, which is beneficial for large-scale production. From the experimental results, it can be seen that the ferric acid can increase the surface area of the HA-Fe and improve the properties of capacity (586 mAh·g^{-1} at 0.1 A·g^{-1}), cycle stability (1000 cycles), and good rate properties (capacity retention ratio of 55.63%).

2. Materials and Methods
2.1. Materials

Humic acid (HA) was purchased from Alfa Aesar Chemical Co., Ltd., Shanghai, China (Purity, 99%), and the typical structure is shown in Figure 1a. Ammonium iron (III) sulfate ($NH_4Fe(SO_4)_2$) was purchased from Damas-beta Chemical Co., Ltd., Shanghai, China (Purity, 99%). Hydrochloric acid (HCl) (Purity, AR) and sodium hydroxide (NaOH) (Purity,

AR) were purchased from Yantai Yuandong Fine Chem. Co., Ltd., Yantai, China. All of the above chemicals were used as received without further processing. The deionized water was made in our laboratory.

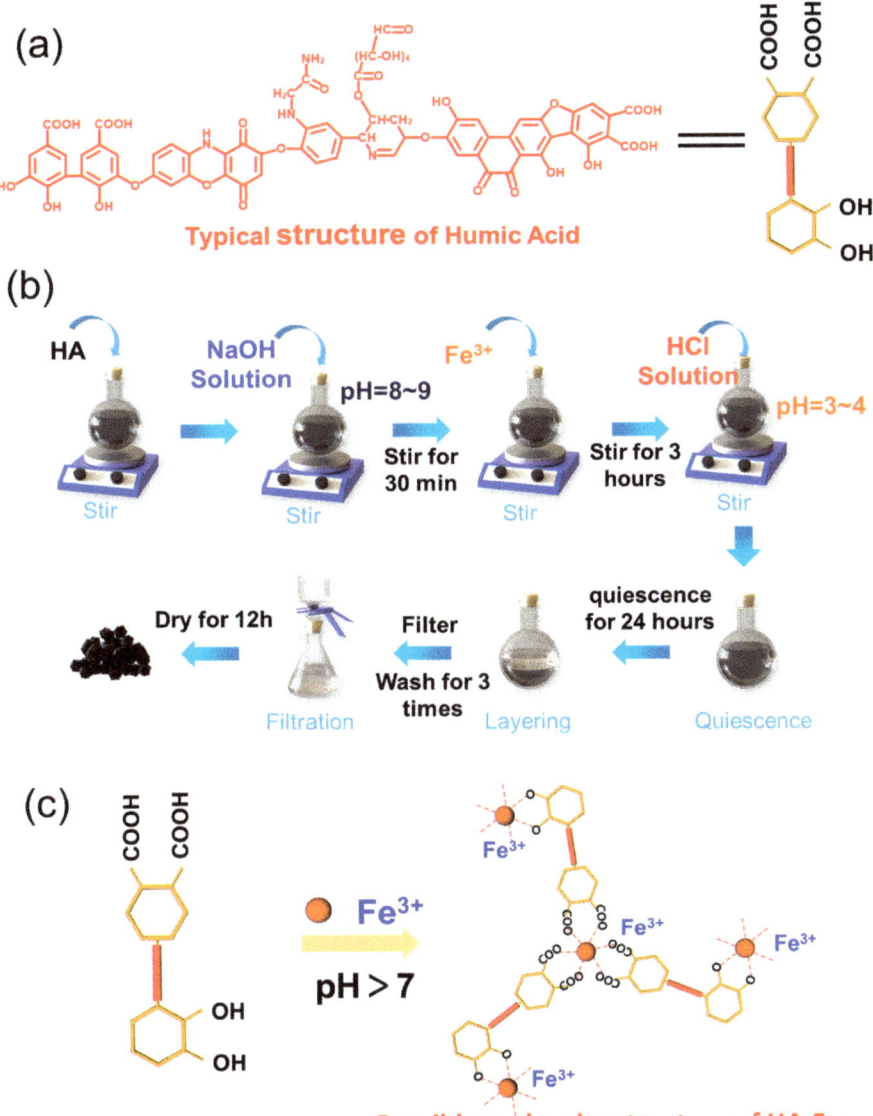

Figure 1. (a) The typical structure of HA molecules. (b) The formation procedures program of HA-Fe material. (c) Possible molecular structure of HA-Fe compound.

2.2. Preparation Procedure of HA-Fe Materials

The experimental set-up is composed of a single mouth bottom flask, a magnetic stirrer, vacuum filter, and a vacuum oven. Figure 1b shows the simple preparation process of HA-Fe. 375 mg of HA was accurately weighed and dispersed in 250 mL deionized water in a

500 mL single mouth bottom flask with continuous stirring by a magnetic stirrer. Following that, 0.1 mol·L^{-1} NaOH solution was used to adjust the pH of the suspension to 8~9 to help dissolve the HA. The solution was stirred for 30 min, and then 1000 mg of $NH_4Fe(SO_4)_2$ was added into the HA solution. The ferric ions were successfully chelated with HA after continuously stirring the mixture for 3 h at room temperature. Adjusting the pH to 3~4 with 0.1 mol·L^{-1} HCl solution, and then a large amount of black powder was precipitated from the solution. Stirring was stopped, and the mixture kept quiescence for 24 h to layer the HA-Fe material. The supernatant liquid was removed, and the underling materials were filtered and washed with deionized water for three times with a vacuum filter. The black HA-Fe powder was obtained after drying at 80 °C for 12 h in the vacuum oven. Figure 1c shows the chelation mechanism diagram and the possible molecular structure of HA-Fe.

2.3. Characterization and Electrochemical Properties of Materials

The Supporting Information (SI) provides the test methods and associated instrumentation information for structure confirmation, morphological characterization, and other physical characterization. The SI also provides a comprehensive introduction of the electrode preparation procedures, electrochemical testing methods, and instrumentation.

3. Results and Discussion

3.1. Characterization Results of HA and HA-Fe Materials

SEM has been used for microstructural and morphological characterization of HA-Fe and HA. As clearly depicted in Figure S1a, the original HA shows large particles, about 5–15 um in size. Interestingly, after chelating with ferric ions, the particles of HA-Fe were much smaller, as shown in Figure S1c. The smaller particle size can always reduce the lithium-ion diffusion distance, which often results in high rates of capability and good electrochemical performance [52]. From the morphology comparison in Figure S1b,d, it can be seen that the surface morphology of HA (Figure S1b) is much clearer and tighter than that of HA-Fe (Figure S1d), which illustrated that after chelating with the ferric ion, the structure of particles was changed to some extent. The homogeneous elemental distribution of oxygen and iron elements in HA-Fe materials is illustrated by the corresponding elemental mapping images in Figure S1f,g.

The TEM of HA-Fe at the dimension of 100 nm and 10 nm are shown in Figure 2a and c, respectively, while the TEM of HA at the dimension of 100 nm and 10 nm are shown in Figure 2b and d, respectively. It can be seen from the areas enclosed by the red boxes in Figure 2a,b that the morphology of HA-Fe and HA were very different. There are many dark pots uniformly distributed in the materials enclosed by the red box in Figure 2a, however, the morphology of HA in the area enclosed by the red box in Figure 2b is very clear. Similarly, the blocky morphology of HA-Fe in Figure 2c, which is enclosed by the red cycles, is very different to the uniform morphology of HA enclosed by the red box in Figure 2d at the 10 nm dimension, as shown in Figure 2c,d. The different morphology illustrated that the ferric ion is chelated with the HA successfully. The mass fractions of C, O, and Fe on the surface spectrum of HA and HA-Fe are shown in Table 1. It can be seen that the mass fraction of Fe increased from <0.1% to 29.2%, which also illustrated the successful formation of HA-Fe.

Table 1. Mass fractions of C, O, and Fe on the surface spectrum of HA and HA-Fe.

Mass Fractions of Different Element on Surface Spectrum (wt%)	C	O	Fe
HA	72.3	27.7	<0.1
HA-Fe	44.0	26.8	29.2

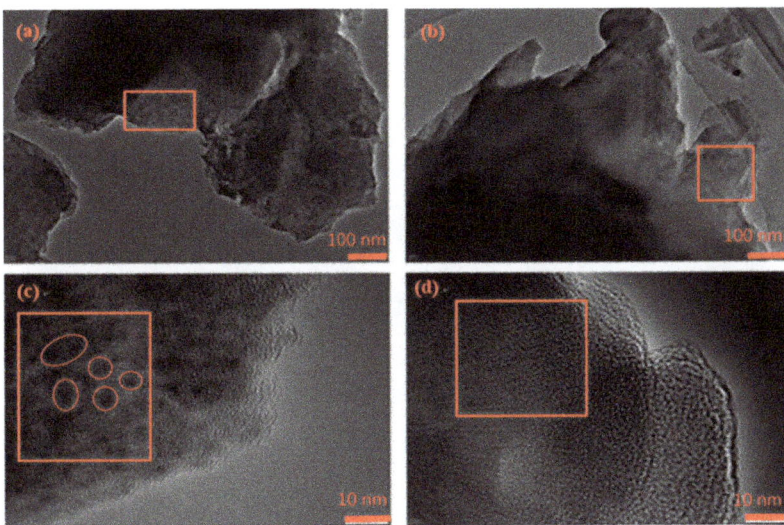

Figure 2. (**a**,**c**) The TEM of HA-Fe at the dimension of 100 nm and 10 nm, respectively; (**b**,**d**) The TEM of HA at the dimension of 100 nm and 10 nm, respectively.

The FTIR spectroscopy (Thermo Fisher, USA) has been used to characterize the typical chemical groups such as carboxyl, hydroxyl groups of HA and HA-Fe. As shown in Figure 3a, the spectroscopy of HA is similar to that reported in the literature [45,53]. Also, it can be seen from Figure 3a that the infrared spectra of HA-Fe and HA are roughly the same, indicating that the structure of HA was not changed obviously after chelating with ferric acid. Specifically, the broad peak around 3429 cm^{-1} is due to the stretching of the O-H bond. The aliphatic C-H stretching of the alkyl groups and the methyl C-H groups can be assigned to the weak shoulders at 2920 cm^{-1} and 2852 cm^{-1}. The peaks at 1580 cm^{-1} can be indexed to aromatic C=C, and 1380 cm^{-1} can be indexed to the symmetric stretching of COO$^-$, C-OH stretching of phenolic OH [53]. However, by observing the infrared spectrum of HA-Fe carefully, the vibration peak positions of phenolic hydroxyl, aromatic C=C and COO$^-$ changed slightly after forming a stable chelating structure with Fe^{3+}. The O-H was transferred from 3429 cm^{-1} to 3439 cm^{-1}, the aromatic C=C was transferred from 1580 cm^{-1} to 1630 cm^{-1}, and the symmetric stretching of COO-, C-OH stretching of phenolic OH was transferred from 1380 cm^{-1} to 1395 cm^{-1}. Similar peak shifts at the sites of OH, C=C, and COO$^-$ were also observed in the spectra of Fe(III)-humic acid complex in the literature [53]. This change in the FTIR spectrum is an indication that the major functional groups chelated by the iron ions are aromatic COOH and hydroxyl or phenolic OH [53]. As is displayed in Figure 3b, XRD was used to characterize the structure of HA, Fe, and the HA-Fe materials. The strong peaks at 10.7° and 37.2° in the XRD patterns of Fe and the strong peaks at 26.6° in the XRD patterns of HA illustrated their highly crystalline nature. In comparison, no significant peaks appeared in the XRD pattern of HA-Fe after chelation with trivalent iron ions, indicating that the synthesized HA-Fe is amorphous [54].

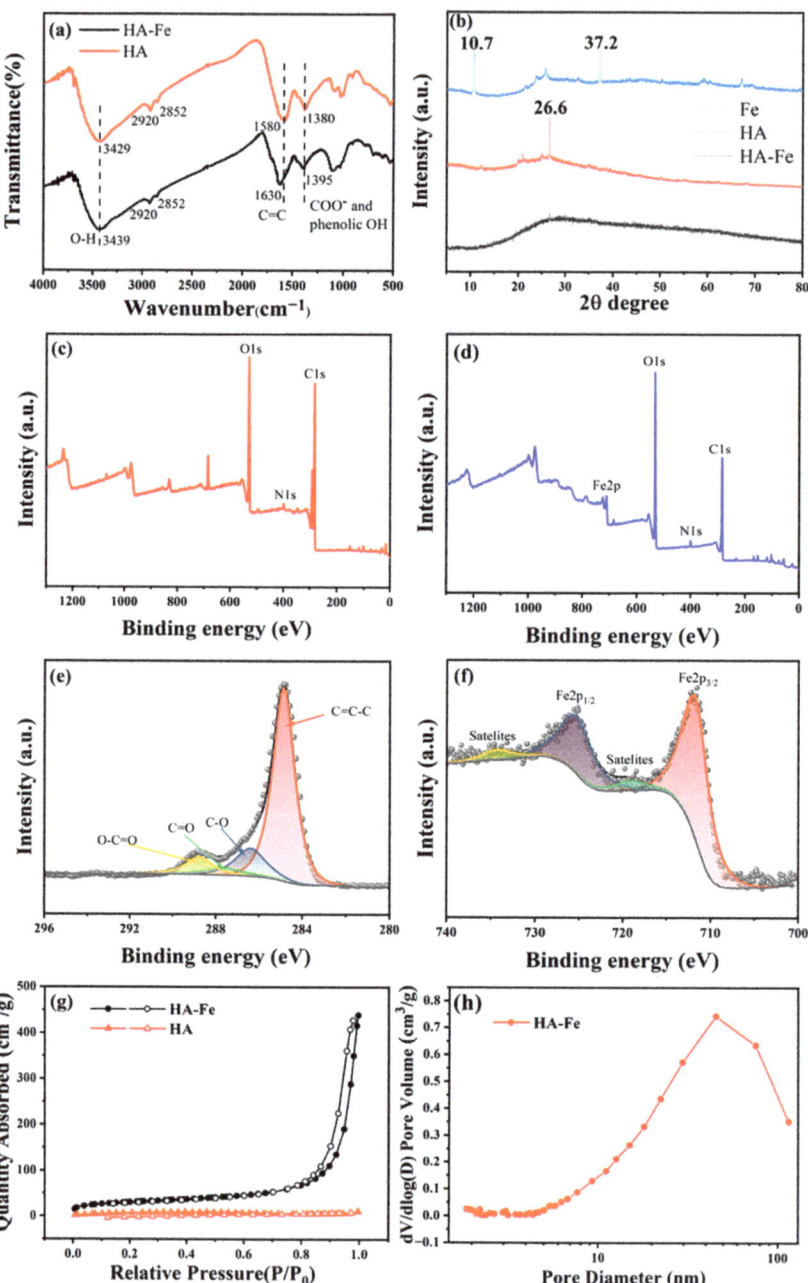

Figure 3. (**a**) FTIR spectra of Fe, HA, and HA-Fe. (**b**) X-ray diffraction pattern of Fe, HA, and HA-Fe. (**c**,**d**) X-ray photoelectron spectroscopy of HA and HA-Fe. (**e**,**f**) C1s and Fe2p XPS spectra of HA-Fe. (**g**) Nitrogen adsorption–desorption isotherm of HA-Fe and HA. (**h**) Pore size distribution of HA-Fe.

To further confirm the chemical compositions of HA and HA-Fe, the XPS measurements of them are shown in Figure 3c–f. As shown in Figure 3c,d, two typical peaks with binding energies (BE) of 285 eV and 531 eV are assigned to the C1s and O1s orbital, respectively, indicating that the main components are carbon and oxygen [45], and the peaks at the BE of 400 eV are easily indexed to the N1s orbital, indicating that both HA and HA-Fe contain N elements [55]. By carefully observing the XPS spectrum of HA-Fe (Figure 3d), it can be seen that the newly emerging peak at the binding energy of 712 eV is indexed to the Fe2p orbital [56], indicating that Fe^{3+} has been successfully chelated to the HA molecule. The C1s spectrum of HA-Fe (Figure 3e) is divided into four different types of carbon species: C=C-C (284.6 eV), C-O (286.4 eV), C=O (287.7 eV), and O-C=O (288.7 eV) groups, which is nearly the same as the XPS spectrum of HA in the literature [45], indicating that Fe ions do not change the chemical composition significantly. Figure 3f shows the XPS spectra of the Fe element in the prepared HA-Fe sample. In the Fe2p XPS spectrum of the HA-Fe material, the two peaks at 712 eV and 725.8 eV are attributed to $Fe2p_{1/2}$ and $Fe2p_{3/2}$, respectively. The weak peaks at 719.3 eV and 734.7 eV are attributed to the satellite peak [57], which can also prove the successful binding of Fe^{3+}.

The N_2 adsorption and desorption at 77 K were used to study the porous structure. As shown in Figure 3g, the isotherms are assigned to the V type. Under these circumstances, the adsorption process initially resembles that of macroporous solids, and the capillaries in the mesopores will be condensed at relatively high pressure, resulting in a sharp increase in adsorption capacity. The adsorption isotherm tends to be stable after these pores have been filled. The condensation of the capillaries and the evaporation of the capillaries usually do not take place at the same pressure, which will lead to the formation of a hysteresis loop [58]. The pore size distribution curve in Figure 3h shows that the pores with various diameters (mainly distributed from 10 nm to 100 nm) are present in HA-Fe. The specific surface areas for HA-Fe and HA are 105.42 $m^2 \cdot g^{-1}$ and 5.01 $m^2 \cdot g^{-1}$, respectively, which illustrates that the ferric ion increased the surface area of HA-Fe materials. The thermal stability of HA and HA-Fe is investigated by the TGA measurement. As shown in Figure S2, the TGA curve shows that HA-Fe has better thermal stability than HA.

3.2. Electrochemical Properties Investigation of HA-Fe Material

In order to illustrate the electrochemical performance of the HA-Fe material as an anode for LIBs, the change–discharge process, the cycle stability, rate capability, CV curves, and electrochemical impedance were investigated systemically.

First, the CV tests were performed to study the redox reaction during charge and discharge. As shown in Figure S3a,b, the CV curves of the HA and HA-Fe materials were measured at the scanning rate of 0.1 $mV \cdot s^{-1}$, and the potential window is from 5 mV to 3 V (vs. Li/Li$^+$). It can be seen from Figure S3a,b that the reduction current of the first charge–discharge cycle is much larger than that of the second and third cycle, which is because of the formation of SEI film [53]. As shown in Figure S3a, HA exhibited a reduction peak centered at 1.0V during the first discharge, indicating the accumulation process of lithium ions in HA materials. Meanwhile, a slightly corresponding anodic peak at 1.05 V (vs. Li/Li$^+$) was detected, indicating the de-intercalation of lithium ions. Similarly, Figure S3b shows a broad oxidation peak at about 1.1 V in the first cycle, which was related to the SEI film formation and the oxidation of Fe^0 to $Fe^{2+/3+}$ [59]. In the subsequent cycles, the reduction peaks and the oxidation peak at 1.1 V both decreased rapidly, which is caused by the end of SEI film formation. In order to compare the response currents of HA and HA-Fe during the same scan rate clearly, the CV plot of HA and HA-Fe at the third cycle is shown in Figure S3c. It can be seen that the response current of HA-Fe is much larger than that of HA, which illustrates that the chelated ferric ion can increase the capacity greatly. Figure S3d showed the CV curves of HA-Fe materials at 3rd and 800th, respectively. It can be seen that the response current and the area enclosed by the current curve increased obviously after 800 times change–discharge process, which means that the change–discharge process can activate the electrochemical process.

In order to better understand the electrochemical properties of HA-Fe materials, we tested the cycle performance of HA-Fe, HA, and ammonium iron (III) sulfate (marked as Fe). The change–discharge voltage range is 5 mV~3.0 V with a constant current of 100 mA·g^{-1}. Figure 4a shows the capacity of the HA, Fe, and HA-Fe at a current density of 0.1 A·g^{-1}. The initial discharge capacities of HA and Fe are about 90 mAh·g^{-1} and 385 mAh·g^{-1}, respectively, which are much smaller than that of HA-Fe (1038 mAh·g^{-1}). As a result, due to the increased impedance caused by the gradual formation of the SEI layer, the capacity of the three materials declines rapidly in the first few cycles. After that, the change–discharge capacities of HA and HA-Fe increased gradually to about 178 mAh·g^{-1} and 530 mAh·g^{-1} after 500 cycles, respectively. The capacity of the HA-Fe gradually leveled off after 500 cycles and stabilized at a value of 586 mAh·g^{-1} after 1000 cycles. The improvement in performance during the change–discharge cycle may be due to the activation process, e.g., the swelling process resulting from the increase in wettability of the polymer in the electrolyte, which allows for more active sites to participate in the battery's change–discharge cycle [60]. Moreover, it can be seen that the activation degree of HA-Fe is more significant than that of HA, which may be attributed to three reasons. Firstly, as shown in the SEM morphology of HA and HA-Fe, the HA-Fe particle is much smaller and looser than that of HA. The loose character is beneficial to the lithium shuttle in the solid HA-Fe material, which makes the reaction efficiency of the lithium ions and the functional groups much higher. Secondly, from the XRD results of HA and HA-Fe, the crystallinity degree of HA materials decreased obviously after chelating with the ferric ions. Thirdly, the chelation between ferric ions and HA can dramatically decrease the solubility in electrolytes. In comparison, the change–discharge capacity of Fe did not increase during the cycling process, which means that Fe did not occur in the activation process.

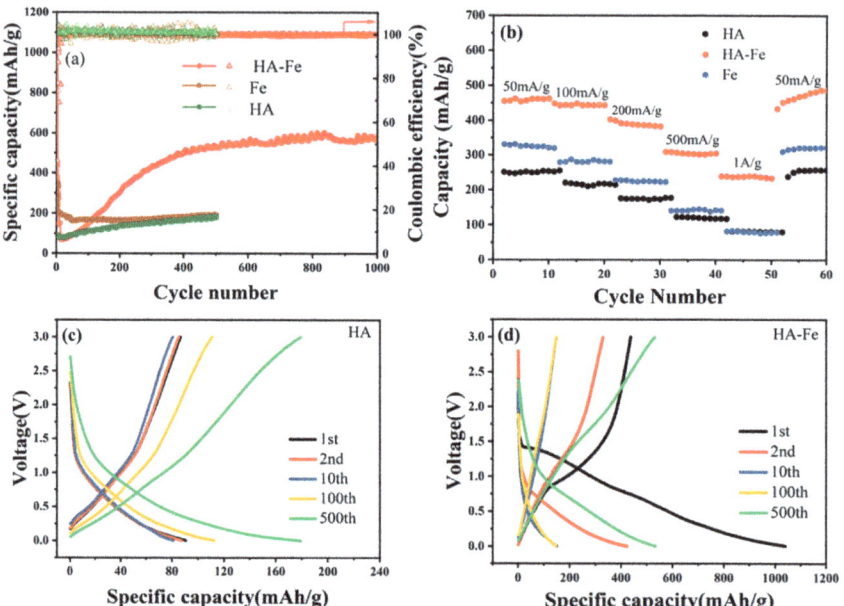

Figure 4. (**a**) Cycle performance diagram of Fe, HA, and HA-Fe at 100mA·g^{-1}. (**b**) Rate performance diagram of Fe, HA, and HA-Fe at different current densities. (**c**,**d**) GDC curves of HA and HA-Fe at a current density of 0.1 A·g^{-1}, respectively.

In order to research the rate ability of Fe, HA, and HA-Fe, the capabilities measurements were conducted at the current densities of 0.05 A·g^{-1}, 0.1 A·g^{-1}, 0.2 A·g^{-1}, 0.5 A·g^{-1}, and 1.0 A·g^{-1}, respectively, and the corresponding data are shown in Figure 4b. The HA-Fe material delivers stable charge/discharge capacities of 462 mAh·g^{-1}, 447 mAh·g^{-1}, 398 mAh·g^{-1}, 309 mAh·g^{-1}, and 252 mAh·g^{-1} at the current densities of 0.05 A·g^{-1},

0.1 A·g^{-1}, 0.2 A·g^{-1}, 0.5 A·g^{-1}, and 1.0 A·g^{-1}, respectively, while the HA material delivers stable charge/discharge capacities of 252 mAh·g^{-1}, 210 mAh·g^{-1}, 175 mAh·g^{-1}, 120 mAh·g^{-1}, and 82 mAh·g^{-1}. It can be seen that at different current densities, the capacities of HA-Fe were all much higher than those of HA and Fe. The capacity retention rates of HA-Fe, HA, and Fe at 1 A·g^{-1} are 55.63%, 32.55%, and 24.85%, respectively, and the results show that the performance of the rate is significantly improved after chelation with ferric ions. In addition, the HA-Fe anode shows a 100% recovery of capacity when the current densities are returned to 0.05 A·g^{-1}, which shows that the high current has not destroyed the molecular structure of the HA-Fe.

Figure 4c,d show the change–discharge curve of HA and HA-Fe, respectively. It can be seen that the main discharge potential of HA and HA-Fe is in the range of 1.2 V~5 mV, which shows that HA and HA-Fe materials are suitable for use as anode material. During the first change–discharge process of the HA-Fe material, additional irreversible charge plateaus (around 1.0 V) and discharge plateaus (around 1.5 V) exist, which may be due to the irreversible formation of the SEI film. The first discharge and charge capacities of HA-Fe material are 1038 mAh·g^{-1} and 437 mAh·g^{-1}, respectively, corresponding to a low coulombic efficiency of 42%. Similar to many other anode materials, the irreversible capacity was also caused by the contribution of SEI formation results in the relatively low coulombic efficiency in the first charge/discharge process. The capacities of both HA and HA-Fe materials all increased with the change–discharge process. However, the increased degree of HA is not as large as that of HA-Fe. In the tenth cycle, the discharge capacity of HA-Fe is 146.9 mAh·g^{-1}, while after 500 change–discharge cycles, the reversible capacity is 530 mAh·g^{-1}, which is much higher than that of HA after 500 cycles (178 mAh·g^{-1}).

To further explain the encouraging electrochemical properties of the synthesized HA-Fe materials, the electrochemical impedance spectroscopy (EIS) of HA and HA-Fe materials were conducted from 0.01 Hz to 100 kHz before cycling, after 100 cycles, and after 400 cycles, respectively. Nyquist plot analysis was performed using an equivalent circuit model, as shown in Figure S4a. The equivalent circuit model is made up of the following six electrical elements (R_s, R_f, R_{ct}, CPE_1, CPE_2, and Q). The R_s corresponded to the ohmic resistance of the electrode system, while R_f and CPE_1 corresponded to the resistance and capacitance attributed to the SEI film, respectively. CPE_2 corresponded to the double-layer electrical capacitance. R_{ct} corresponded to the change–transfer resistance, and Q corresponded to the diffusion resistance in the solid phase, respectively [61]. This equivalent circuit model produced fitted lines as shown in Figure S4b–f, and the electrical element values are given in Table 2. From Table 2, it can be seen that before the cycling process, the resistance of HA (1403.6 Ω) was much larger than that of HA-Fe (389.6 Ω), which means that the addition of ferric ions can decrease the change–transfer resistance. After 100 cycles and 400 cycles of change–discharge processes, the change–transfer resistance of HA decreased to 240.7 Ω and 162.0 Ω, respectively, while the change–transfer resistance of HA-Fe decreased to 21.3 Ω and 13.7 Ω, respectively. The decreased change–transfer resistance during the change–discharge cycles can explain the obvious activation process of the HA-Fe material during the change–discharge process in Figure 4a.

Table 2. Values of electrical elements in the equivalent circuit model.

Sample	Rs (Ω)			Rct (Ω)		
	Before Cycling	After 100 Cycles	After 400 Cycles	Before Cycling	After 100 Cycles	After 400 Cycles
HA	9.39	2.08	2.55	1403.6	240.7	162.0
HA-Fe	4.96	9.06	1.39	389.6	21.3	13.7

Table 3 shows the electrochemical properties comparison between some previously reported HA or biomass macromolecular composites and the material reported in this work. It can be seen that the capacity of HA-Fe composites can rival most of the composites reported in the literature. Considering the renewability, low cost, the simple preparation process, and environment friendliness, HA-Fe has great commercial application potential.

Table 3. Electrochemical properties of several HA or biomass macromolecular composites.

Sample	Current Density (A·g^{-1})	Capacity (mAh·g^{-1})	Cycles	Reference
HA-Fe	0.1	586	1000	This work
HA	0.04	180	200	[45]
H-CF	0.1	249	100	[62]
PTA-700	0.1	535	100	[63]
L-900	0.1C	433	100	[64]
PCS-CaCl2	0.2C	546	100	[65]

3.3. Lithium Storage Mechanism Investigation

Ex-situ XPS was performed to characterize the electrodes at different electrochemical states in the first cycle to investigate the redox mechanism of the HA-Fe material during the change–discharge process, as shown in Figure 5a. The Li1s spectra of the HA-Fe electrode in different electrochemical states are shown in Figure 5b. In the case of the original electrode (A point), almost no Li1s signal could be detected in the XPS spectra. During the discharging process (A→B→C), the intensity of the Li1s peak increases continuously, which indicates that the continuous lithiation process is taking place in the HA-Fe. When the electrode is recharged (C→D→E), a gradual reduction of Li peak was detected by ex-situ X-ray photoelectron spectroscopy, indicating that the Li ions were gradually eliminated from the HA-Fe electrode. Nevertheless, when the electrode was recharged to 3.0 V, the Li1s peak could still be detected, indicating that not all lithium ions released from the HA-Fe material during the charge. The formation of SEI in the first change–discharge cycle and the irreversible reaction between HA-Fe and Li$^+$ can explain this irreversible phenomenon.

As shown in Figure 5c–f, the peak intensity of the C1s peak during the change–discharge process was investigated in detail. The peaks at 284.05 eV, 284.5 eV, 286.31 eV, and 287.95 eV were assigned to the C=C, C-C, C-O, and C=O bonds, respectively. The intensity of the C=O peak dropped to its lowest value and the intensity of the C-O peak increased to its highest value when the voltage decreased from 0.8 V to 0.005 V, which illustrated that the C=O bond participated in the lithiation process and transferred to C-O bond. When recharged from 0.005 V to 1.5 V, the C-O peak decreased and the C=O peak increased, which illustrated the reversible redox transformation between the C=O double bond and the C-O single bond. However, as shown in Figure 5d,e, the intensities of C-C and C=C did not change obviously, which illustrated that the redox transfer from C-C to C=C nearly did not occur in the voltage range of 0.005 V–1.5 V. During the discharge process from 1.5 V to 3.0 V, the intensity of the C-C peak decreased and the C=C peak increased, which means that the C=C also participated in the lithiation process and transferred to the C-C bond.

Figure S5a–d showed the O1s characterization of the HA-Fe electrode at different states during the discharge–charge processes. As shown in Figure S5a,b, the C=O peak decreased during the discharge process from 0.8 V to 0.005 V, while the C-O peak increased accordingly, which is consistent with that from the C=O changing process in the C1s peak. In contrast, the C=O peak increased while the C-O peak decreased again during the charged process from 0.005 V to 1.5 V due to the reversible redox transformation between the C=O bond and the C-O bond (from Figure S5b,c). The obvious transformation process illustrated that the C=O is a very important active site for Li storage. In addition, it can be seen that the C=O bond nearly did not change from 1.5 V to 3.0 V, which illustrated that the C=O bond redox voltage range is 0.005 V~1.5 V but not 1.5 V~3.0 V. From the above discussion, it can be known that the voltage ranges of redox conversion between C-O and C=O are lower than the voltage ranges of redox conversion between C=C and C-C.

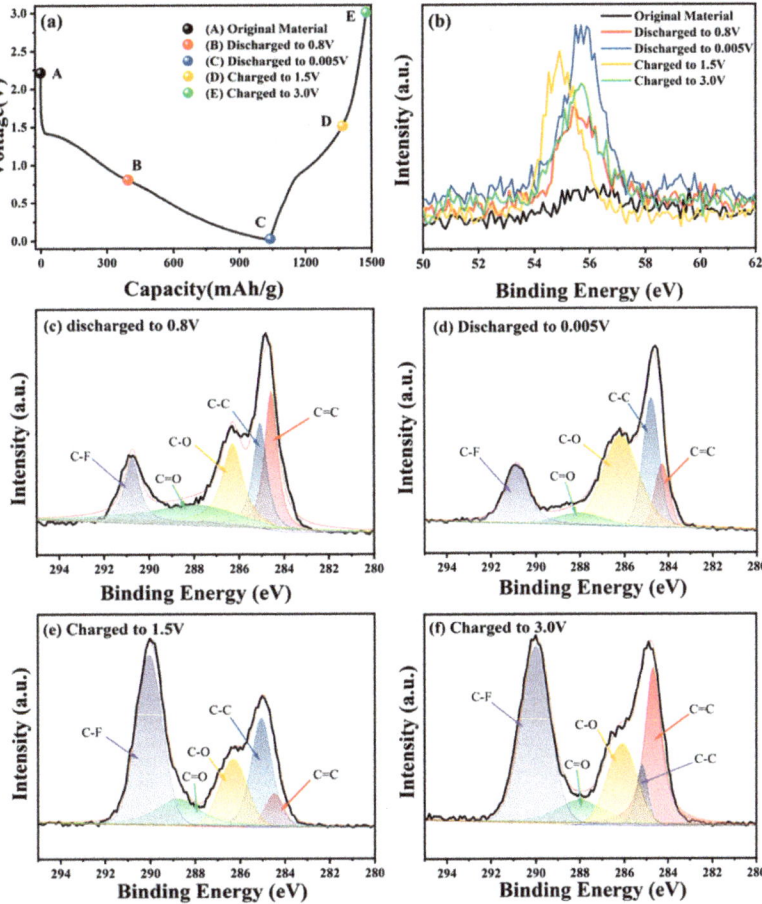

Figure 5. (a) The selected points for XPS measurements in the first discharge/charge cycle. (b) XPS profile of Li1s of the HA-Fe electrodes during the first discharge/charge cycle. (c,d) Ex-situ XPS spectra of C1s when discharged to 0.8 V, 0.005 V in the tenth discharge/charge cycle, respectively. (e,f) Ex-situ XPS spectra of C1s when recharged to 1.5 V, 3.0 V in the tenth discharge/charge cycle, respectively.

The total stored charge comes from two processes: a diffusion controlled faradaic process and a capacitive process, which mainly includes a pseudocapacitance process that refers to the redox reaction taking place at the surface locations [66–69]. In order to study the lithium storage mechanism, the faradaic and capacitive contribution of HA and HA-Fe materials were calculated by CV curves at different scanning rates (0.1 mV·s^{-1}, 0.3 mV·s^{-1}, 0.5 mV·s^{-1}, 0.7 mV·s^{-1}, 1.0 mV·s^{-1}), as shown in Figure 6a,b. The charge storage process can be characterized according to the equation: $log(i) = log(a) \times b\, log(v)$. Where i represents the peak current in CV curve, v denotes the corresponding scan rate, and a and b are constants [66,70]. When the b value is close to 0.5, it indicates that the electrochemical process is controlled by the internal solid-phase diffusion process. And when the b value is close to 1, it indicates that the electrochemical process is controlled by the capacitive process [71]. As shown in Figure 6c, the b values of HA and HA-Fe were 0.87 and 0.93, respectively, which indicates that HA-Fe has more capacitive contribution than HA, which can illustrate that the surface area of the HA-Fe materials is larger than that of HA material. Furthermore, for a charge storage process combined with diffusion-controlled faradaic process and capacitive process, its current response i at a given potential is the sum of the above two contributions, which can be presented as: $i = k_1 v + k_2 v^{0.5}$ or $\frac{i}{v^{0.5}} = k_1 v^{0.5} + k_2$. Where $k_1 v$ and $k_2 v^{0.5}$ correspond to the pseudocapacitive process and diffusion-controlled faradaic process, respectively [67,68,72]. Therefore, by obtaining k_1 and k_2 values at a constant potential, it is possible to determine the quantitative contributions of capacitive and diffusive processes at a given voltage. Figure 6d showed the total specific capacity with contributions from capacitive and diffusive processes of HA and HA-Fe at 0.1 mV·s^{-1}. The HA-Fe material has both higher capacitive and diffusion capacities (241 mAh·g^{-1} and 345 mAh·g^{-1}) than that of HA material (62 mAh·g^{-1} and 116 mAh·g^{-1}), which means that the diffusion-controlled faradaic contribution capability and surface redox sites all increased after chelating with Fe ions. Figure 6e,f showed the capacitive contribution at different scan rates of the HA and HA-Fe, respectively. It can be seen that the capacitive contribution of HA-Fe is higher than that of HA at different scan rates, which is mainly due to the fact that the high specific surface area can provide a large number of active sites for the rapid embedding and de-embedding of lithium ions, thus improving the capacitance of the electrode.

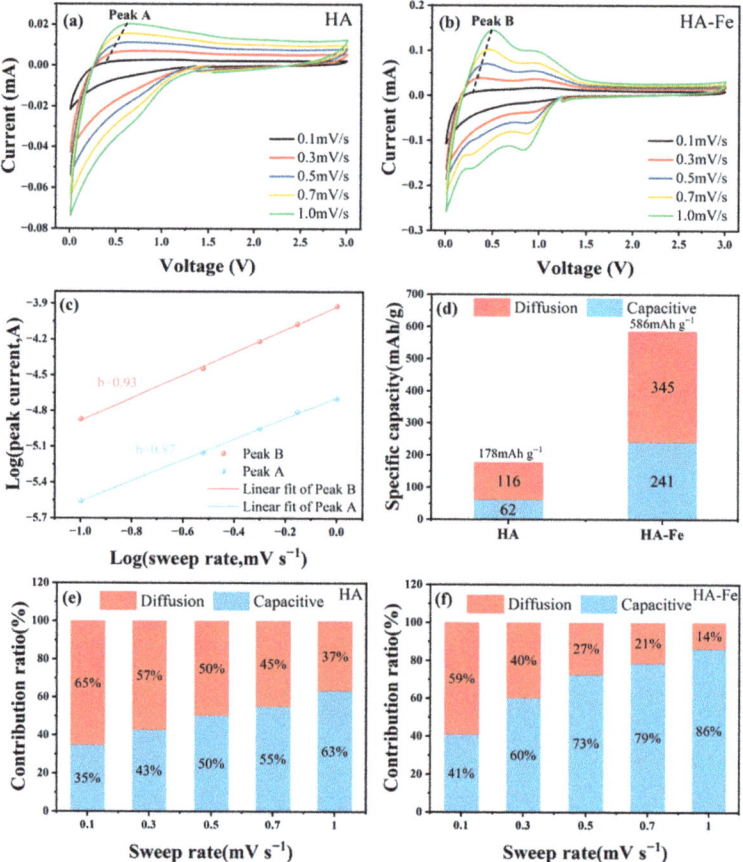

Figure 6. (**a**,**b**) CV curves at various scan rates ranging from 0.1 to 1.0 mV·s^{-1} of HA and HA-Fe. (**c**) Log(i) versus log (v) plots of HA and HA-Fe. (**d**) Total specific capacity with contributions from capacitive and diffusive processes of HA and HA-Fe at 0.1 mV·s^{-1}. (**e**,**f**) Capacitive contribution at different scan rates of HA and HA-Fe, respectively.

4. Conclusions

In summary, a novel material, named HA-Fe, by chelating ferric ion with humic acid was prepared in this work and used as anode materials of LIBs. The HA-Fe materials were synthesized by a simple way without any heating or emission of toxic substances. According to the SEM, TEM, XPS, XRD, and nitrogen adsorption–desorption experimental results, it was found that the ferric ions can chelate with humic acid successfully under the mild conditions, and the ferric ions can increase the surface area of materials and transform HA into amorphous structure. HA-Fe has excellent electrochemical properties, including long cycle stability and excellent rate performance. A high specific capacity of 586 mAh·g^{-1} was maintained at 0.1 A·g^{-1} after 1000 cycles. In addition, when the current density increased from 0.05 A·g^{-1} to 1 A·g^{-1}, the capacity of HA-Fe changed from 462 mAh·g^{-1} to 252 mAh·g^{-1} with the capacity retention rate of 55.63%, which is much higher than that of HA (32.55%) and Fe (24.85%). Moreover, the chelation between ferric ion and humic acid can improve the activation degree of HA-Fe material (improved about eight times) compared with pure HA material (improved about 2 times) during the change–discharge process. In the end, the intensified mechanism was also investigated with ex-situ XPS measurements. It was found that C=O and C=C bonds are activation sites for storing lithium ions, but with different redox voltages. In our opinion, the simple synthesis conditions and favorable electrochemical performance of this HA-Fe anodes make it a promising material for the development of truly powerful "green" LIBs.

Supplementary Materials: The following supporting information can be downloaded at: https://www.mdpi.com/article/10.3390/ma16196477/s1, Figure S1. (a–d) SEM images with different magnifications of HA and HA-Fe. (e–g) SEM image of HA-Fe and the corresponding elemental mapping images of O (purple) and Fe (yellow). Figure S2. TGA curves of HA and HA-Fe. Figure S3. (a,b) Cyclic voltammetry measurements of HA and HA-Fe during the first three cycles in the voltage range of 0.005 V~3.0 V at 0.1 mV·s^{-1}; (c) Cyclic voltammetry comparison plot of HA and HA-Fe at the 3rd cycle; (d) Cyclic voltammetry comparison plot of HA-Fe at the 3rd cycle and the 800th cycle. Figure S4. (a) The equivalent circuit model used to analyze the nyquist plots. (b,c) Electrochemical impedance plots of the HA-Fe, HA before cycling, after 100th cycling and after 400th cycling. (d–f) Electrochemical impedance comparison of HA and HA-Fe materials before cycling, after 100th cycling and after 400th cycling. Figure S5. (a) XPS spectra of O1s when discharged to 0.8 V, (b) O1s when discharged to 0.005 V, (c) O1s when charged to 1.5 V, (d) O1s when charged to 3.0 V.

Author Contributions: Conceptualization, H.Z. and Q.Y.; methodology, H.Z., S.J. and Y.W.; validation, M.K., M.G. and K.X.; formal analysis, M.K. and X.W.; investigation, H.Z., Y.W. and R.Z.; resources, Q.Y. and J.Z.; data curation, M.G., K.X. and G.T.; writing—original draft preparation, H.Z.; writing—review and editing, Q.Y. and J.Z.; visualization, G.T., X.W. and S.J.; supervision, J.Z.; project administration, Q.Y. and J.Z.; funding acquisition, Q.Y. All authors have read and agreed to the published version of the manuscript.

Funding: This work is supported by the National Natural Science Foundation of China (No. 21978126) and Shandong Province Science and Technology Small and Medium Enterprises Innovation Ability Enhancement Project (2022TSGC1370).

Institutional Review Board Statement: Not applicable.

Informed Consent Statement: Not applicable.

Data Availability Statement: The data are unavailable due to privacy.

Conflicts of Interest: The authors declare that they have no conflict of interest.

References

1. Peng, L.; Fang, Z.; Zhu, Y.; Yan, C.; Yu, G. Holey 2D nanomaterials for electrochemical energy storage. *Adv. Energy Mater.* **2018**, *8*, 1702179. [CrossRef]
2. Abdah, M.A.A.M.; Mokhtar, M.; Khoon, L.T.; Sopian, K.; Dzulkurnain, N.A.; Ahmad, A.; Sulaiman, Y.; Bella, F.; Su'ait, M.S. Synthesis and electrochemical characterizations of poly (3,4-ethylenedioxythiophene/manganese oxide coated on porous carbon nanofibers as a potential anode for lithium-ion batteries. *Energy Rep.* **2021**, *7*, 8677–8687. [CrossRef]
3. Kim, T.; Song, W.; Son, D.-Y.; Ono, L.K.; Qi, Y. Lithium-ion batteries: Outlook on present, future, and hybridized technologies. *J. Mater. Chem. A* **2019**, *7*, 2942–2964. [CrossRef]
4. Liang, F.; Wu, D.; Jiang, L.; Zhang, Z.; Zhang, W.; Rui, Y.; Tang, B.; Liu, F. Layered niobium oxide hydrate anode with excellent performance for lithium-ion batteries. *ACS Appl. Mater. Interfaces* **2021**, *13*, 51057–51065. [CrossRef] [PubMed]
5. Fei, H.; Liu, X.; Li, Z.; Feng, W. Synthesis of manganese coordination polymer microspheres for lithium-ion batteries with good cycling performance. *Electrochim. Acta* **2015**, *174*, 1088–1095. [CrossRef]
6. Dai, Y.; Mo, D.-C.; Qu, Z.-T.; Wang, W.-K.; Lyu, S.-S. Organic—Inorganic Hybrid Interfaces Enable the Preparation of Nitrogen-Doped Hollow Carbon Nanospheres as High-Performance Anodes for Lithium and Potassium-Ion Batteries. *Materials* **2023**, *16*, 4936. [CrossRef]
7. Armand, M.; Tarascon, J.-M. Building better batteries. *Nature* **2008**, *451*, 652–657. [CrossRef]
8. Lin, F.; Markus, I.M.; Nordlund, D.; Weng, T.-C.; Asta, M.D.; Xin, H.L.; Doeff, M.M. Surface reconstruction and chemical evolution of stoichiometric layered cathode materials for lithium-ion batteries. *Nat. Commun.* **2014**, *5*, 3529. [CrossRef]
9. Gauthier, M.; Carney, T.J.; Grimaud, A.; Giordano, L.; Pour, N.; Chang, H.-H.; Fenning, D.P.; Lux, S.F.; Paschos, O.; Bauer, C. Electrode-electrolyte interface in Li-ion batteries: Current understanding and new insights. *J. Phys. Chem. Lett.* **2015**, *6*, 4653–4672. [CrossRef]
10. Liu, W.; Zong, K.; Li, Y.; Deng, Y.; Hussain, A.; Cai, X. Nano-Graphite Prepared by Rapid Pulverization as Anode for Lithium-Ion Batteries. *Materials* **2022**, *15*, 5148. [CrossRef]
11. Lee, S.W.; Gallant, B.M.; Byon, H.R.; Hammond, P.T.; Shao-Horn, Y. Nanostructured carbon-based electrodes: Bridging the gap between thin-film lithium-ion batteries and electrochemical capacitors. *Energy Environ. Sci.* **2011**, *4*, 1972–1985. [CrossRef]
12. Wang, D.; Yu, Y.; He, H.; Wang, J.; Zhou, W.; Abruna, H.D. Template-free synthesis of hollow-structured Co_3O_4 nanoparticles as high-performance anodes for lithium-ion batteries. *ACS Nano* **2015**, *9*, 1775–1781. [CrossRef] [PubMed]
13. Bhosale, M.E.; Chae, S.; Kim, J.M.; Choi, J.-Y. Organic small molecules and polymers as an electrode material for rechargeable lithium ion batteries. *J. Mater. Chem. A* **2018**, *6*, 19885–19911. [CrossRef]
14. Zhuang, J.; Xu, X.; Peleckis, G.; Hao, W.; Dou, S.X.; Du, Y. Silicene: A promising anode for lithium-ion batteries. *Adv. Mater.* **2017**, *29*, 1606716. [CrossRef]
15. Chae, S.; Kim, N.; Ma, J.; Cho, J.; Ko, M. One-to-one comparison of graphite-blended negative electrodes using silicon nanolayer-embedded graphite versus commercial benchmarking materials for high-energy lithium-ion batteries. *Adv. Energy Mater.* **2017**, *7*, 1700071. [CrossRef]
16. Min, X.; Xu, G.; Xie, B.; Guan, P.; Sun, M.; Cui, G. Challenges of prelithiation strategies for next generation high energy lithium-ion batteries. *Energy Storage Mater.* **2022**, *47*, 297–318. [CrossRef]
17. Nzabahimana, J.; Chang, P.; Hu, X. Porous carbon-coated ball-milled silicon as high-performance anodes for lithium-ion batteries. *J. Mater. Sci.* **2019**, *54*, 4798–4810. [CrossRef]
18. Croguennec, L.; Palacin, M.R. Recent achievements on inorganic electrode materials for lithium-ion batteries. *J. Am. Chem. Soc.* **2015**, *137*, 3140–3156. [CrossRef]
19. Li, W.; Song, B.; Manthiram, A. High-voltage positive electrode materials for lithium-ion batteries. *Chem. Soc. Rev.* **2017**, *46*, 3006–3059. [CrossRef]
20. Lupi, C.; Pasquali, M.; Dell'Era, A. Nickel and cobalt recycling from lithium-ion batteries by electrochemical processes. *Waste Manag.* **2005**, *25*, 215–220. [CrossRef]
21. Yu, J.; Tang, W.; Hu, Y.; Gao, J.; Wang, M.; Liu, S.; Lai, H.; Xu, L.; Fan, C. Novel low-cost, high-energy-density (>700 Wh kg^{-1}) Li-rich organic cathodes for Li-ion batteries. *Chem. Eng. J.* **2021**, *415*, 128509. [CrossRef]
22. Yang, Y.; Yuan, J.; Huang, S.; Chen, Z.; Lu, Y.; Yang, C.; Zhai, G.; Zhu, J.; Zhuang, X. Porphyrinic conjugated microporous polymer anode for Li-ion batteries. *J. Power Sources* **2022**, *531*, 231340. [CrossRef]
23. Yu, J.; Li, N.; Wang, H.-G.; Gao, B.; Wang, B.; Li, Z. Unraveling the superior anodic lithium storage behavior in the redox-active porphyrinic triazine frameworks. *Chem. Eng. J.* **2023**, *463*, 142434. [CrossRef]
24. Xu, Z.; Yang, J.; Hou, S.; Lin, H.; Chen, S.; Wang, Q.; Wei, H.; Zhou, J.; Zhuo, S. Thiophene-diketopyrrolopyrrole-based polymer derivatives/reduced graphene oxide composite materials as organic anode materials for lithium-ion batteries. *Chem. Eng. J.* **2022**, *438*, 135540. [CrossRef]
25. Araujo, R.B.; Banerjee, A.; Panigrahi, P.; Yang, L.; Strømme, M.; Sjödin, M.; Araujo, C.M.; Ahuja, R. Designing strategies to tune reduction potential of organic molecules for sustainable high capacity battery application. *J. Mater. Chem. A* **2017**, *5*, 4430–4454. [CrossRef]
26. Yu, J.; Chen, X.; Wang, H.-G.; Gao, B.; Han, D.; Si, Z. Conjugated ladder-type polymers with multielectron reactions as high-capacity organic anode materials for lithium-ion batteries. *Sci. China Mater.* **2022**, *65*, 2354–2362. [CrossRef]

27. Sang, P.; Chen, Q.; Wang, D.-Y.; Guo, W.; Fu, Y. Organosulfur materials for rechargeable batteries: Structure, mechanism, and application. *Chem. Rev.* **2023**, *123*, 1262–1326. [CrossRef]
28. Chen, Q.; Li, L.; Wang, W.; Li, X.; Guo, W.; Fu, Y. Thiuram Monosulfide with Ultrahigh Redox Activity Triggered by Electrochemical Oxidation. *J. Am. Chem. Soc.* **2022**, *144*, 18918–18926. [CrossRef]
29. Wang, D.-Y.; Guo, W.; Fu, Y. Organosulfides: An emerging class of cathode materials for rechargeable lithium batteries. *Acc. Chem. Res.* **2019**, *52*, 2290–2300. [CrossRef]
30. Wu, M.; Cui, Y.; Bhargav, A.; Losovyj, Y.; Siegel, A.; Agarwal, M.; Ma, Y.; Fu, Y. Organotrisulfide: A high capacity cathode material for rechargeable lithium batteries. *Angew. Chem.* **2016**, *128*, 10181–10185. [CrossRef]
31. Du, W.; Du, X.; Ma, M.; Huang, S.; Sun, X.; Xiong, L. Polymer electrode materials for lithium-ion batteries. *Adv. Funct. Mater.* **2022**, *32*, 2110871. [CrossRef]
32. Choi, W.; Ohtani, S.; Oyaizu, K.; Nishide, H.; Geckeler, K.E. Radical polymer-wrapped SWNTs at a molecular level: High-rate redox mediation through a percolation network for a transparent charge-storage material. *Adv. Mater.* **2011**, *23*, 4440–4443. [CrossRef] [PubMed]
33. Guo, W.; Yin, Y.-X.; Xin, S.; Guo, Y.-G.; Wan, L.-J. Superior radical polymer cathode material with a two-electron process redox reaction promoted by graphene. *Energy Environ. Sci.* **2012**, *5*, 5221–5225. [CrossRef]
34. Ou, Y.; Zhang, Y.; Xiong, Y.; Hu, Z.; Dong, L. Three-dimensional porous radical polymer/reduced graphene oxide composite with two-electron redox reactions as high-performance cathode for lithium-ion batteries. *Eur. Polym. J.* **2021**, *143*, 110191. [CrossRef]
35. Mao, P.; Fan, H.; Zhou, G.; Arandiyan, H.; Liu, C.; Lan, G.; Wang, Y.; Zheng, R.; Wang, Z.; Bhargava, S.K. Graphite-like structured conductive polymer anodes for high-capacity lithium storage with optimized voltage platform. *J. Colloid Interface Sci.* **2023**, *634*, 63–73. [CrossRef]
36. Zhang, B.; Dong, Y.; Han, J.; Zhen, Y.; Hu, C.; Liu, D. Physicochemical Dual Crosslinking Conductive Polymeric Networks Combining High Strength and High Toughness Enable Stable Operation of Silicon Microparticles Anodes. *Adv. Mater.* **2023**, *35*, 2301320. [CrossRef] [PubMed]
37. Wang, C.; Dong, H.; Jiang, L.; Hu, W. Organic semiconductor crystals. *Chem. Soc. Rev.* **2018**, *47*, 422–500. [CrossRef]
38. Acker, P.; Rzesny, L.; Marchiori, C.F.; Araujo, C.M.; Esser, B. π-conjugation enables ultra-high rate capabilities and cycling stabilities in phenothiazine copolymers as cathode-active battery materials. *Adv. Funct. Mater.* **2019**, *29*, 1906436. [CrossRef]
39. Li, S.; Lin, J.; Zhang, Y.; Zhang, S.; Jiang, T.; Hu, J.; Liu, J.; Wu, D.Y.; Zhang, L.; Tian, Z. Eight-electron redox cyclohexanehexone anode for high-rate high-capacity lithium storage. *Adv. Energy Mater.* **2022**, *12*, 2201347. [CrossRef]
40. Ba, Z.; Wang, Z.; Zhou, Y.; Li, H.; Dong, J.; Zhang, Q.; Zhao, X. Electrochemical Properties of a Multicarbonyl Polyimide Superstructure as a Hierarchically Porous Organic Anode for Lithium-Ion Batteries. *ACS Appl. Energy Mater.* **2021**, *4*, 13161–13171. [CrossRef]
41. Wang, J.; Shen, Z.; Yi, M.; Zhang, X. Muconic acid as high-performance organic anode for lithium ion batteries. *J. Alloys Compd.* **2021**, *865*, 158573. [CrossRef]
42. Li, K.; Yu, J.; Si, Z.; Gao, B.; Wang, H.-G.; Wang, Y. One-dimensional π-d conjugated coordination polymer with double redox-active centers for all-organic symmetric lithium-ion batteries. *Chem. Eng. J.* **2022**, *450*, 138052. [CrossRef]
43. Wu, Y.; Zhang, Y.; Chen, Y.; Tang, H.; Tang, M.; Xu, S.; Fan, K.; Zhang, C.; Ma, J.; Wang, C. Heterochelation boosts sodium storage in π-d conjugated coordination polymers. *Energy Environ. Sci.* **2021**, *14*, 6514–6525. [CrossRef]
44. Yang, H.; Cui, W.; Han, Y.; Wang, B. Porous nanocomposite derived from Zn, Ni-bimetallic metal-organic framework as an anode material for lithium-ion batteries. *Chin. Chem. Lett.* **2018**, *29*, 842–844. [CrossRef]
45. Zhu, H.; Yin, J.; Zhao, X.; Wang, C.; Yang, X. Humic acid as promising organic anodes for lithium/sodium ion batteries. *Chem. Commun.* **2015**, *51*, 14708–14711. [CrossRef] [PubMed]
46. Li, J.; Luo, M.; Ba, Z.; Wang, Z.; Chen, L.; Li, Y.; Li, M.; Li, H.-B.; Dong, J.; Zhao, X. Hierarchical multicarbonyl polyimide architectures as promising anode active materials for high-performance lithium/sodium ion batteries. *J. Mater. Chem. A* **2019**, *7*, 19112–19119. [CrossRef]
47. Wang, Y.; Liu, Z.; Liu, H.; Liu, H.; Li, B.; Guan, S. A Novel High-Capacity Anode Material Derived from Aromatic Imides for Lithium-Ion Batteries. *Small* **2018**, *14*, 1704094. [CrossRef]
48. Lu, Y.; Zhang, Q.; Li, L.; Niu, Z.; Chen, J. Design strategies toward enhancing the performance of organic electrode materials in metal-ion batteries. *Chem* **2018**, *4*, 2786–2813. [CrossRef]
49. Ma, C.; Wang, Z.; Zhao, Y.; Li, Y.; Shi, J. A novel raspberry-like yolk-shell structured Si/C micro/nano-spheres as high-performance anode materials for lithium-ion batteries. *J. Alloys Compd.* **2020**, *844*, 156201. [CrossRef]
50. Ravikumar, M.M.; Shetty, V.R.; Suresh, G.S. Synthesis and Applications of Aurin Tricarboxylic Acid-Copper Metal Organic Framework for Rechargeable Lithium-Ion Batteries. *J. Electrochem. Soc.* **2020**, *167*, 100533. [CrossRef]
51. Zhang, G.; Wang, H.; Deng, X.; Yang, Y.; Zhang, T.; Zeng, H.; Wang, C.; Deng, Y. Metal chelation based supramolecular self-assembly enables a high-performance organic anode for lithium ion batteries. *Chem. Eng. J.* **2021**, *413*, 127525. [CrossRef]
52. Wang, J.; Yao, H.; Du, C.; Guan, S. Polyimide schiff base as a high-performance anode material for lithium-ion batteries. *J. Power Sources* **2021**, *482*, 228931. [CrossRef]
53. Xie, L.; Shang, C. Role of humic acid and quinone model compounds in bromate reduction by zerovalent iron. *Environ. Sci. Technol.* **2005**, *39*, 1092–1100. [CrossRef] [PubMed]

54. Zhang, G.; Yang, Y.; Zhang, T.; Xu, D.; Lei, Z.; Wang, C.; Liu, G.; Deng, Y. Fe^{III} chelated organic anode with ultrahigh rate performance and ultra-long cycling stability for lithium-ion batteries. *Energy Storage Mater.* **2020**, *24*, 432–438. [CrossRef]
55. Wang, X.; Lv, L.; Cheng, Z.; Gao, J.; Dong, L.; Hu, C.; Qu, L. High-Density Monolith of N-Doped Holey Graphene for Ultrahigh Volumetric Capacity of Li-Ion Batteries. *Adv. Energy Mater.* **2016**, *6*, 1502100. [CrossRef]
56. Zhang, C.; Chen, Z.; Wang, H.; Nie, Y.; Yan, J. Porous Fe_2O_3 nanoparticles as lithium-ion battery anode materials. *ACS Appl. Nano Mater.* **2021**, *4*, 8744–8752. [CrossRef]
57. Yao, J.; Jin, T.; Li, Y.; Xiao, S.; Huang, B.; Jiang, J. Electrochemical performance of $Fe_2(SO_4)_3$ as a novel anode material for lithium-ion batteries. *J. Alloys Compd.* **2021**, *886*, 161238. [CrossRef]
58. Kruk, M.; Jaroniec, M. Gas adsorption characterization of ordered organic-inorganic nanocomposite materials. *Chem. Mater.* **2001**, *13*, 3169–3183. [CrossRef]
59. Wu, S.; Lu, M.; Tian, X.; Jiang, C. A facile route to graphene-covered and carbon-encapsulated $CoSO_4$ nanoparticles as anode materials for lithium-ion batteries. *Chem. Eng. J.* **2017**, *313*, 610–618. [CrossRef]
60. Yuan, Q.; Li, C.; Guo, X.; Zhao, J.; Zhang, Y.; Wang, B.; Dong, Y.; Liu, L. Electrochemical performance and storage mechanism study of conjugate donor-acceptor organic polymers as anode materials of lithium-ion battery. *Energy Rep.* **2020**, *6*, 2094–2105. [CrossRef]
61. Yang, H.; Zhang, W.; Yuan, Q.; Zhao, J.; Li, Y.; Xie, Y. The fabrication of hierarchical porous nano-SnO_2@carbon@humic acid ternary composite for enhanced capacity and stability as anode material for lithium ion battery. *Colloids Surf. A* **2022**, *650*, 129560. [CrossRef]
62. Zhao, P.-Y.; Yu, B.-J.; Sun, S.; Guo, Y.; Chang, Z.-Z.; Li, Q.; Wang, C.-Y. High-performance anode of sodium ion battery from polyacrylonitrile/humic acid composite electrospun carbon fibers. *Electrochim. Acta* **2017**, *232*, 348–356. [CrossRef]
63. Huang, G.; Kong, Q.; Yao, W.; Wang, Q. Poly tannic acid carbon rods as anode materials for high performance lithium and sodium ion batteries. *J. Colloid Interface Sci.* **2023**, *629*, 832–845. [CrossRef] [PubMed]
64. Wu, Z.; Li, Z.; Chou, S.; Liang, X. Novel Biomass-derived Hollow Carbons as Anode Materials for Lithium-ion Batteries. *Chem. Res. Chin. Univ.* **2023**, *39*, 283–289. [CrossRef]
65. Yu, K.; Wang, J.; Wang, X.; Liang, J.; Liang, C. Sustainable application of biomass by-products: Corn straw-derived porous carbon nanospheres using as anode materials for lithium ion batteries. *Mater. Chem. Phys.* **2020**, *243*, 122644. [CrossRef]
66. Zhang, C.; He, Y.; Mu, P.; Wang, X.; He, Q.; Chen, Y.; Zeng, J.; Wang, F.; Xu, Y.; Jiang, J.X. Toward high performance thiophene-containing conjugated microporous polymer anodes for lithium-ion batteries through structure design. *Adv. Funct. Mater.* **2018**, *28*, 1705432. [CrossRef]
67. Wang, J.; Polleux, J.; Lim, J.; Dunn, B. Pseudocapacitive contributions to electrochemical energy storage in TiO_2 (anatase) nanoparticles. *J. Phys. Chem. C* **2007**, *111*, 14925–14931. [CrossRef]
68. Sathiya, M.; Prakash, A.; Ramesha, K.; Tarascon, J.M.; Shukla, A.K. V_2O_5-anchored carbon nanotubes for enhanced electrochemical energy storage. *J. Am. Chem. Soc.* **2011**, *133*, 16291–16299. [CrossRef]
69. Brezesinski, T.; Wang, J.; Tolbert, S.H.; Dunn, B. Ordered mesoporous α-MoO_3 with iso-oriented nanocrystalline walls for thin-film pseudocapacitors. *Nat. Mater.* **2010**, *9*, 146–151. [CrossRef]
70. Guo, Y.; Yuan, Q.; Li, C.; Du, H.; Zhao, J.; Liu, L.; Li, Y.; Xie, Y.; Vaidya, V. The synthesis of alternating donor–acceptor polymers based on pyrene-4, 5, 9, 10-tetraone and thiophene derivatives, their composites with carbon, and their lithium storage performances as anode materials. *RSC Adv.* **2021**, *11*, 15044–15053. [CrossRef]
71. Lian, L.; Li, K.; Ren, L.; Han, D.; Lv, X.; Wang, H.-G. Imine-linked triazine-based conjugated microporous polymers/carbon nanotube composites as organic anode materials for lithium-ion batteries. *Colloids Surf. A* **2023**, *657*, 130496. [CrossRef]
72. Li, C.; Kong, L.; Zhao, J.; Liang, B. Preparation of DAD conjugated polymers based on [1,2,5] thiadiazolo[3,4-c]pyridine and thiophene derivatives and their electrochemical properties as anode materials for lithium-ion batteries. *Colloids Surf. A* **2022**, *651*, 129707. [CrossRef]

Disclaimer/Publisher's Note: The statements, opinions and data contained in all publications are solely those of the individual author(s) and contributor(s) and not of MDPI and/or the editor(s). MDPI and/or the editor(s) disclaim responsibility for any injury to people or property resulting from any ideas, methods, instructions or products referred to in the content.

Article

Three-Dimensional Flower-like MoS₂ Nanosheets Grown on Graphite as High-Performance Anode Materials for Fast-Charging Lithium-Ion Batteries

Yeong A. Lee [1,2,†], Kyu Yeon Jang [1,3,†], Jaeseop Yoo [2,†], Kanghoon Yim [4], Wonzee Jung [4,5], Kyu-Nam Jung [1], Chung-Yul Yoo [6], Younghyun Cho [7], Jinhong Lee [1], Myung Hyun Ryu [1], Hyeyoung Shin [2,*], Kyubock Lee [2,*] and Hana Yoon [1,*]

[1] Korea Institute of Energy Research (KIER), Daejeon 34129, Republic of Korea; yeonga1902@kier.re.kr (Y.A.L.); kyuyeonjang@gmail.com (K.Y.J.); mitamire@kier.re.kr (K.-N.J.); jinhong02@kier.re.kr (J.L.); nicengood@kier.re.kr (M.H.R.)
[2] Graduate School of Energy Science and Technology (GEST), Chungnam National University, Daejeon 34134, Republic of Korea; yjs567@naver.com
[3] Department of Advanced Energy Technologies and System Engineering, Korea University of Science and Technology (UST), Daejeon 34113, Republic of Korea
[4] Computational Science and Engineering Laboratory, Korea Institute of Energy Research (KIER), Daejeon 34129, Republic of Korea; khyim@kier.re.kr (K.Y.); kyjung1020@kier.re.kr (W.J.)
[5] Department of Physics, Chungnam National University, Daejeon 34134, Republic of Korea
[6] Department of Chemistry, Mokpo National University, Muan-gun 58554, Republic of Korea; chungyulyoo@mokpo.ac.kr
[7] Department of Energy Systems, Soonchunhyang University, Asan 31538, Republic of Korea; yhcho@sch.ac.kr
* Correspondence: shinhy@cnu.ac.kr (H.S.); kyubock.lee@cnu.ac.kr (K.L.); hanayoon@kier.re.kr (H.Y.)
† These authors contributed equally to this work.

Abstract: The demand for fast-charging lithium-ion batteries (LIBs) with long cycle life is growing rapidly due to the increasing use of electric vehicles (EVs) and energy storage systems (ESSs). Meeting this demand requires the development of advanced anode materials with improved rate capabilities and cycling stability. Graphite is a widely used anode material for LIBs due to its stable cycling performance and high reversibility. However, the sluggish kinetics and lithium plating on the graphite anode during high-rate charging conditions hinder the development of fast-charging LIBs. In this work, we report on a facile hydrothermal method to achieve three-dimensional (3D) flower-like MoS₂ nanosheets grown on the surface of graphite as anode materials with high capacity and high power for LIBs. The composite of artificial graphite decorated with varying amounts of MoS₂ nanosheets, denoted as MoS₂@AG composites, deliver excellent rate performance and cycling stability. The 20−MoS₂@AG composite exhibits high reversible cycle stability (~463 mAh g^{-1} at 200 mA g^{-1} after 100 cycles), excellent rate capability, and a stable cycle life at the high current density of 1200 mA g^{-1} over 300 cycles. We demonstrate that the MoS₂-nanosheets-decorated graphite composites synthesized via a simple method have significant potential for the development of fast-charging LIBs with improved rate capabilities and interfacial kinetics.

Keywords: graphite; molybdenum disulfide; fast charging; high rate capability; hydrothermal synthesis; lithium-ion battery; anode materials

1. Introduction

Among various energy storage technologies, lithium-ion batteries (LIBs) have been widely investigated as power sources for portable electronics and electric vehicles (EVs) due to their high energy density and long lifespan [1–4]. The rapid expansion of the global EV and energy storage systems (ESSs) market has led to significant demand for fast-charging battery technology that can support the high power and long cycle life requirements of these applications. Graphite, a commercial anode material used in LIBs, is still considered the most promising material because of its excellent cycle reversibility, stable cycle life, and superior electronic conductivity. Despite these advantages, due to its low theoretical capacity (~372 mAh g^{-1}) and limited rate capability under fast charging conditions, graphite cannot meet the growing performance requirements of LIBs. The slow kinetics of lithium intercalation at the graphite–electrolyte interface during rapid charging conditions can lead to an undesirable anode voltage drop below 0 V vs. Li/Li$^+$, resulting in the formation of lithium plating on the graphite surface. This phenomenon can cause capacity decay, dendrite growth, short circuits, and even serious safety issues [5–8]. To overcome this challenge, various approaches have been proposed to develop high-power lithium-ion battery (LIB) anodes with improved interfacial kinetics [9–15]. Among these approaches, surface modification using functional materials has emerged as an effective strategy to enhance the rate capability and cycle stability of graphite anodes [7,8,12–15].

Transition metal sulfides (TMDs) have been extensively investigated in various energy storage applications, such as batteries, supercapacitors, and electrocatalysts [16–18]. Among various TMDs, molybdenum disulfide (MoS$_2$), a typical two-dimensional (2D) layered material, has emerged as a promising anode material for LIBs due to its high theoretical capacity (~670 mAh g^{-1}), which is much higher than that of graphite [19]. Furthermore, the interlayer spacing of MoS$_2$ (~0.62 nm) is larger than that of graphite (0.34 nm), providing ample space for fast Li$^+$ diffusion paths [20,21]. However, MoS$_2$ suffers from rapid capacity decay and inferior rate capability due to its low intrinsic electronic conductivity, large volume variation during cycling, and the production of polysulfide dissolution during the charge and discharge process [22–26].

Herein, we report a facile and efficient approach to synthesizing a 3D hierarchical MoS$_2$/artificial graphite (MoS$_2$@AG) composite for use as an anode material in LIBs. The composite was prepared using a hydrothermal reaction to decorate 3D hierarchically aligned flower-like MoS$_2$ nanosheets directly on the surface of graphite. The resulting composite exhibits improved specific capacity and rate capability compared to commercial graphite anodes. The graphite matrix provides a conductive pathway for fast electron transfer through the electrode, while the 3D hierarchically aligned MoS$_2$ nanosheets form a stable interface with the graphite, effectively preventing structural degradation and providing excellent electron transport. Moreover, the 3D hierarchical morphology of the MoS$_2$ nanosheets enhances the electrode/electrolyte contact area, facilitating charge transfer kinetics. As a result, the MoS$_2$-decorated graphite composite demonstrates exceptional rate capability, achieving charging times of less than 20 min for approximately 84% of its capacity. The composite also exhibits superior cycle stability under fast charging conditions compared to commercial graphite anodes. These results highlight the potential of the MoS$_2$/graphite composite as a promising candidate for high-energy and high-power LIBs.

2. Materials and Methods

2.1. Preparation of MoS$_2$- and MoS$_2$-Decorated Graphite Composites

The MoS$_2$-decorated graphite composite was synthesized via a facile hydrothermal reaction. To prepare the precursor solution, hexaammonium molybdate tetrahydrate ((NH$_4$)$_6$Mo$_7$O$_{24}$·4H$_2$O) and thiourea (NH$_2$CSNH$_2$), both purchased from Sigma Aldrich, were dissolved in a molar ratio of 1:14 in 50 mL of water dispersion solution containing artificial graphite (Hitachi Chemical Co., Japan) at a concentration of 0.0342 g mL^{-1}. The mixture was vigorously stirred for 1 h. Subsequently, the mixture was transferred into a Teflon-lined stainless autoclave and reacted at 220 °C for 16 h. After cooling naturally, the black precipitates were collected by centrifugation, washed with distilled water and ethanol, and dried at 80 °C for 24 h. The amount of thiourea and (NH$_4$)$_6$Mo$_7$O$_{24}$·4H$_2$O added to the graphite solution varied to achieve mass ratios of 2.5:100, 5:100, 10:100, 20:100, or 30:100 and the corresponding products were denoted as 2.5−MoS$_2$@AG, 5−MoS$_2$@AG, 10−MoS$_2$@AG, 20−MoS$_2$@AG, or 30−MoS$_2$@AG. The preparation process for the MoS$_2$-decorated graphite composite is illustrated in Figure 1a. The synthesis procedure for a pristine flower-like MoS$_2$ was similar to that of MoS$_2$-decorated graphite composite but without the presence of graphite.

Figure 1. (**a**) Schematic illustration and microstructural analysis of one-step hydrothermal synthesis of the 3D hierarchically aligned flower-like MoS$_2$ nanosheets grown directly on the surface of graphite. The MoS$_2$@AG composite was formed through spontaneous nucleation and growth of few-layer MoS$_2$ nanosheets on the surface of graphite. Representative images of the composite obtained by (**b**) SEM, (**c**) TEM, and (**d**) high-resolution TEM (HR-TEM), revealing the morphology and structure of the composite at different scales.

2.2. Characterizations

Powder X-ray diffraction (XRD) was measured on an X-ray diffractometer (Rigaku SmartLab High Resolution, Tokyo, Japan) using Cu-Kα radiation (λ = 1.5406 Å). Raman spectroscopy measurements were taken using a Raman spectrometer (HORIBA, LabRAM HR Evolution, Lyon, France) with an excitation wavelength of 514 nm. X-ray photoelectron microscopy (XPS) analysis was conducted using a Thermo VG Scientific Sigma Probe instrument with a micro-focused monochromatized Al Kα X-ray source (1486.6 eV). Field-emission scanning electron microscopy (FE-SEM) images were acquired using a FEI NovaNano SEM 450, and transmission electron microscopy (TEM) with energy-dispersive X-ray spectroscopy (EDS) was performed using a JEOL F200 instrument to observe the surface morphologies and microstructures. The specific surface areas of pristine graphite (AG) and x−MoS$_2$@AG composites were obtained by using the Brunauer–Emmett–Teller (BET) method, and pore size distribution was calculated using Barrett–Joyner–Halenda (BJH) method (BELSORP Max, Microtrac MRB, Osaka, Japan). Thermogravimetric analysis (TGA) was carried out on a TGA2 (Mettler Toledo) at a heating rate of 10 °C min^{-1} from 30 to 700 °C under an air atmosphere.

2.3. Electrochemical Measurements

The electrochemical properties of the MoS$_2$@AG composites were evaluated using a CR2032 coin-type half-cell assembled in an argon-filled glove box. A slurry was prepared by mixing the active material (MoS$_2$@AG composites), conducting agent (Super P), and polyvinylidene fluoride (PVDF) binder at a weight ratio of 8:1:1 in a solvent of N-methyl-2-pyrrolidone (NMP). The resulting slurry was coated onto a copper foil and dried in a vacuum at 120 °C for 12 h. The average loading density of the anode electrodes was about 2 mg cm^{-2}. The anode was composed of the MoS$_2$@AG-composite-coated copper foil, while a lithium metal foil was used as the counter/reference electrode. A polypropylene membrane was used as the separator. The electrolyte solution comprised a 1M LiPF$_6$ solution in a mixture of ethylene carbonate (EC)/ethylene methyl carbonate (EMC)/diethyl carbonate (DEC) (=3:4:3, $v/v/v$) containing 1% vinylene carbonate (VC). The electrochemical cell was fabricated using 160 µL of electrolyte. The galvanostatic charge and discharge curves of the cells were obtained using a battery testing system (Biologic, BCS) in a voltage range from 0.01 to 2.5 V vs. Li/Li$^+$ with various current densities. The electrochemical impedance spectroscopy (EIS) measurements were performed at an AC amplitude of 10 mV over the frequency range from 200 kHz to 50 mHz (Biologic, VSP) and the equivalent circuit fitting was conducted using ZView Software 3.2 (Scribner Associates, Inc., Southern Pines, NC, USA).

3. Results and Discussion

Figure 1a illustrates the one-step hydrothermal synthesis process for forming 3D hierarchical flower-like MoS$_2$ nanosheets directly grown on the surface of graphite. Figure 1b–d show representative SEM, TEM, and high-resolution TEM (HR-TEM) images of the MoS$_2$@AG composites, respectively. The x−MoS$_2$@AG composites with different weight percentages of sulfur and molybdenum sources relative to graphite were synthesized. In this process, MoS$_2$ nanosheets were spontaneously nucleated and grown on the surface of graphite. Ultimately, a 3D composite of few-layer MoS$_2$ nanosheets with a low loading on graphite was obtained. The unique structure of the MoS$_2$-decorated graphite composite is anticipated to significantly enhance the capacity and rate capability of LIBs. Graphite acts as a support for MoS$_2$ nucleation, thereby providing an excellent electron transfer channel. Additionally, 3D hierarchically aligned flower-like MoS$_2$ nanosheets grown on the surface of graphite improve the rate characteristics by facilitating lithium diffusion and increase the specific capacity by offering abundant exposed active sites.

The morphologies of pristine graphite and MoS$_2$-decorated graphite composites were examined by SEM and SEM-backscattered electron (SEM-BSE) images to investigate the effect of the amounts of thiourea and (NH$_4$)$_6$Mo$_7$O$_{24}$·4H$_2$O on the formation process (Figures 2 and S1). Three-dimensionally grown plate-like MoS$_2$ nanosheets were observed on the surface of 2.5−MoS$_2$@AG and 5−MoS$_2$@AG composites, as shown in Figure 2a–c and Figure 2d–f, respectively. In contrast, flower-shaped MoS$_2$ nanosheets were observed on the surfaces of 10−MoS$_2$@AG and 20−MoS$_2$@AG composites (Figure 2g–l). The increase in the amounts of thiourea and (NH$_4$)$_6$Mo$_7$O$_{24}$·4H$_2$O provided more nucleation sites, resulting in the formation of more flower-like MoS$_2$ nanosheets. Figure S2 shows the enlarged SEM images of the flower-like MoS$_2$ nanosheets. Furthermore, Figure 2c,f,i,l show the SEM-BSE images corresponding to Figure 2b,e,h,k, respectively. The brighter areas indicate higher average atomic numbers, confirming the successful formation of MoS$_2$ on the graphite surface.

The specific surface area and pore structure of pristine graphite (AG) and 20−MoS$_2$@AG composite were examined by the N$_2$ adsorption and desorption isotherm measurements (Figure S3). The Brunauer–Emmett–Teller (BET) specific surface area of the 20−MoS$_2$@AG composite was calculated to be 2.98 m^2 g^{-1}, which is almost 2.6 times larger than that of pristine AG (1.15 m^2 g^{-1}). As shown in the inset of Figure S3, the mean pore diameters of AG and 20−MoS$_2$@AG were 29.28 and 21.48 nm, respectively, and 20−MoS$_2$@AG showed narrower average pores. The increased specific surface area and decreased pore diameter of the 20−MoS$_2$@AG composite can be attributed to the formation of 3D hierarchically aligned MoS$_2$ nanosheets on the surface of graphite. These structural features are expected to improve the rate characteristic by facilitating lithium diffusion and increase the specific capacity by offering abundant exposed active sites.

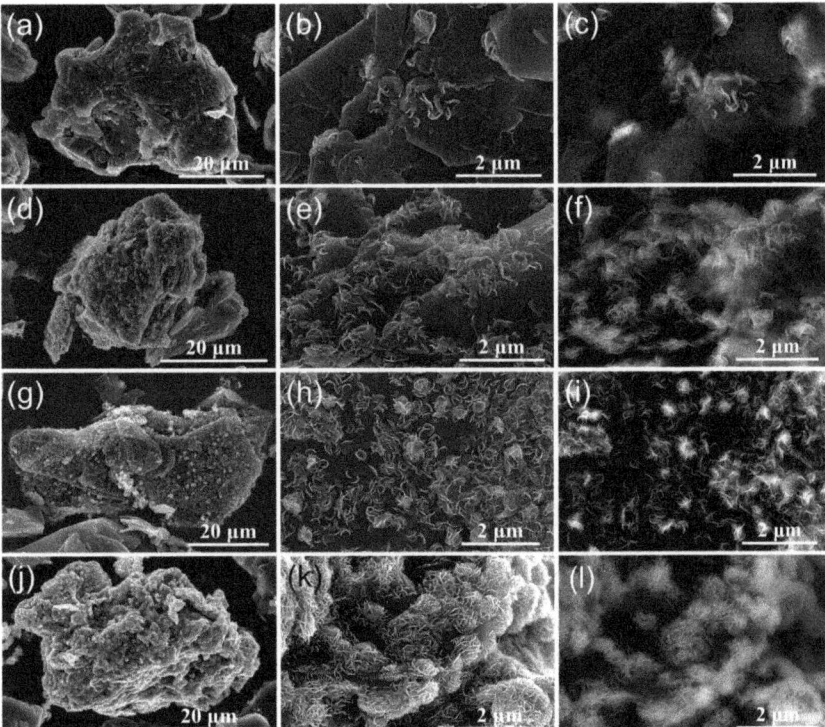

Figure 2. SEM and SEM-BSE images of MoS_2-nanosheets-decorated graphite composites with varying amounts of thiourea and $(NH_4)_6Mo_7O_{24} \cdot 4H_2O$: (**a**–**c**) 2.5−$MoS_2$@AG, (**d**–**f**) 5−$MoS_2$@AG, (**g**–**i**) 10−$MoS_2$@AG, and (**j**–**l**) 20−MoS_2@AG. The images show vertically aligned plate-like MoS_2 nanosheets on 2.5−MoS_2@AG and 5−sMoS_2@AG composites, and flower-shaped MoS_2 nanosheets on 10−MoS_2@AG and 20−MoS_2@AG composites. The SEM-BSE images in (**c**,**f**,**i**,**l**) confirm the successful formation of MoS_2 nanosheets on the graphite surface by brighter contrast indicating a higher atomic number.

The amount of MoS_2 grown on the graphite surface was calculated through thermogravimetric analysis (TGA), as shown in Figure S4. In the case of the pristine MoS_2, the weight loss from 300 to 500 °C is indicative of the oxidation of MoS_2 to MoO_3. As for the 20−MoS_2@AG composites, the TGA profiles display a two-step weight decrease, which can be attributed to the consecutive oxidations of MoS_2 and carbon, respectively. The MoS_2 content in the 20−MoS_2@AG was estimated to be <5 wt% through the TGA results.

The low-magnification TEM image (Figure 3a) reveals the MoS_2 nanosheets directly grown on the graphite surface. High-resolution TEM (HR-TEM) images in Figure 3b,c show that the few-layered MoS_2 nanosheets are grown on the surface of graphite. The MoS_2 nanosheets consist of 9–14 layers, and the interlayer distance (002) of MoS_2 was found to be 0.64~0.67 nm, which is slightly larger than that of conventional bulk MoS_2 (0.62 nm). This expanded interlayer spacing could contribute to enhanced kinetics and a low energy barrier for ion intercalation [25–27]. Furthermore, high-angle annular dark-field scanning transmission electron microscopy (HAADF-STEM) images of MoS_2-decorated graphite are shown in Figure 3d and energy dispersive spectroscopic (EDS) mapping images of MoS_2-decorated graphite reveal the existence of C, Mo, and S elements in the composite, as shown in Figure 3e–g.

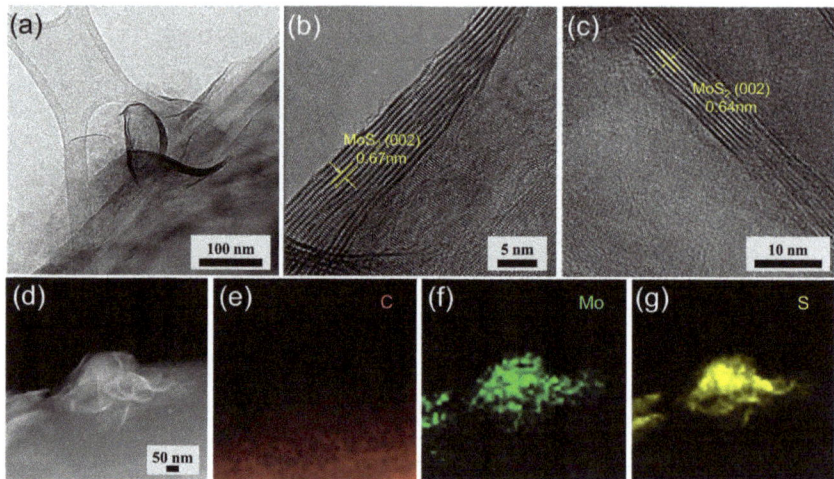

Figure 3. TEM and STEM images of MoS$_2$-nanosheets-decorated graphite composites. (**a**) Low-magnification TEM image showing 3D architectural MoS$_2$ nanosheets formed on the graphite surface. (**b**,**c**) HR-TEM images displaying few-layered MoS$_2$ nanosheets grown on the graphite surface. (**d**) STEM-HAADF image and (**e**–**g**) TEM-EDS mapping images indicating the distribution of C, Mo, and S elements in the composite.

Figure 4a shows the powder XRD pattern of AG, 2.5−MoS$_2$@AG, 5−MoS$_2$@AG, 10−MoS$_2$@AG, and 20−MoS$_2$@AG. The diffraction peaks observed in all MoS$_2$-decorated graphite composites at 26.44°, 42.30°, 44.48°, and 54.58° were consistent with the (002), (100), (101), and (004) planes of hexagonal graphite (JCPDS 41-1487). Additionally, the (002) diffraction peak shifted to a lower angle compared to that of pristine graphite (26.58°), indicating an increased interlayer spacing (Figure S5). The three-dimensionally aligned few-layer MoS$_2$ nanosheets grown on the surface of graphite are expected to provide more space for Li$^+$ ion transport and reduce the kinetic barriers for their movement. Figure 4b shows an enlarged XRD pattern of the low-angle region in Figure 4a, and the peaks at 14.22° and 33.24° match well with the (002) and (101) planes of hexagonal MoS$_2$ (JCPDS 37-1492). The average interlayer spacing (002) of MoS$_2$ calculated from Bragg's law is about 0.622 nm, which is slightly longer than that of highly crystalline MoS$_2$ (0.615 nm).

The Raman spectra of MoS$_2$-nanosheets-decorated graphite are presented in Figure 4c, showing two sharp peaks at 380 and 407 cm^{-1} attributed to the E$^1_{2g}$ and A$_{1g}$ vibration modes of MoS$_2$, respectively [28]. Two characteristic bands were also observed at 1358 and 1580 cm^{-1}, corresponding to the D band and G band of graphite, respectively.

The XPS measurements were used to analyze the elemental composition of MoS$_2$-decorated graphite. The XPS survey spectra in Figure 4d show that MoS$_2$-decorated graphite contains Mo, S, C, and O elements. In the high-resolution Mo 3d spectrum (Figure 4e), two peaks are observed at 229.7 and 232.8 eV, which can be attributed to the Mo 3d$_{5/2}$ and Mo 3d$_{3/2}$ binding energies, respectively, and are characteristic peaks of Mo^{4+} in MoS$_2$. The peaks at 232.9 and 236.1 eV are related to the Mo 3d$_{5/2}$ and 3d$_{3/2}$ of Mo^{6+} (typical of the Mo-O bond). Furthermore, a small S 2s peak is displayed at 226.7 eV. The high-resolution S 2p spectrum is consistent with S 2p$_{1/2}$ at 163.6 eV and S 2p$_{3/2}$ at 162.5 eV, respectively (Figure 4f) [22].

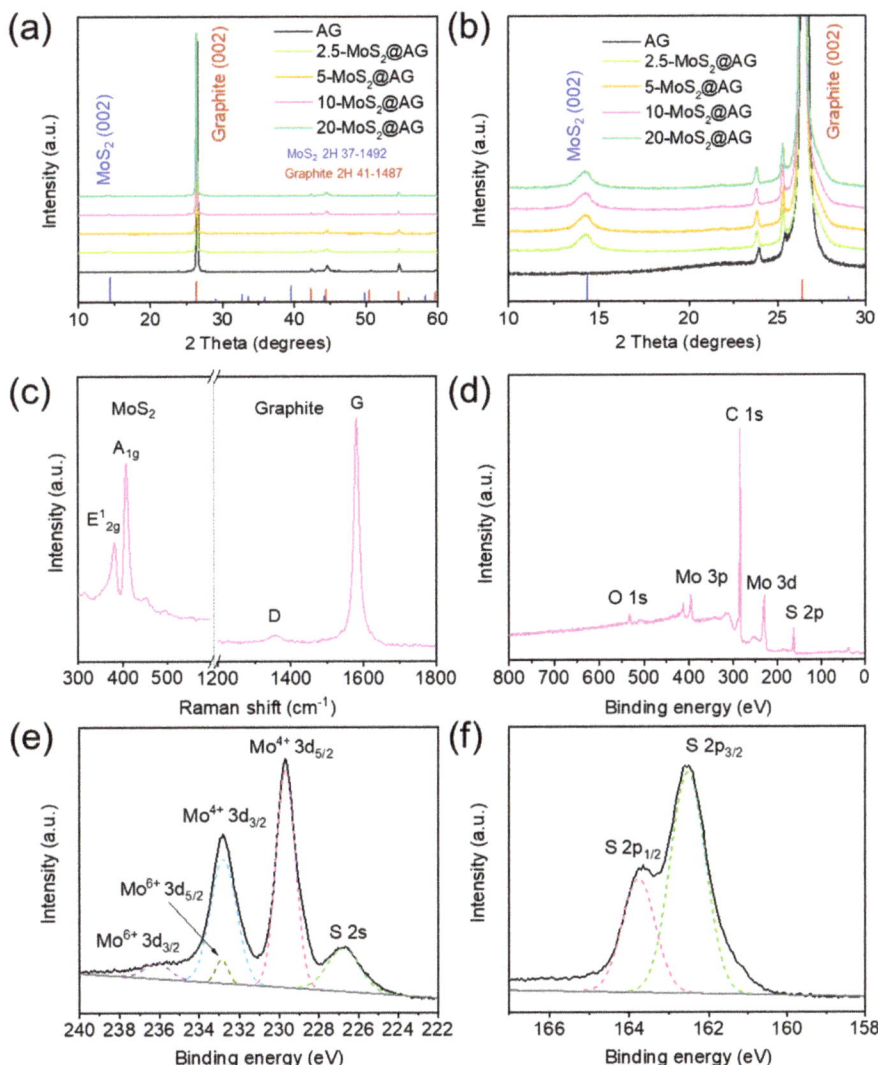

Figure 4. (**a**,**b**) XRD patterns, (**c**) Raman spectra, (**d**) XPS survey spectrum, and high-resolution XPS spectra of (**e**) Mo 3d and (**f**) S 2p for MoS$_2$@AG composites.

The electrochemical performance of the MoS$_2$-decorated graphite composite was evaluated as an LIB anode material. Figure 5a shows the galvanostatic charge and discharge (GCD) profiles of the 20−MoS$_2$@AG electrode during the initial first cycle in a voltage range of 0.01–2.5 V vs. Li/Li$^+$ at a current density of 35 mA g^{-1}. The 20−MoS$_2$@AG electrode exhibited an initial charge capacity of 494.1 mAh g^{-1} with a coulombic efficiency of 86.1%. The capacity of 20−MoS$_2$@AG is higher than that of pristine graphite (373.3 mAh g^{-1}) due to the formation of MoS$_2$ nanosheets on the graphite surface. However, the initial coulombic efficiency is slightly lower than that of graphite (~91.2%), indicating that electrochemically active MoS$_2$ nanosheets contribute to the total capacity of the 20−MoS$_2$@AG composite [12,14].

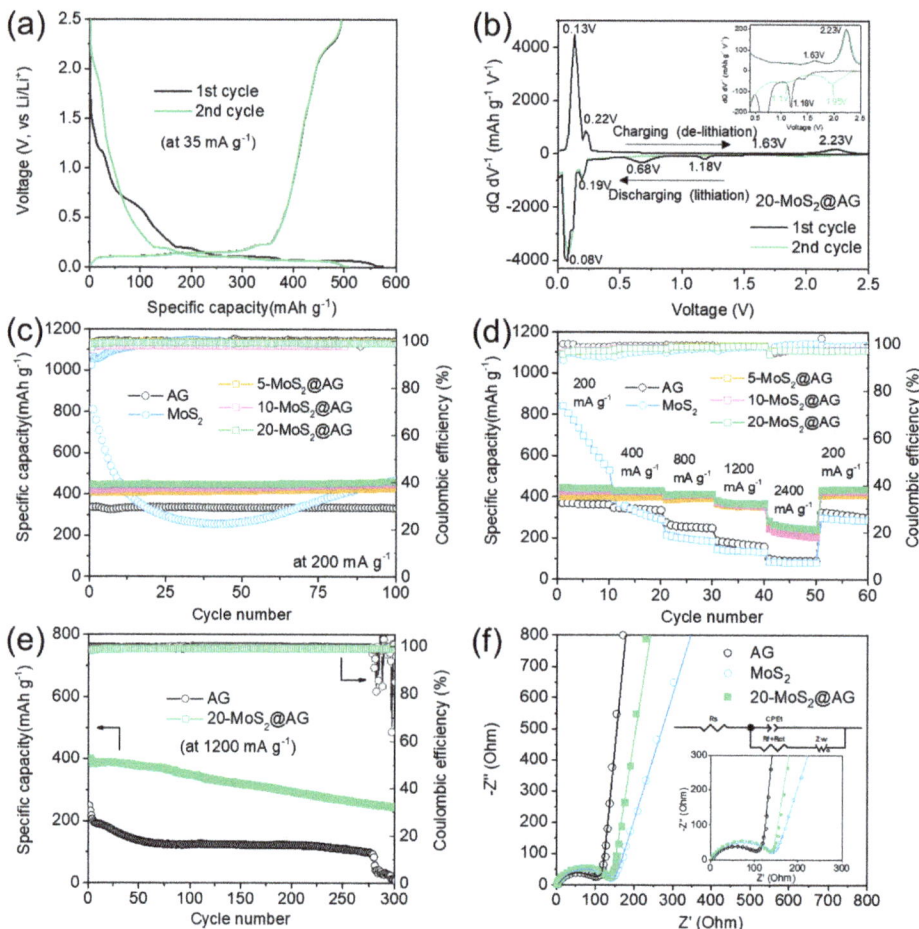

Figure 5. (**a**) Initial galvanostatic charge and discharge curves at a current density of 35 mA g^{-1} and (**b**) corresponding differential voltage profiles of the 20−MoS$_2$@AG electrode, (**c**) cycle performance at a current density of 200 mA g^{-1}, (**d**) rate performance comparison of pristine AG, pristine MoS$_2$, and MoS$_2$@AG composite electrodes at different current densities ranging from 200 to 2400 mA g^{-1}, (**e**) cycle performance at a high current density of 1200 mA g^{-1} for pristine AG and 20−MoS$_2$@AG electrodes, and (**f**) Nyquist plots from EIS data (symbols) and fitting results (solid lines) of pristine AG, pristine MoS$_2$, and 20−MoS$_2$@AG electrodes before the cycle.

Figure 5b shows the differential voltage profile of the 20−MoS$_2$@AG. In the first cycle, the cathodic peaks at 0.08 V and 0.19 V in the discharge (lithiation) process are attributed to the gradual intercalation of Li$^+$ into the interlayers of graphite [15,29]. Additional discharge contributions corresponding to the MoS$_2$ nanosheets on the graphite surface were observed at higher voltages of ~0.68 and 1.18 V vs. Li/Li$^+$. The peak at 1.18 V is ascribed to Li$^+$ insertion into the structure of MoS$_2$ to form Li$_x$MoS$_2$. Another peak at 0.68 V corresponds to the reduction/decomposition of MoS$_2$ to Mo metal particles and Li$_2$S through a conversion reaction (MoS$_2$ + 4Li$^+$ → Mo + 2Li$_2$S), and the formation of the solid electrolyte interphase (SEI) layers [25,30,31]. MoS$_2$ reacts with Li$^+$ ions at a higher operating voltage compared to graphite. Therefore, the lithiated 20−MoS$_2$@AG electrode can lower the Li$^+$ adsorption energy on the surface of graphite and form a stable interface between MoS$_2$ and graphite, promoting Li$^+$ migration during cycling [12,14,29]. Additionally, the slightly higher anodic potential in the MoS$_2$@AG composite would prevent the formation of lithium dendrites during repeated cycling, which typically occurs in conventional graphite anodes [29]. In the subsequent reverse anodic scan, two peaks at 1.63 V and 2.23 V are associated with the incomplete oxidation of Mo metal into MoS$_2$ and the partial de-lithiation of Li$_2$S into S, respectively. In the next cycle, the reduction peaks at 0.68 V and 1.18 V are disappeared and two new peaks at 1.1 V and 1.95 V are observed, assigned to the following reactions: 2Li$^+$ + S + 2e$^-$ → Li$_2$S and MoS$_2$ + xLi$^+$ + xe$^-$ → Li$_x$MoS$_2$, respectively [25,32]. These peak shifts could be explained by the structural change and phase transformation during the first cycling [28]. Even at the 100th cycle, it was confirmed that two weak peaks at 1.3 V and 1.88 V were still observed (Figure S6a,b). As shown in Figure S6c,d, the pristine MoS$_2$ electrode shows a similar differential voltage profile to the 20−MoS$_2$@AG electrode. This means that the 20−MoS$_2$@AG composite electrode has an energy storage mechanism combined with those of graphite and MoS$_2$. Moreover, the presence of two main pairs of redox peaks during cycling indicated the stable cycle stability and reversibility of the 20−MoS$_2$@AG electrode.

Figure 5c and Figure S7a show the cycle stabilities of pristine AG, pristine MoS$_2$, 5−MoS$_2$@AG, 10−MoS$_2$@AG, 20−MoS$_2$@AG, and 30−MoS$_2$@AG electrodes, with their initial charge (de-lithiation) capacities at a current density of 200 mA g^{-1} being 337.1, 812.0, 407.0, 418.0, 448.0, and 523.2 mAh g^{-1}, respectively. The highest specific capacity of the 30-MoS$_2$@AG can be ascribed to the greater loading mass of MoS$_2$. The pristine flower-like MoS$_2$ nanosheets electrode exhibited a rapid capacity decay during initial cycles, likely due to its low conductivity and severe structural damage caused by volume changes, leading to the pulverization of the active material [33]. In contrast, the 20−MoS$_2$@AG composite electrode displayed excellent cycle performance with an average coulombic efficiency of 99.3% and delivered the highest reversible capacity of 463.2 mAh g^{-1} after 100 cycles. The reversible capacity of the 20−MoS$_2$@AG electrode is much larger than the theoretical capacity of 386.9 mAh g^{-1} (C$_{20-MoS2@AG}$ = C$_{graphite}$ × wt% of graphite (95%) + C$_{MoS2}$ × wt% of MoS$_2$ (5%) = 372 × 0.95 + 670 × 0.05 = 386.9 mAh g^{-1}). This can be attributed to the formation of 3D hierarchical MoS$_2$ nanosheets on the graphite surface, resulting in abundant active sites for Li$^+$ diffusion, expanded interlayer spacing, and a large electrode/electrolyte contact area [25,31].

The galvanostatic charge and discharge curves of pristine MoS$_2$ and 20−MoS$_2$@AG electrodes at current densities of 35 and 200 mA g^{-1} are presented in Figure S8. Interestingly, both pristine MoS$_2$ nanosheets and x−MoS$_2$@AG composite electrodes exhibited a capacity climbing phenomenon during cycles, which is likely due to the increasing electrochemically accessible surface area, resulting from the gradual appearance of cracks on the (002) basal planes of MoS$_2$ [33–37].

The rate capabilities of pristine AG, pristine MoS$_2$, 5−MoS$_2$@AG, 10−MoS$_2$@AG, 20−MoS$_2$@AG, and 30−MoS$_2$@AG electrodes were evaluated at current densities of 200, 400, 800, 1200, and 2400 mA g^{-1} (1C = ~400 mA g^{-1}), as shown in Figures 5d and S7b. Among them, the 20−MoS$_2$@AG electrode demonstrated the best rate performance compared to pristine AG and pristine MoS$_2$. At current densities of 200, 400, 800, 1200, and 2400 mA g^{-1}, the average charge capacities of 20−MoS$_2$@AG delivered reversible capacities of 441.7, 431.4, 413.1, 371.0, and 255.8 mAh g^{-1}, respectively, which were superior to those of pristine MoS$_2$ (699.5, 349.2, 198.8, 139.7, and 84.1 mAh g^{-1}) and pristine AG (365.5, 342.7, 256.8, 172.6, and 92.5 mAh g^{-1}). The 20−MoS$_2$@AG electrode displayed a high reversible capacity of 371.0 mAh g^{-1} at a high current density of 1200 mA g^{-1}. Notably, the 20−MoS$_2$@AG electrode maintained a high-capacity value even under high-rate conditions, retaining approximately 84.0% of the capacity compared to that at a relatively low current density of 200 mA g^{-1} (Table S1). Even when the current density recovered to 200 mA g^{-1} after 50 cycles, the capacity of the 20−MoS$_2$@AG electrode remained at 439.5 mAh g^{-1}, indicating excellent reversibility and rate cycle stability. On the other hand, the 30-MoS$_2$@AG electrode with a higher amount of MoS$_2$ showed more capacity fading as the current density increased (71.0% retention of the capacity at 1200 mA g^{-1}). These results showed that the higher the MoS$_2$ content, the lower the rate characteristics due to its low intrinsic electronic conductivity and large volume change upon cycling. Therefore, it was determined that the 20−MoS$_2$@AG composite was the most optimal condition. Additionally, all MoS$_2$-decorated graphite electrodes exhibited enhanced rate performance and reversibility. This can be attributed to the combination of graphite's excellent conductivity and the flower-like MoS$_2$ nanosheets grown on its surface, which increases the contact/accessible area with the electrolyte, provides more reaction sites, and facilitates Li$^+$ diffusion, thereby contributing to the excellent electrochemical performance [26].

A long-term cycling performance at a high current density is essential for practical applications of LIBs [25]. Figure 5e illustrates the cycle performance of graphite and 20−MoS$_2$@AG electrodes at a high current density of 1200 mA g^{-1} over 300 cycles. The 20−MoS$_2$@AG electrode exhibits an initial capacity of ~400 mAh g^{-1}, higher than that of AG (~248 mAh g^{-1}), and shows stable cycle performance over 300 cycles. The significant improvement in the overall performance of x−MoS$_2$@AG composites compared to pristine graphite is mainly due to the surface decoration with MoS$_2$ nanosheets. Furthermore, the 20−MoS$_2$@AG electrode exhibited stable cycle characteristics over 400 cycles at a high current density of 1200 mA g^{-1}, while the 30−MoS$_2$@AG electrode showed a rapid decrease in capacity from around the 315th cycle (Figure S7c).

The electrochemical performance of our MoS$_2$@AG as anode materials of LIBs was compared with those of the previously reported transition metal sulfides (MS$_x$) or transition metal oxides (MO$_x$)-based carbon composites (Table S2) [38–42]. Upon comparing the specific capacity of our MoS$_2$@AG composite anode to that of carbon composites incorporating transition metal sulfides or transition metal oxides, we observed that the capacity of our composite was relatively lower than carbon-based composites containing more than 50 wt% of transition metal sulfides or oxides. However, in comparison to anode materials that contained small amounts (<10 wt%) of transition metal sulfides or oxides, our composite demonstrated superior capacity characteristics under not only low current density conditions, but also high current density conditions. These findings are expected to provide the MoS$_2$@AG composite with a small amount of MoS$_2$ surface modification on the graphite surface as a potential anode material for high-performance and cost-effective lithium-ion batteries.

Electrochemical impedance spectroscopy (EIS) was used to obtain further insight into the electrochemical reaction kinetics of AG, pristine MoS$_2$ nanosheets, and 20−MoS$_2$@AG electrodes (Figure 5f). EIS plots were analyzed based on the fitting results obtained using an equivalent circuit model (inset of Figure 5f). The Nyquist plots of the electrodes can be divided into four components. The intercept at the real axis in the high-frequency region represents the internal resistance (R_s), a semicircle in the high-frequency region corresponds to the surface film resistance (R_f), a semicircle in the medium-frequency region represents charge transfer resistance (R_{ct}), and a straight line in the low-frequency region represents the Warburg impedance (Z_w) related to the diffusion resistance of Li$^+$ within the bulk of the electrode materials [15,21,29]. CPE represents a constant phase element, corresponding to the charge-transfer reaction. The values for the resistance parameters obtained from fitting using the equivalent circuit are summarized in Tables S3 and S4. The diameter of the semicircle ($R_f + R_{ct}$) of the 20−MoS$_2$@AG electrode (128 Ω) in the high/medium frequency is similar to that of AG (101 Ω). Moreover, the slopes of the straight lines at low frequencies for AG and 20−MoS$_2$@AG are larger than those for pristine MoS$_2$ nanosheets, indicating that the MoS$_2$@AG electrode has a faster diffusion rate at the interface. The equivalent circuit fitting result confirms that both Warburg resistance (Z_wR (Ω)) and Warburg time constant (Z_wT (s)) values of AG and MoS$_2$@AG electrodes are about 2–5 times smaller than those of MoS$_2$. The ion diffusion coefficient is inversely proportional to the inverse ratio values of Z_wT and Z_wR^2 [43]. In addition, the Nyquist plots of the pristine AG, pristine MoS$_2$, and 20−MoS$_2$@AG electrodes were obtained after two cycles at a current density of 35 mA g^{-1}, as shown in Figure S9. The pristine AG and 20−MoS$_2$@AG after two cycles show a decrease in both surface film and charge-transfer resistance ($R_f + R_{ct}$) showing 101 → 76.3 Ω for AG and 128 → 76.8 Ω for MoS$_2$@AG. This observation is attributed to the improved infiltration of electrolyte into electrode materials and charge transfer kinetics. However, the $R_f + R_{ct}$ of cycled MoS$_2$ electrode increases slightly (128 → 139 Ω), which corresponds mainly to the rapid capacity decay due to the pulverization of active materials.

Examining the changes in surface and thickness of the electrode before and after cycling provides a visual means to understand the mechanism underlying the improved performance of the MoS$_2$ nanosheets decorated on the graphite composite in comparison to pristine graphite. In light of this, we investigated the structural stability of the 20−MoS$_2$@AG electrode before and after cycling by performing SEM measurements, as shown in Figure S10. Prior to cycling, the MoS$_2$@AG electrode had a rough surface covered with numerous nanosheets and its thickness was approximately 36.2 μm. After 300 cycles at a current density of 200 mA g^{-1}, the 3D hierarchically aligned MoS$_2$ nanosheets synthesized on the surface of the graphite had transformed into nanoparticles and aggregated with each other, but some MoS$_2$ nanosheets still maintained their original shape. In addition, we observed that the thickness of the 20−MoS$_2$@AG electrode after cycling was about 36.5 μm, which was almost unchanged compared to the initial thickness of 36.2 μm before cycling. These ex situ SEM images provide evidence that the 20−MoS$_2$@AG electrode maintains a stable structure without significant volume changes even after long-term cycling, indicating that high reversible capacity can be maintained even after prolonged cycling.

4. Conclusions

In conclusion, we have successfully synthesized a 3D hierarchical MoS$_2$@AG composite that exhibits high performance as an anode material for LIBs. The use of graphite as a substrate for the nucleation and growth of three-dimensionally aligned few-layer MoS$_2$ nanosheets provided a conductive matrix and exposed surface area, resulting in a significant improvement in the rate capability and capacity of the composite. The hydrothermal reaction facilitated the growth of MoS$_2$ nanosheets on the graphite surface, forming a 3D network with excellent structural stability during the charge/discharge process. The three-dimensionally grown few-layer MoS$_2$ nanosheets on the surface of graphite facilitated the diffusion of Li$^+$ and reduced the diffusion resistance, leading to improved rate performance. The 20−MoS$_2$@AG electrode exhibited excellent cyclability and rate capability with outstanding stability over 300 cycles, even at a high current density of 1200 mA g^{-1}. Notably, at this current density, the 20−MoS$_2$@AG composite displayed an initial capacity of around 400 mAh g^{-1}, which is approximately 60% higher than that of the graphite electrode. These results demonstrate that the MoS$_2$-nanosheets-decorated graphite composite developed through a simple hydrothermal method holds great promise as a high-power anode material for LIBs in various practical applications.

Supplementary Materials: The following supporting information can be downloaded at: https://www.mdpi.com/article/10.3390/ma16114016/s1, Figure S1: SEM images of pristine artificial graphite (AG) obtained at (a) low magnification (scale bar: 100 μm) and (b) enlarged magnification (scale bar: 30 μm); Figure S2: SEM images of the MoS$_2$@AG composite obtained at various enlarged magnifications; Figure S3: N$_2$ adsorption and desorption isotherms of pristine graphite (AG) and 20−MoS$_2$@AG composite. The inset shows their pore size distribution curves; Figure S4: TGA curves of pristine MoS$_2$, 5−MoS$_2$@AG, 10−MoS$_2$@AG, and 20−MoS$_2$@AG composites with a heating rate of 10 °C min^{-1} in air; Figure S5: XRD patterns of MoS$_2$@AG composites; Figure S6: The differential voltage profiles of (a,b) the 20−MoS$_2$@AG electrode and (c,d) pristine MoS$_2$; Figure S7: (a) Cycle performance at a current density of 200 mA g^{-1}, (b) rate performance comparison of pristine AG, pristine MoS$_2$, 20−MoS$_2$@AG, and 30−MoS$_2$@AG electrodes at different current densities ranging from 200 to 2400 mA g^{-1}, and (c) cycle performance at a high current density of 1200 mA g^{-1} for pristine AG, 20−MoS$_2$@AG, and 30−MoS$_2$@AG electrodes; Figure S8: Galvanostatic charge and discharge curves of (a) pristine MoS$_2$ and (b) 20−MoS$_2$@AG electrodes; Figure S9: Nyquist plots of pristine AG, pristine MoS$_2$, and 20−MoS$_2$@AG obtained after 2 cycles at a current density of 35 mA g^{-1}; Figure S10: Ex situ SEM images of the 20−MoS$_2$@AG electrodes before and after 300 cycles at a current density of 200 mA g^{-1}; (a–c) before cycling and (d–f) after cycling. Panels (c) and (f) of Figure S10 show cross-sectional SEM images of the electrode before and after cycling, respectively; Table S1: Comparison of the initial capacities and rate performance for pristine graphite, pristine MoS$_2$, 5−MoS$_2$@AG, 10−MoS$_2$@AG, 20−MoS$_2$@AG, and 30−MoS$_2$@AG composites; Table S2: Comparison of LIB electrochemical performance for transition metal sulfides (MS$_x$) and transition metal oxides (MO$_x$)-based composites; Table S3: EIS fitting parameters for AG, MoS$_2$, and 20−MoS$_2$@AG before the cycling tests; Table S4: EIS fitting parameters for AG, MoS$_2$, and 20−MoS$_2$@AG after 2 cycles. References [38–42] are cited in the supplementary file.

Author Contributions: Conceptualization, Y.A.L., H.S., K.L. and H.Y.; methodology, Y.A.L., K.Y.J., C.-Y.Y. and H.Y.; investigation, Y.A.L., K.Y.J., J.Y., K.Y., W.J., K.-N.J., J.L. and M.H.R.; resources, H.Y.; data curation, Y.A.L., K.Y.J. and J.Y.; writing—original draft preparation, Y.A.L., K.Y.J., J.Y. and H.Y.; writing—review and editing, Y.A.L., K.Y.J., K.Y., K.-N.J., C.-Y.Y., Y.C., J.L., M.H.R., H.S., K.L. and H.Y.; visualization, Y.A.L.; supervision, H.S., K.L. and H.Y.; project administration, H.Y.; funding acquisition, H.S. and H.Y. All authors have read and agreed to the published version of the manuscript.

Funding: This research was funded by framework of the research and development program of the Korea Institute of Energy Research (KIER, Project No. C3-2408, C3-8601, and C3-2420), the R&D Program for Forest Science Technology (Project No. 2020229B10-2222-AC01) provided by Korea Forest Service (Korea Forestry Promotion Institute), and Korea Institute of Planning and Evaluation for Technology in Food, Agriculture and Forestry (IPET) funded by Ministry of Agriculture, Food and Rural Affairs (MAFRA) (321077022HD020). This work was also supported by research fund of Chungnam National University.

Institutional Review Board Statement: Not applicable.

Informed Consent Statement: Not applicable.

Data Availability Statement: The data are available upon request from the corresponding authors.

Acknowledgments: SEM analysis was supported by Go-Woon Lee and TEM analysis was supported by Byung-Sun Ahn, both from the Analytical Center of Energy Research (ACER) at KIER.

Conflicts of Interest: The authors declare no conflict of interest.

References

1. Xie, J.; Lu, Y.-C. A retrospective on lithium-ion batteries. *Nat. Commun.* **2020**, *11*, 2499. [CrossRef] [PubMed]
2. Goodenough, J.B.; Park, K.-S. The Li-Ion Rechargeable Battery: A Perspective. *J. Am. Chem. Soc.* **2013**, *135*, 1167–1176. [CrossRef]
3. Etacheri, V.; Marom, R.; Elazari, R.; Salitra, G.; Aurbach, D. Challenges in the development of advanced Li-ion batteries: A review. *Energy Environ. Sci.* **2011**, *4*, 3243–3262. [CrossRef]
4. Tarascon, J.-M.; Armand, M. Issues and challenges facing rechargeable lithium batteries. *Nature* **2001**, *414*, 359–367. [CrossRef]
5. Liu, Q.; Du, C.; Shen, B.; Zuo, P.; Cheng, X.; Ma, Y.; Yin, G.; Gao, Y. Understanding undesirable anode lithium plating issues in lithium-ion batteries. *RSC Adv.* **2016**, *6*, 88683–88700. [CrossRef]
6. Waldmann, T.; Hogg, B.-I.; Wohlfahrt-Mehrens, M. Li plating as unwanted side reaction in commercial Li-ion cells—A review. *J. Power Sources* **2018**, *384*, 107–124. [CrossRef]
7. Li, S.; Wang, K.; Zhang, G.; Li, S.; Xu, Y.; Zhang, X.; Zhang, X.; Zheng, S.; Sun, X.; Ma, Y. Fast Charging Anode Materials for Lithium-Ion Batteries: Current Status and Perspectives. *Adv. Funct. Mater.* **2022**, *32*, 2200796. [CrossRef]
8. Li, L.; Zhang, D.; Deng, J.; Gou, Y.; Fang, J.; Cui, H.; Zhao, Y.; Cao, M. Carbon-based materials for fast charging lithium-ion batteries. *Carbon* **2021**, *183*, 721–734. [CrossRef]
9. Cheng, Q.; Yuge, R.; Nakahara, K.; Tamura, N.; Miyamoto, S. KOH etched graphite for fast chargeable lithium-ion batteries. *J. Power Sources* **2015**, *284*, 258–263. [CrossRef]
10. Chen, K.-H.; Goel, V.; Namkoong, M.J.; Wied, M.; Müller, S.; Wood, V.; Sakamoto, J.; Thornton, K.; Dasgupta, N.P. Enabling 6C Fast Charging of Li-Ion Batteries with Graphite/Hard Carbon Hybrid Anodes. *Adv. Energy Mater.* **2021**, *11*, 2003336. [CrossRef]
11. Kim, N.; Chae, S.; Ma, J.; Ko, M.; Cho, J. Fast-charging high-energy lithium-ion batteries via implantation of amorphous silicon nanolayer in edge-plane activated graphite anodes. *Nat. Commun.* **2017**, *8*, 812. [CrossRef]
12. Rhee, D.Y.; Kim, J.; Moon, J.; Park, M.S. Off-stoichiometric TiO$_{2-x}$-decorated graphite anode for high-power lithium-ion batteries. *J. Alloys Compd.* **2020**, *843*, 156042. [CrossRef]
13. Lee, S.-M.; Kim, J.; Moon, J.; Jung, K.-N.; Kim, J.H.; Park, G.-J.; Choi, J.-H.; Rhee, D.Y.; Kim, J.-S.; Park, M.-S. A cooperative biphasic MoOx–MoPx promoter enables a fast-charging lithium-ion battery. *Nat. Commun.* **2021**, *12*, 39. [CrossRef] [PubMed]
14. Lee, J.W.; Kim, S.Y.; Rhee, D.Y.; Park, S.; Jung, J.Y.; Park, M.-S. Tailoring the Surface of Natural Graphite with Functional Metal Oxides via Facile Crystallization for Lithium-Ion Batteries. *ACS Appl. Mater. Interfaces* **2022**, *14*, 29797–29805. [CrossRef] [PubMed]

15. Yan, X.; Jiang, F.; Sun, X.; Du, R.; Zhang, M.; Kang, L.; Han, Q.; Du, W.; You, D.; Zhou, Y. A simple, low-cost and scale-up synthesis strategy of spherical-graphite/Fe_2O_3 composites as high-performance anode materials for half/full lithium ion batteries. *J. Alloys Compd.* **2020**, *822*, 153719. [CrossRef]
16. Song, Y.; Wang, Z.; Yan, Y.; Zhao, W.; Bakenov, Z. $NiCo_2S_4$ nanoparticles embedded in nitrogen-doped carbon nanotubes networks as effective sulfur carriers for advanced Lithium–Sulfur batteries. *Micro. Meso. Mater.* **2021**, *316*, 110924. [CrossRef]
17. Qi, J.-Q.; Huang, M.-Y.; Ruan, C.-Y.; Zhu, D.-D.; Zhu, L.; Wei, F.-X.; Sui, Y.-W.; Meng, Q.-K. Construction of $CoNi_2S_4$ nanocubes interlinked by few-layer $Ti_3C_2T_x$ MXene with high performance for asymmetric supercapacitors. *Rare Metals* **2022**, *41*, 4116–4126. [CrossRef]
18. Xiao, Y.; Shen, Y.; Su, D.; Zhang, S.; Yang, J.; Yan, D.; Fang, S.; Wang, X. Engineering $Cu_{1.96}S/Co_9S_8$ with sulfur vacancy and heterostructure as an efficient bifunctional electrocatalyst for water splitting. *J. Mater. Sci. Technol.* **2023**, *154*, 1–8. [CrossRef]
19. Yu, S.-H.; Zachman, M.J.; Kang, K.; Gao, H.; Huang, X.; Disalvo, F.J.; Park, J.; Kourkoutis, L.F.; Abruña, H.D. Atomic-Scale Visualization of Electrochemical Lithiation Processes in Monolayer MoS_2 by Cryogenic Electron Microscopy. *Adv. Energy Mater.* **2019**, *9*, 1902773. [CrossRef]
20. Zhan, J.; Wu, K.; Yu, X.; Yang, M.; Cao, X.; Lei, B.; Pan, D.; Jiang, H.; Wu, M. α-Fe_2O_3 Nanoparticles Decorated C@MoS_2 Nanosheet Arrays with Expanded Spacing of (002) Plane for Ultrafast and High Li/Na-Ion Storage. *Small* **2019**, *15*, 1901083. [CrossRef]
21. Tian, H.; Yu, M.; Liu, X.; Qian, J.; Qian, W.; Chen, Z.; Wu, Z. Plant-cell oriented few-layer MoS_2/C as high performance anodes for lithium-ion batteries. *Electrochim. Acta* **2022**, *424*, 140685. [CrossRef]
22. Wang, G.; Zhang, J.; Yang, S.; Wang, F.; Zhuang, X.; Müllen, K.; Feng, X. Vertically Aligned MoS_2 Nanosheets Patterned on Electrochemically Exfoliated Graphene for High-Performance Lithium and Sodium Storage. *Adv. Energy Mater.* **2018**, *8*, 1702254. [CrossRef]
23. Tiwari, A.P.; Yoo, H.; Lee, J.; Kim, D.; Park, J.H.; Lee, H. Prevention of sulfur diffusion using MoS_2-intercalated 3D-nanostructured graphite for high-performance lithium-ion batteries. *Nanoscale* **2015**, *7*, 11928–11933. [CrossRef] [PubMed]
24. Wang, B.; Zhang, Y.; Zhang, J.; Xia, R.; Chu, Y.; Zhou, J.; Yang, X.; Huang, J. Facile Synthesis of a MoS_2 and Functionalized Graphene Heterostructure for Enhanced Lithium-Storage Performance. *ACS Appl. Mater. Interfaces* **2017**, *9*, 12907–12913. [CrossRef] [PubMed]
25. Zhen, M.; Wang, J.; Guo, S.-Q.; Shen, B. Vertically aligned nanosheets with MoS_2/N-doped-carbon interfaces enhance lithium-ion storage. *Appl. Surf. Sci.* **2019**, *487*, 285–294. [CrossRef]
26. Hu, S.; Chen, W.; Zhou, J.; Yin, F.; Uchaker, E.; Zhang, Q.; Cao, G. Preparation of carbon coated MoS_2 flower-like nanostructure with self-assembled nanosheets as high-performance lithium-ion battery anodes. *J. Mater. Chem. A* **2014**, *2*, 7862–7872. [CrossRef]
27. Hu, Z.; Wang, L.; Zhang, K.; Wang, J.; Cheng, F.; Tao, Z.; Chen, J. MoS_2 Nanoflowers with Expanded Interlayers as High-Performance Anodes for Sodium-Ion Batteries. *Angew. Chem. Int. Ed.* **2014**, *53*, 12794–12798. [CrossRef]
28. Jiao, Y.; Mukhopadhyay, A.; Ma, Y.; Yang, L.; Hafez, A.M.; Zhu, H. Ion Transport Nanotube Assembled with Vertically Aligned Metallic MoS_2 for High Rate Lithium-Ion Batteries. *Adv. Energy Mater.* **2018**, *8*, 1702779. [CrossRef]
29. Xiang, F.; Hou, W.; Gu, X.; Wen, L.; Sun, Y.; Lu, W. One-pot synthesis of MnO-loaded mildly expanded graphite composites as high-performance lithium-ion battery anode materials. *J. Alloys Compd.* **2022**, *897*, 163202. [CrossRef]
30. Hu, L.; Ren, Y.; Yang, H.; Xu, Q. Fabrication of 3D Hierarchical MoS_2/Polyaniline and MoS_2/C Architectures for Lithium-Ion Battery Applications. *ACS Appl. Mater. Interfaces* **2014**, *6*, 14644–14652. [CrossRef]
31. Zhao, J.; Ren, H.; Gu, C.; Guan, W.; Song, X.; Huang, J. Synthesis of hierarchical molybdenum disulfide microplates consisting of numerous crosslinked nanosheets for lithium-ion batteries. *J. Alloys Compd.* **2019**, *781*, 174–185. [CrossRef]
32. Luo, X.; Li, N.; Guo, X.; Wu, K. One-pot hydrothermal synthesis of MoS_2 anchored corncob-derived carbon nanospheres for use as a high-capacity anode for reversible Li-ion battery. *J. Solid State Chem.* **2021**, *296*, 122020. [CrossRef]
33. Teng, Y.; Zhao, H.; Zhang, Z.; Li, Z.; Xia, Q.; Zhang, Y.; Zhao, L.; Du, X.; Du, Z.; Lv, P.; et al. MoS_2 Nanosheets Vertically Grown on Graphene Sheets for Lithium-Ion Battery Anodes. *ACS Nano* **2016**, *10*, 8526–8535. [CrossRef]
34. Liu, M.; Li, N.; Wang, S.; Li, Y.; Liang, C.; Yu, K. 3D nanoflower-like MoS_2 grown on wheat straw cellulose carbon for lithium-ion battery anode material. *J. Alloys Compd.* **2023**, *933*, 167689. [CrossRef]
35. Liu, Y.; Zhang, L.; Zhao, Y.; Shen, T.; Yan, X.; Yu, C.; Wang, H.; Zeng, H. Novel plasma-engineered MoS_2 nanosheets for superior lithium-ion batteries. *J. Alloys Compd.* **2019**, *787*, 996–1003. [CrossRef]
36. Zhang, L.; Fan, W.; Tjiu, W.W.; Liu, T. 3D porous hybrids of defect-rich MoS_2/graphene nanosheets with excellent electrochemical performance as anode materials for lithium ion batteries. *RSC Adv.* **2015**, *5*, 34777. [CrossRef]
37. Chang, K.; Geng, D.; Li, X.; Yang, J.; Tang, Y.; Cai, M.; Li, R.; Sun, X. Ultrathin MoS_2/Nitrogen-Doped Graphene Nanosheets with Highly Reversible Lithium Storage. *Adv. Energy Mater.* **2013**, *3*, 839. [CrossRef]
38. Gao, M.; Liu, B.; Zhang, X.; Zhang, Y.; Li, X.; Han, G. Ultrathin MoS_2 nanosheets anchored on carbon nanofibers as free-standing flexible anode with stable lithium storage performance. *J. Alloys Compd.* **2022**, *894*, 162550. [CrossRef]
39. Sun, H.; Wang, J.-G.; Zhang, X.; Li, C.; Liu, F.; Zhu, W.; Hua, W.; Li, Y.; Shao, M. Nanoconfined Construction of MoS_2@C/MoS_2 Core−Sheath Nanowires for Superior Rate and Durable Li-Ion Energy Storage. *ACS Sustain. Chem. Eng.* **2019**, *7*, 5346–5354. [CrossRef]
40. Kim, H.; Nguyen, Q.H.; Kim, I.T.; Hur, J. Scalable synthesis of high-performance molybdenum diselenide-graphite nanocomposite anodes for lithium-ion batteries. *Appl. Surf. Sci.* **2019**, *481*, 1196–1205. [CrossRef]
41. Liu, X.; Wang, Q.; Zhang, J.; Guan, H.; Zhang, C. One-Step Preparation of MoS_2/Graphene Nanosheets via Solid-State Pan-Milling for High Rate Lithium-Ion Batteries. *Ind. Eng. Chem. Res.* **2020**, *59*, 16240–16248. [CrossRef]

42. Choi, J.H.; Kim, M.-C.; Moon, S.-H.; Kim, H.; Kim, Y.-S.; Park, K.-W. Enhanced electrochemical performance of MoS_2/graphene nanosheet nanocomposites. *RSC Adv.* **2020**, *10*, 19077. [CrossRef] [PubMed]
43. Yoon, H.; Kim, H.-J.; Yoo, J.J.; Yoo, C.-Y.; Park, J.H.; Lee, Y.A.; Cho, W.K.; Han, Y.-K.; Kim, D.H. Pseudocapacitive slurry electrodes using redox-active quinone for high-performance flow capacitors: An atomic-level understanding of pore texture and capacitance enhancement. *J. Mater. Chem. A* **2015**, *3*, 23323–23332. [CrossRef]

Disclaimer/Publisher's Note: The statements, opinions and data contained in all publications are solely those of the individual author(s) and contributor(s) and not of MDPI and/or the editor(s). MDPI and/or the editor(s) disclaim responsibility for any injury to people or property resulting from any ideas, methods, instructions or products referred to in the content.

Article

Facile Synthesis of Mesoporous Nanohybrid Two-Dimensional Layered Ni-Cr-S and Reduced Graphene Oxide for High-Performance Hybrid Supercapacitors

Ravindra N. Bulakhe [1,2], Anh Phan Nguyen [3], Changyoung Ryu [1], Ji Man Kim [2] and Jung Bin In [1,3,*]

1. Soft Energy Systems and Laser Applications Laboratory, School of Mechanical Engineering, Chung-Ang University, Seoul 06974, Republic of Korea; bulakhern@gmail.com (R.N.B.); asd24zx@cau.ac.kr (C.R.)
2. Department of Chemistry, Sungkyunkwan University, Suwon 16419, Republic of Korea; jimankim@skku.edu
3. Department of Intelligent Energy and Industry, Chung-Ang University, Seoul 06974, Republic of Korea; anhpn@cau.ac.kr
* Correspondence: jbin@cau.ac.kr

Abstract: This study describes the single-step synthesis of a mesoporous layered nickel-chromium-sulfide (NCS) and its hybridization with single-layered graphene oxide (GO) using a facile, inexpensive chemical method. The conductive GO plays a critical role in improving the physicochemical and electrochemical properties of hybridized NCS/reduced GO (NCSG) materials. The optimized mesoporous nanohybrid NCSG is obtained when hybridized with 20% GO, and this material exhibits a very high specific surface area of 685.84 m^2/g compared to 149.37 m^2/g for bare NCS, and the pore diameters are 15.81 and 13.85 nm, respectively. The three-fold superior specific capacity of this optimal NCSG (1932 C/g) is demonstrated over NCS (676 C/g) at a current density of 2 A/g. A fabricated hybrid supercapacitor (HSC) reveals a maximum specific capacity of 224 C/g at a 5 A/g current density. The HSC reached an outstanding energy density of 105 Wh/kg with a maximum power density of 11,250 W/kg. A 4% decrement was observed during the cyclic stability study of the HSC over 5000 successive charge–discharge cycles at a 10 A/g current density. These results suggest that the prepared nanohybrid NCSG is an excellent cathode material for gaining a high energy density in an HSC.

Keywords: electrochemical impedance spectroscopy; energy density; hybrid supercapacitor; nickel-chromium-sulfide (Ni-Cr-S); specific surface area

1. Introduction

Supercapacitors (SCs) are classified into three types: electric double-layer capacitors (EDLCs), pseudocapacitors, and hybrid SCs (HSCs) [1,2]. The EDLC [3] has limited specific energy due to a simple electrostatic charge storage mechanism and pseudocapacitor [4]. The charge transfer is limited at the electrode surface and electrolyte interface, compromising SCs' capability of ultra-high power and a long cycle life [2,5]. Therefore, the combination of high specific power and energy and long cycle life in one cell has been a research goal for energy storage devices. In addition, HSCs are one of the best alternative solutions to increasing specific energy and maintaining high specific power without sacrificing the cyclic stability of SC devices [5–7]. The HSC integrates the advantages of the EDLC and pseudocapacitor, minimizing their disadvantages, and is classified into three categories: asymmetric, composite, and battery-type, according to the nature of their anode and cathode materials.

Among them, battery-type hybrid materials display the best supercapacitive performance, offering a high charge storage capacity, operating voltage, and capacity retention; a long lifetime; and outstanding specific energy and power due to their unique charge storage mechanism of deep and diffusion-limited intercalation [8–10]. The charge storage mechanism of SCs depends on the composition and physicochemical properties of the electrode materials (e.g., conductivity, porosity, wettability, specific surface area, redox properties, etc.). Therefore, the charge stored in the SC is directly proportional to multiple faradic and non-faradic reactions occurring at a) the interface of the electrode surface with electrolytes and b) the interplanar surface of the layered materials [11,12].

Many researchers have used various materials and methods, such as carbon materials and their derivatives [13], metal hydroxide [14,15], metal oxides [16], metal sulfides [4,17,18], metal phosphates [11], and polymers [19] to fabricate HSCs. Among them, layered ternary metal sulfides are promising candidates due to their unique intrinsic properties, which include a high conductive nature, high theoretical capacity, low cost, and high specific power and energy with excellent cyclic stability due to the low electronegativity of sulfur [20,21]. In addition, the layered structure provides large interlayered spacing, which stores a large capacity using deep intercalation and limited diffusion, confirming a battery-type material [11].

Very few reports are available on Ni- and Cr-based ternary metal sulfides for SC applications. For example, Xu et al. synthesized mesoporous $NiCo_2S_4$ microspheres using a solvothermal method and obtained a specific capacity of 857 C/g at the current density of 1 A/g [22]. Similarly, Ni et al. reported a maximum specific capacity of 803.08 C/g at a current density of 1 A/g for the core-cell structure of $NiMn_2O_4@NiMn_2S_4$ nanoflowers@nansheets synthesized using a hydrothermal method [23]. Du et al. prepared $NiMoS_4$ and nickel-copper-sulfide for HSCs and achieved maximum specific capacities of 313 and 422.37 C/g at a current density of 1 A/g, respectively [24,25]. Hai et al. reported Cr-doped $(Co, Ni)_3S_4/Co_9S_8/NiS_2$ nanowires/nanoparticles using the hydrothermal method and achieved a high capacity of 1117 C/g [26]. Bulakhe et al. reported on a polyhedron-structured $CuCr_2S_4$ cathode using the hydrothermal method and obtained a capacity of 1536 C/g at 1 A/g current density [17]. Thus far, there have been no reports on $NiCrS_4$ and its hybrid/composite materials, so we focused this research on this topic. The literature in Table S1 concludes that the hydrothermal method efficiently synthesizes ternary metal disulfides and their hybrid/composite materials.

We synthesized $NiCr_2S_4$ (NCS) and hybridized it using single-layered graphene oxide (GO) and prepared nanohybrid $NiCr_2S_4/rGO$ (NCSG) materials. GO has been successfully applied to supercapacitors and batteries owing to its low production cost and outstanding physicochemical and electrochemical properties [27–29]. The conducting GO plays an important role in hybridization, and a reduced GO (rGO) amount was successfully engineered to enhance the performance of the NCSG. The optimized nanohybrid NCSG-2 provides a high specific surface area of 685.84 m^2/g compared to bare NCS at 149.37 m^2/g, resulting in a mesoporous nature with a pore diameter of around 15 nm. The NCSG nanohybrid material displays excellent capacity and retention with outstanding cyclic stability. An HSC device was devised using NCSG-2 as the cathode and rGO as the anode, denoted as NCSG//rGO. The HSC demonstrated outstanding specific energy and power with excellent cyclic stability. While submitting the manuscript, we found that this is the first report on an HSC study.

2. Experimental Section

2.1. Chemicals

All chemicals were purchased from Sigma-Aldrich (St. Louis, MO, USA). Nickel chloride ($NiCl_2$), chromium chloride ($CrCl_3$), hexamethylenetetramine ($C_6H_{12}N_4$), sodium sulfide (Na_2S), and GO ink were used prior to purification. Nickel foam (NF) with >99.99 purity was purchased from MTI Korea (Seoul, Korea).

2.2. Experimental Details on $NiCr_2S_4$ and $NiCr_2S_4/rGO$

The synthesis process for NCSG is described in Scheme 1. A chemical bath was prepared using 0.17 g of $NiCl_2$, 0.38 g of $CrCl_3$, and 0.22 g of Na_2S, serving as nickel, chromium, and sulfur precursors, respectively. Additionally, 0.99 g of HMT was introduced as a complexing agent during reactions. The chemical bath was prepared using water as an aqueous solvent. The chemical bath solution was stirred for 30 min at a ramping speed of 100 rpm. The prepared homogeneous chemical bath was poured into a 100 mL Teflon liner. The Teflon liner was closed and assembled in a stainless-steel case. The autoclave was kept in a convection oven for 12 h at 120 °C. After completing the reaction, the autoclave was cooled to the atmospheric temperature. The detailed reaction mechanism is explained in Equations S1–S6 in Note S1. The obtained product was repeatedly cleaned and filtered with deionized water and ethanol. The prepared NCS powder was dried for 12 h at 60 °C. A similar experiment was repeated with the addition of a single-layered GO ink of 10%, 20%, and 30% volume ratios to prepare hybrid NCSG materials. The prepared samples were labeled NCSG-1, NCSG-2, and NCSG-3, respectively. All of the NCS and NCSG samples were annealed at 450 °C for 2 h in an inert atmosphere. All prepared samples were used for further physicochemical and electrochemical characterizations.

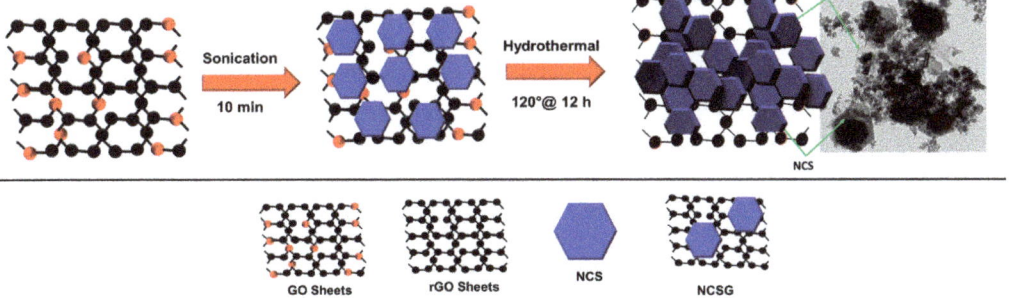

Scheme 1. Synthesis process of Ni-Cr-S/reduced graphene oxide (NCSG).

2.3. Material Characterizations

All physicochemical characterizations for NCS, NCSG-1, NCSG-2, and NCSG-3 samples were investigated. Structural characterizations were performed using X-ray powder diffraction (XRD, wavelength: 0.15406 nm) and X-ray photoelectron spectroscopy (XPS) techniques. Morphological characterizations were investigated using field-emission scanning electron microscopy (FE-SEM) and high-resolution transmission electron microscopy (HR-TEM). A porosity study was performed using the Brunauer–Emmett–Teller (BET) technique. The electrochemical studies were conducted using Bio-Logic science instruments.

2.4. Electrochemical Characterizations

The synthesized NCS, NCSG-1, NCSG-2, and NCSG-3 samples were investigated for electrochemical studies. Prior to the experiment, NF was cleaned with 2 M HCl to remove the oxide layer on the surface of the NF. The cleaned NF was used as a current collector. In an appropriate proportion of active materials, conductive carbon black and polyvinylidene difluoride (85%, 10%, and 5%) were mixed to prepare a homogeneous slurry. The obtained slurry was uniformly coated onto the NF, and the prepared electrodes were dried at 60 °C for 12 h. Afterward, the prepared electrodes were used as working, platinum chips, and Ag and AgCl were used as the counter and reference electrodes, respectively. The aqueous 2 M KOH electrolyte was used for the entire electrochemical performance. The cyclic voltammetry (CV), charge–discharge (CD), electrochemical impedance spectroscopy (EIS), and cyclic stability were performed for three-electrode systems. Afterward, the HSC was devised using NCSG-2 as the cathode, with rGO as the anode, a porous glass fiber separator, and an aqueous KOH electrolyte. The electrochemical cell (EC-CELL) setup was used to determine the performance regarding the CV, GCD, EIS, and cyclic stability.

3. Results and Discussion

The structural analysis of the prepared NCS and NCSG samples was performed based on XRD. The two major broad and intense characteristic peaks are found at the angles of 2θ: 17.20°, 30.28°, 34.81°, 35.87°, 44.63°, 49.42°, 56.18°, 59.41°, 65.07°, and 67.75°, which correspond to the (101), (110), (202), (013), (−114), (006), (−312), (215), (017), and (008) planes, respectively, belonging to the NCS (PDF card no. 01-088-0659). It confirms the formation of crystalline monoclinic $NiCr_2S_4$. The weak peaks marked with a star symbol in the plot belong to nickel sulfide. The peaks belonging to rGO in the NCSG XRD spectra are absent, which is possibly ascribed to the small percentage of rGO, so they are masked by the diffraction signal for NCS or the destruction of the regular GO during the synthesis processes through the interaction of NCS [30].

Further, the result was confirmed with the XPS study. The elemental composition and valance states of the NCS and NCSG-2 samples were investigated using the XPS analysis. Figure S1 depicts the full survey spectrum of the NCS and NCSG-2 samples. The XPS survey confirms the coexistence of Ni, Cr, and S in NCS and the additional C element in the NCSG-2 spectrum. Figure 1a' presents the C1s spectrum deconvoluted in three peaks at 284.7, 286.3, and 288.4 eV, belonging to the C-C/C=C, C-O, and C=O functional groups, respectively [31]. The small peaks for C-O and C=O confirm rGO formation during synthesis. The C-C/C=C functional group indicates the same carbon skeleton structure before and after the reaction; hence, rGO may provide a large specific surface area of the nanohybrid NCSG-2 material [11,32,33].

In Figure 1b,b', the Ni 2p spectra are comparable between the NCS and NCSG-2 samples. Spin–orbit doublets of $2p_{3/2}$ and $2p_{1/2}$ are revealed; in order, Ni^{2+} peaks are located at 855.73 and 873.48 eV, Ni^{3+} at 857.33 eV and 875.1 eV, and satellite peaks at 861.83 eV and 879.68 eV, respectively [34–36]. The spin–orbit splitting of doublet pairs is ~17.75 eV, typical of Ni 2p doublets [35,36]. These results confirm the coexistence of Ni^{2+} and Ni^{3+} [35]. Figure 1c,c' depict the resembling Cr 2p spectra of NCS and NCSG-2 samples. They are deconvoluted into two spin–orbit doublet pairs of $2p_{3/2}$ and $2p_{1/2}$. Cr^{3+} peaks are located at 576.73 and 586.16 eV, Cr^{4+} at 578.58 and 587.88 eV, respectively [31,37], and the spin–orbit splitting is ~9.3 eV, which is typical of Cr 2p [37], confirming the coexistence of Cr^{3+} and Cr^{4+} [26]. The S 2p spectra are depicted in Figure 1d,d'. The S 2p spectra were also deconvoluted into $2p_{3/2}$ and $2p_{1/2}$ pairs. For these pairs, S peaks are placed at 163.43 and 164.88 eV, while metal sulfate peaks are found at 169.28 and 170.38 eV, respectively [34,35,38–40]. The 1.15 eV spin–orbit is typical of S 2p [41].

The morphological study of the NCS and NCSG samples was performed using FE-SEM to determine the crystal shape, size, and morphology. The NCS sample displayed hexagonal nanosheets with a 400 nm diameter with nanoparticles in Figure 2a. Figure 2b demonstrates that the rGO sheets are covered with NCS hexagonal nanosheets with nanoparticles and form an NCSG-2 nanohybrid. Similar results for the NCSG-1 and NCSG-3 samples are depicted in Figure S2. Similar results have also been reported for nanohybrid materials in the literature [42,43]. The elemental mapping and energy-dispersive spectroscopy (EDS) spectra of the NCSG-2 sample are depicted in Figure S3. Figure S3a–e presents the uniform distribution of all individual elements (Ni, Cr, S, and C) throughout the scanned area. Furthermore, the NCSG-2 structure was examined using HR-TEM. Figure 2c,d reveals that the NCS hexagonal nanosheets and nanoparticles completely cover the rGO sheets. Figure 2e–g illustrates the high-resolution image of the nanosheets. The HR-TEM image (Figure 2e,f) depicts the parallel lattice fringes with 0.27 and 0.25 nm 'd'-spacing values, corresponding to the (004) and (202) planes of the monoclinic structure of NCS. Figure 2h displays the selected area electron diffraction pattern and the resulting polycrystalline nature of the NCSG-2 material. Figure S4 provides the elemental mapping analysis of the NCSG-2 sample employed using the HR-TEM connected to the EDS for all constituent elements. Figure S4 suggests that the uniform distribution for Ni, Cr, S, and C is observed throughout the NCSG-2 structure.

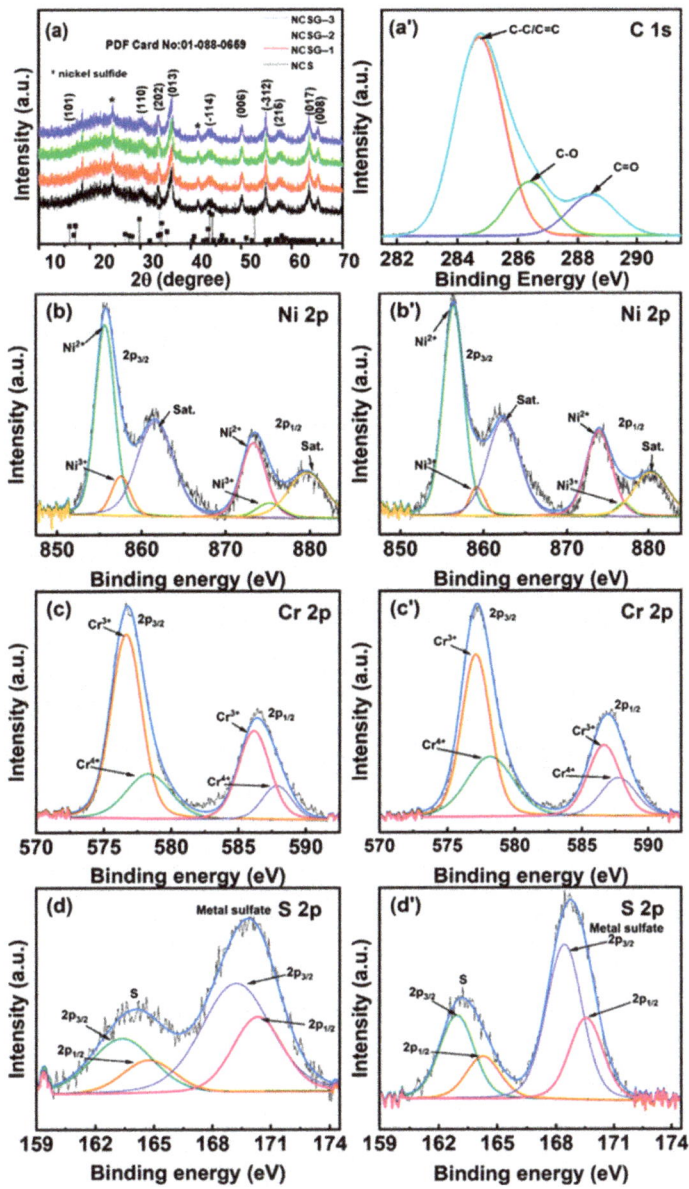

Figure 1. (**a**) X-ray powder diffraction spectra for NCS and all NCSG samples; high-resolution X-ray photoelectron spectroscopy spectra of (**a′**) C, (**b,b′**) Ni, (**c,c′**) Cr, and (**d,d′**) S elements of NCS and NCSG-2 samples, respectively.

The effects of rGO on the specific surface area, pore diameter, and pore volume of NCS were studied using the N_2 adsorption–desorption isotherm using BET and Barret–Joyner–Halenda (BJH) measurements. The N_2 adsorption–desorption isotherms of bare NCS and nanohybrid NCSG-2 are presented in Figure 3a. The obtained isotherms are categorized as BDDT type-IV isotherms and H_3-type hysteresis loops, suggesting the mesoporous nature of the materials [44]. The measured specific surface areas using the BET measurements are 149.37 m^2/g for bare NCS and 685.84 m^2/g for NCSG-2 materials, much higher than for bare NCS. Using the BJH method, the pore-size and pore-volume distributions of bare NCS and nanohybrid NCSG-2 were investigated. The observed pore sizes of the NCS and NCSG-2 materials are 13.85 and 15.81 nm (Figure 3b), confirming a mesoporous nature. In addition, Figure 3b displayed pore volumes of 1.50 and 0.29 cm^3/g for the NCSG-2 and NCS materials, respectively. The obtained pore sizes are <20 nm, which may reduce the electrical resistance and enhance the electrochemical activity of the materials [45]. The presence of plenty of mesopores within the nanohybrid materials facilitates the fast movement of electrolyte ions and works as ion storage, possibly enhancing electrochemical performance [45].

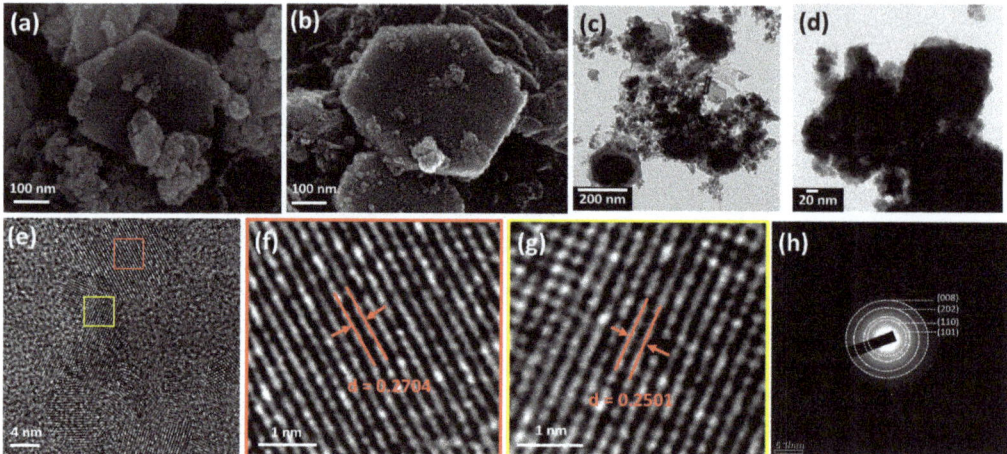

Figure 2. (**a**) Field-emission scanning electron microscopy images of bare NiCr$_2$S$_4$ (NCS), (**b**) nanohybrid NCS with 20% graphene oxide (NCSG-2), (**c**,**d**) transmission electron microscopy (TEM) images, (**e**) high-resolution TEM image, (**f**,**g**) additional high-resolution TEM images corresponding to the areas shown in Figure 2e marked in red and yellow, respectively, and (**h**) selected area electron diffraction pattern of the NCSG-2 nanohybrid sample.

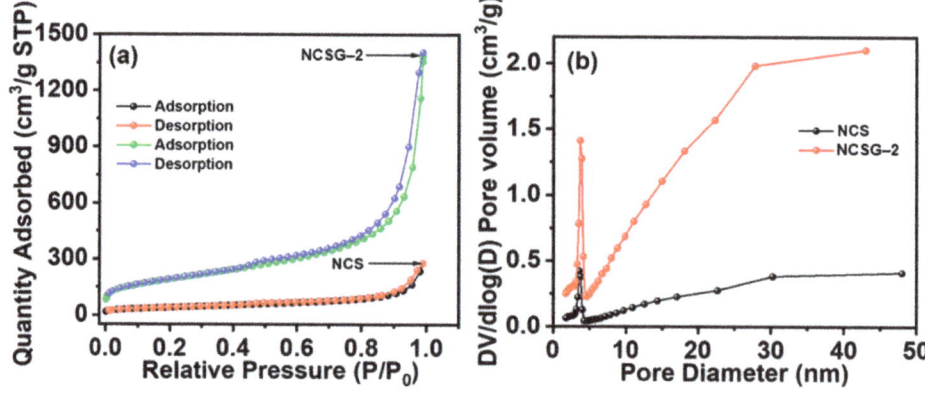

Figure 3. (**a**) Adsorption-desorption isotherm plot and (**b**) Barret–Joyner–Halenda (BJH) desorption pore-size distribution plot for NiCr$_2$S$_4$ (NCS) and NCS with 20% graphene oxide (NCSG-2).

4. Electrochemical Studies

The supercapacitive study was investigated using a three-electrode system consisting of NCS/NCSG electrodes serving as the working electrode and platinum chip and Ag/AgCl as the counter and reference electrodes, respectively. The aqueous 2 M KOH was used as an electrolyte. Figure 4 depicts all graphs of SC studies for NCS and NCSG electrodes. Initially, the comparative CV measurements of the bare NCS and nanohybrid NCSG-1, NCSG-2, and NCSG-3 electrodes were performed at a 20 mV/s scan rate in Figure 4a. Among them, NCSG-2 revealed an excellent capacity due to the high current value (large area under the curve). All CV curves exhibited similar shapes, including a couple of intense redox peaks. This result suggests that the specific capacity attributes, due to a fast and reversible redox reaction of Ni^{2+}/Ni^{3+} and Cr^{2+}/Cr^{3+}, indicate battery-type faradic reactions, strongly aligning with the literature reports [14,46]. The possible redox reaction may be executed as follows:

$$NiCr_2S_4 + 4OH^- \Leftrightarrow NiSOH + 2CrSOH + SOH^- + 3e^- \tag{1}$$

$$CrSOH + OH^- \Leftrightarrow CrSO + H_2O + e^- \tag{2}$$

$$NiSOH + OH^- \Leftrightarrow CrSO + H_2O + e^- \tag{3}$$

Figure 4b,c combined the CV curves of NCS and NCSG-2 electrodes, and Figure S5a,b depicts all CV curves for the NCSG-1 and NCSG-3 electrodes at various scan rates from 2–20 mV/s, respectively. The specific capacities of NCS and all NCSG electrodes were estimated using Equation S7. The specific capacities of NCS, NCSG-1, NCSG-2, and NCSG-3 are 764, 1109, 1960, and 1500 C/g, respectively. Battery-type faradic reactions suggest that the total stored charge contributes to diffusion-controlled (Q_d) and capacitive-type (Q_s) mechanisms. The following equation can be used to separate the total stored charge analysis:

$$Q_t = Q_s + Q_d, \tag{4}$$

where $Q_d = cv^{-1/2}$, Q_t represents the total charge, Q_s denotes the capacitive surface charge, and c is constant. In addition, Q_s can be obtained using the graph Q_t vs. $v^{-1/2}$ after extrapolating v on the y-axis. Figure S6a plots $v^{-1/2}$ vs. the total charge for all electrodes. All calculated Q_s and Q_d value fragments are presented in Figure S6b for various scan rates from 2–20 mV/s for the NCS and all NCSG electrodes. The Q_s values for NCS, NCSG-1, NCSG-2, and NCSG-3 electrodes are 86%, 83%, 84%, and 90%, respectively, at the 20 mV/s scan rate.

Furthermore, the charge storage ability of the NCS and NCSG electrodes was investigated using a CD study at various current density values. Figure S7a combines the curves of the CD for the NCS, NCSG-1, NCSG-2, and NCSG-3 electrodes at a 2 A/g current density. The NCSG-2 electrode displayed the maximum discharge time compared with the other electrodes, suggesting a high charge-storing ability. Similarly, Figure 4d,e and Figure S7b,c depict the CD curves of the NCS, NCSG-2, NCSG-1, and NCSG-3 electrodes at various current density values from 2–6 A/g. The specific capacities of NCS and all NCSG electrodes were estimated using the following Equation.

$$Specific\ capacity\ (C/g) = \frac{\int i(t)dt}{m} \tag{5}$$

where i, t, and m represent the applied current density in mA, the discharge time in seconds, and the active mass of the electrode material in mg, respectively. The specific capacity values for NCS, NCSG-1, NCSG-2, and NCSG-3 are 676, 920, 1932, and 1269 C/g, respectively. The NCSG-2 electrode displayed the highest capacity among them because the layered morphology supports a large, effective specific surface area, providing numerous active electrochemical sites.

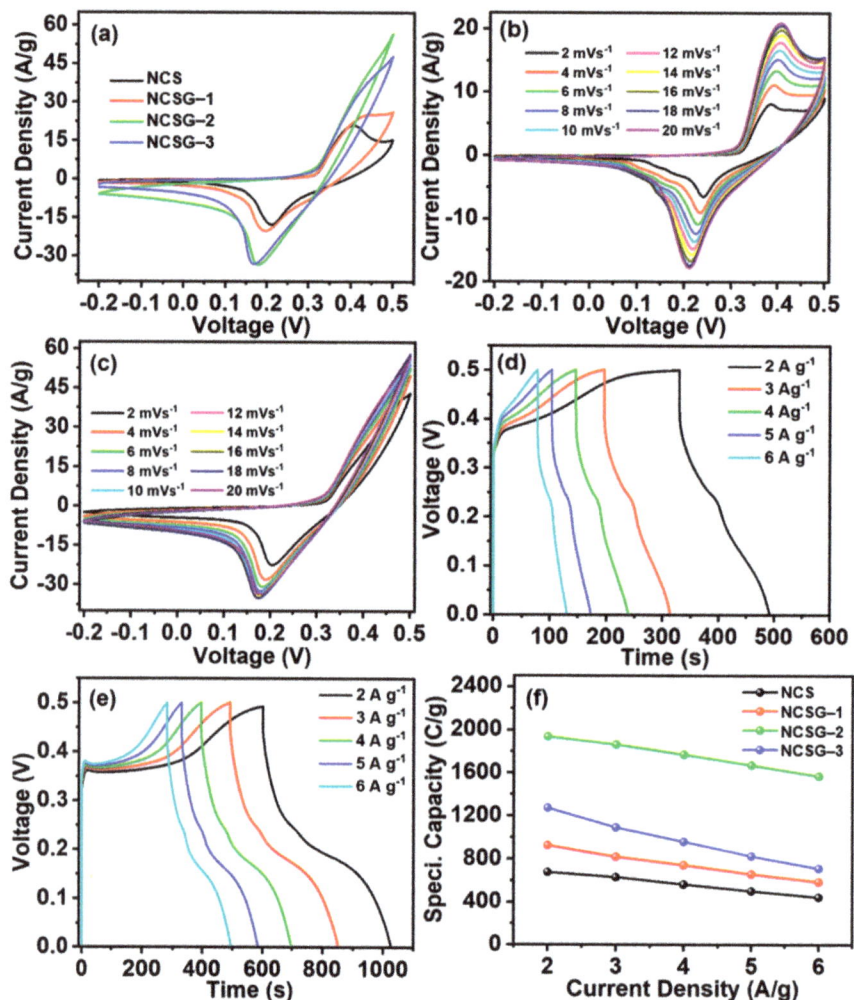

Figure 4. (**a**) Combined cyclic voltammetry (CV) graphs of NiCr$_2$S$_4$ (NCS) and NCS with 10%, 20%, and 30% graphene oxide (NCSG-1, NCSG-2, and NCSG-3) at a 20 mV/s scan rate; (**b**) CV graph of the NCS electrode (2–20 mV/s); (**c**) CV graphs of the NCSG-2 electrode (2–20 mV/s); (**d**) charge–discharge (CD) curves of the NCS electrode (2–6 A/g); (**e**) CD curves of the NCSG-2 electrode (2–6 A/g); and (**f**) graph of the current density vs. specific capacity of the NCS, NCSG-1, NCSG-2, and NCSG-3 electrodes, respectively.

Additionally, NCSG-2 exhibits mesoporosity (~16 nm), accelerating the diffusion of the electrolyte ions through pores and contributing to more redox reactions [12]. Based on the CD measurements, the specific capacity values for NCS and NCSG are plotted in Figure 4f. The NCSG-2 electrode delivers a much better rate capability than the NCS, NCSG-1, and NCSG-3 electrodes. The rate capabilities of the NCS, NCSG-1, NCSG-2, and NCSG-3 electrodes are 60.6%, 65.4%, 81%, and 69%, respectively. Figure 4f reveals that the bare NCS electrode had a smaller capacity than all NCSG electrodes because the conductive rGO plays a crucial role in the hybridization. Table S1 provides a comparative SC study of the Ni- and Cr-based ternary metal dichalcogenides. The specific capacity reported in the literature suggests that NCS and NCSG perform better [17,24,25].

The charge storage capabilities are directly proportional to the amount of GO hybridized in the NCSG compositions. The NCSG-2 offers a high capacity value (1932 C/g) compared to NCSG-1 (920 C/g) and NCSG-3 (1269 C/g) because the amount of GO is optimum for NCSG-2. If the GO amount is low, the probability of forming a conductive channel is low. In another case, when the amount of GO is too high, the GO sheets are likely to restack, reducing the charge storage performance [11].

Moreover, EIS is an important tool providing an intrinsic resistance of the active materials and the solution resistance (electrode and electrolyte interface). A comparative EIS study for NCS and NCSG electrodes was investigated. Figure S8a depicts all combined Nyquist plots for the NCS and NCSG electrodes. All hybrid NCSG electrodes displayed a semicircle in the decreasing trend compared to the bare NCS electrodes, suggesting that incorporating the GO sheets minimized the charge-transfer resistance (R_{ct}). The low-frequency region of the Nyquist plots is ascribed to the diffusion resistance of the electrolyte (W). All NCSG electrodes displayed an enlarged slope, indicating that the porous structure enhanced the electrolyte ion's kinetic transport in the hybrid electrodes. Table S2 provides the comparative values of all parameters from the fitting of the Nyquist plots for all NCS and NCSG electrodes. The NCSG-2 electrode exhibited better R_s, R_{ct}, C_{dl}, and W values, respectively.

The study of the long-term stability of SCs is important for NCS and NCSG electrodes (Figure S8b). The continuous 10,000 GCD cycles were performed at a 10 A/g current density for each electrode. During the CD processes, swelling and shrinkage while intercalation and deintercalation processes result in stress and cracks within the material and interface [47]. The NCSG-1, NCSG-2, and NCSG-3 electrodes retain 90.8%, 95.2%, and 92.9%, and the NCS electrode retains 89.1% of their original values over continuous 10,000 GCD cycles. The results indicate that NCSG-2 is an excellent electrode among the studied options.

Hybrid Supercapacitor

To investigate the potential of the nanohybrid NCSG-2 electrode in the full cell assembly, the HSC was devised using NCSG-2 as the positive electrode and rGO as the negative electrode. The performance of rGO (the EDLC) is studied in an aqueous electrolyte of 2 M KOH using a three-electrode assembly, and all results (CV, CD, and EIS) are provided in Figure S9. The mass balance of the anode and cathode materials was optimized using the equation provided in Note S2 to obtain the high electrochemical performance of the HSC. In the HSC, we combined two potential materials as separate electrodes to achieve a higher cell potential than individual electrodes, resulting in a high energy density. In this study, the 0 to −1 V potential for the anode is coupled with the −0.2 to 0.5 V of the cathode, providing a full cell voltage of 1.5 V (Figure S10). The CV and CD curves were examined at various potential ranges from 0 to 1.6 V to investigate the optimum operating potential range of NCSG-2//rGO HSC (Figure 5a,b) [48]. The specific capacity was calculated using CD curves for various potential ranges (Figure 5c), indicating that the capacity increases linearly with potential ranges. Similarly, a linear energy density increase was observed from 2.13 to 105 Wh/kg as the potential range increases in the range of 0.8 to 1.5 V, resulting in the effectiveness of assembling the HSC (Figure 5c).

The studies of the HSC were performed using an optimum 1.5 V potential window at various scan rates (5 to 100 mV/s). Figure 5d depicts the semi-rectangular shapes of all CV curves, suggesting a hybrid-type charge storage mechanism (EDLC and battery-type) contributed to the capacitance as a result of the positive (cathode) and negative (anode) materials. As the scan rates increased from 5 to 100 mV/s, the shapes of the CV curves continued with the same enlarged area, indicating the high performance of the HSC device. Afterward, the CD performance of the HSC device was investigated at various current density values (Figure 5e). The specific capacities were estimated using discharge curves for each current density (Figure 5f). At the lowest current density of 5 A/g, the HSC delivers 224 C/g, and for a high current density of 10 A/g, it remains at 140 C/g. The capacity retention remains at 62.5% of its original value after increasing the current density to 10 A/g. The energy and power density values of the HSC device were calculated and plotted in Figure 5g. The HSC delivers the maximum energy density of 105 Wh/kg at the power density of 5625 W/kg and the maximum power density of 11,250 W/kg at the lowest energy density of 65.62 Wh/kg. Table S3 provides a comparative study of the electrochemical performance of Ni- and Cr-based HSC devices reported earlier and presents the working NCSG-2//rGO device. Table S3 suggests that the HSC device displays superior performance compared to the reported HSC devices [23–26,35].

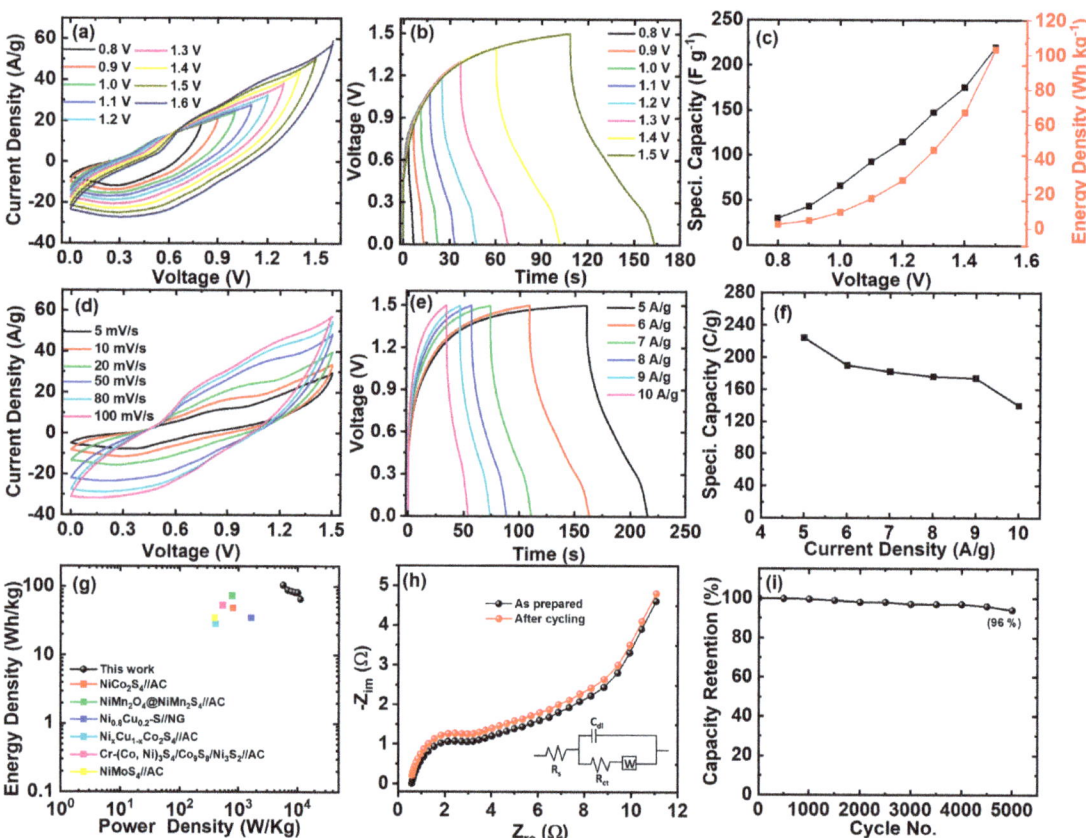

Figure 5. Complete hybrid supercapacitor (HSC) study: (**a**) combined cyclic voltammetry (CV) graphs from 0.8 to 1.6 V potential windows with a 100 mV/s scan rate; (**b**) combined charge–discharge (CD) curves from 0.8 to 1.6 V potential window with a 6 A/g current density; (**c**) graph of potential variation with specific capacity (black color) and energy density (red color); (**d**) CV curves from 5 to 100 mV/s of the scan rate; (**e**) CD curves from 5 to 10 A/g of the current density; (**f**) graph of the current density vs. specific capacity; (**g**) Ragone plot; (**h**) Nyquist plot; and (**i**) Capacity retention obtained from continuous 5000 CD cycles for HSC cell.

The electrochemical impedance measurements of the NCSG-2//rGO HSC device have been investigated from a lower frequency of 100 mHz to a higher frequency of 1 MHz (Figure 5h). The HSC delivers a smaller solution resistance (R_s) and charge-transfer resistance (R_{ct}) of 0.47 and 1.03 Ω, respectively. After completing 5000 continuous CD cycles, R_s and R_{ct} increased to 0.71 and 1.70 Ω, respectively. The enhancement in the EIS parameters is very small, suggesting the excellent performance of the electrode materials. The equivalent circuit was fitted to the Nyquist plot and depicted in the Figure 5h inset. All values of the circuit components are given in Table S4. The capacitance retention against the cycle number graph is plotted in Figure 5i. The HSC device reduced capacity retention by 4% over 5000 CD successive cycles at a 10 A/g current density. A similar study was reported for Ni- and Cr-based HSCs (Table S3). The performance of the NCSG electrodes can also be evaluated based on specific capacitance (F/g), and the converted results are shown in Figure S11 of the Supplementary Materials.

5. Conclusions

The mesoporous layered nanohybrid novel NCSG was successfully synthesized in a single step using a simple and inexpensive chemical route. The physicochemical and electrochemical properties of the NCSG were tuned using various amounts of single-layer GO during the nanohybrid material preparation. Mesoporous NCSG provides a fourfold greater specific surface area of 658.84 m^2/g compared to bare NCS (149.37 m^2/g). The conductive rGO enhanced the three-fold higher capacity of the optimized NCSG-2 (1932 C/g) compared to the bare NCS (676 C/g) materials. The NCSG-2 was used as the cathode, and rGO was used as the anode during the fabrication of the HSC. The HSC delivered a maximum specific capacity of 224 C/g at a 5 A/g density. The HSC had a maximum energy density of 105 Wh/kg and a maximum power density of 11,250 W/kg. The HSC obtained 96% cyclic stability over successive 5000 CD cycles. The robust and outstanding supercapacitive performance of the NCSG nanohybrid material suggests that it is a promising candidate for high-energy HSCs.

Supplementary Materials: The following supporting information can be downloaded at: https://www.mdpi.com/article/10.3390/ma16196598/s1, Figure S1: Entire X-ray photoelectron spectroscopy survey spectrum; Figure S2: Field-emission scanning electron microscopy images; Figure S3: Field-emission scanning electron microscopy elemental mapping; Figure S4: Energy dispersive spectroscopy mapping for each element; Figure S5: Cyclic voltammetry curves; Figure S6: Plot of the reciprocal square root of the scan rate vs. the total charge; Figure S7: Combined charge–discharge (CD) curves of $NiCr_2S_4$ (NCS) and NCS with 10%, 20%, and 30% graphene oxide; Figure S8: Nyquist plots and stability study at a 10 A/g current density for $NiCr_2S_4$ (NCS) and NCS; Figure S9: Supercapacitor study of reduced graphene oxide; Figure S10: Comparative cyclic voltammetry graphs of reduced graphene oxide (rGO; black) and $NiCr_2S_4$; Figure S11: Graph of specific capacitance vs. current density for NCS, NCSG 1, NCSG-2, and NCSG-3 electrodes; Table S1: Detailed description of Ni- and Cr-based TMD's for supercapacitive performance using chemical methods; Table S2: Comparative data from Nyquist plot fittings; Table S3: The Ni- Cr-based HSC device study for supercapacitive parameters; Table S4: Fitting parameters. References [49–53] are cited in the supplementary materials.

Author Contributions: Conceptualization, R.N.B. and J.B.I.; methodology, R.N.B. and J.B.I.; validation, R.N.B., C.R. and A.P.N.; formal analysis, R.N.B., A.P.N., C.R. and J.B.I.; investigation, R.N.B. and C.R.; writing—original draft preparation, R.N.B., A.P.N. and J.B.I.; writing—review and editing, R.N.B. and J.B.I.; visualization, R.N.B.; supervision, J.M.K. and J.B.I.; funding acquisition, J.B.I. All authors have read and agreed to the published version of the manuscript.

Funding: This research was supported by the National Research Foundation of Korea (NRF) grant funded by the Korean government (MSIT) (No. NRF-2022R1A2C1010296). Additionally, this work was supported by another grant from the National Research Foundation of Korea (NRF) funded by the Korean government (MSIT, NRF-2022R1A4A1032832). C. Ryu was supported by the Chung-Ang University Research Scholarship Grants in 2021.

Informed Consent Statement: Not applicable.

Data Availability Statement: The data presented in this study are available on request from the corresponding author.

Conflicts of Interest: The authors declare no conflict of interest.

References

1. Lim, E.; Jo, C.; Lee, J. A mini review of designed mesoporous materials for energy-storage applications: From electric double-layer capacitors to hybrid supercapacitors. *Nanoscale* **2016**, *8*, 7827–7833.
2. Olabi, A.G.; Abbas, Q.; Al Makky, A.; Abdelkareem, M.A. Supercapacitors as next generation energy storage devices: Properties and applications. *Energy* **2022**, *248*, 123617.
3. Chodankar, N.R.; Pham, H.D.; Nanjundan, A.K.; Fernando, J.F.S.; Jayaramulu, K.; Golberg, D.; Han, Y.-K.; Dubal, D.P. True Meaning of Pseudocapacitors and Their Performance Metrics: Asymmetric versus Hybrid Supercapacitors. *Small* **2020**, *16*, 2002806. [CrossRef]
4. Bulakhe, R.N.; Alfantazi, A.; Rok Lee, Y.; Lee, M.; Shim, J.-J. Chemically synthesized copper sulfide nanoflakes on reduced graphene oxide for asymmetric supercapacitors. *J. Ind. Eng. Chem.* **2021**, *101*, 423–429. [CrossRef]
5. Gao, D.; Luo, Z.; Liu, C.; Fan, S. A survey of hybrid energy devices based on supercapacitors. *Green Energy Environ.* **2022**, *8*, 972–988.
6. Dubal, D.P.; Ayyad, O.; Ruiz, V.; Gómez-Romero, P. Hybrid energy storage: The merging of battery and supercapacitor chemistries. *Chem. Soc. Rev.* **2015**, *44*, 1777–1790. [CrossRef]
7. Devi, M.; Moorthy, B.; Thangavel, R. Recent developments in zinc metal anodes, cathodes, and electrolytes for zinc-ion hybrid capacitors. *Sustain. Energy Fuels* **2023**, *7*, 3776–3795.
8. Pal, B.; Yang, S.; Ramesh, S.; Thangadurai, V.; Jose, R. Electrolyte selection for supercapacitive devices: A critical review. *Nanoscale Adv.* **2019**, *1*, 3807–3835.
9. Huang, B.; Yao, D.; Yuan, J.; Tao, Y.; Yin, Y.; He, G.; Chen, H. Hydrangea-like $NiMoO_4$-Ag/rGO as Battery-type electrode for hybrid supercapacitors with superior stability. *J. Colloid Interf. Sci.* **2022**, *606*, 1652–1661. [CrossRef]
10. Shi, C.; Sun, J.; Pang, Y.; Liu, Y.; Huang, B.; Liu, B.-T. A new potassium dual-ion hybrid supercapacitor based on battery-type $Ni(OH)_2$ nanotube arrays and pseudocapacitor-type V_2O_5-anchored carbon nanotubes electrodes. *J. Colloid Interf. Sci.* **2022**, *607*, 462–469. [CrossRef]
11. Bulakhe, R.N.; Lee, J.; Tran, C.V.; In, J.B. Mesoporous nanohybrids of 2D layered Cu–Cr phosphate and rGO for high-performance asymmetric hybrid supercapacitors. *J. Alloy Compd.* **2022**, *926*, 166864.
12. Sadavar, S.V.; Padalkar, N.S.; Shinde, R.B.; Patil, A.S.; Patil, U.M.; Magdum, V.V.; Chitare, Y.M.; Kulkarni, S.P.; Kale, S.B.; Bulakhe, R.N.; et al. Lattice engineering exfoliation-restacking route for 2D layered double hydroxide hybridized with 0D polyoxotungstate anions: Cathode for hybrid asymmetric supercapacitors. *Energy Storage Mater.* **2022**, *48*, 101–113.
13. Yang, W.; Ni, M.; Ren, X.; Tian, Y.; Li, N.; Su, Y.; Zhang, X. Graphene in Supercapacitor Applications. *Curr. Opin. Colloid Interface Sci.* **2015**, *20*, 416–428.
14. Padalkar, N.S.; Sadavar, S.V.; Shinde, R.B.; Patil, A.S.; Patil, U.M.; Magdum, V.V.; Chitare, Y.M.; Kulkarni, S.P.; Bulakhe, R.N.; Parale, V.G.; et al. 2D-2D nanohybrids of Ni–Cr-layered double hydroxide and graphene oxide nanosheets: Electrode for hybrid asymmetric supercapacitors. *Electrochim. Acta* **2022**, *424*, 140615.
15. Li, Y.; Huang, B.; Zhao, X.; Luo, Z.; Liang, S.; Qin, H.; Chen, L. Zeolitic imidazolate framework-L-assisted synthesis of inorganic and organic anion-intercalated hetero-trimetallic layered double hydroxide sheets as advanced electrode materials for aqueous asymmetric super-capacitor battery. *J. Power Sources* **2022**, *527*, 231149. [CrossRef]
16. Jadhav, S.B.; Malavekar, D.B.; Bulakhe, R.N.; Patil, U.M.; In, I.; Lokhande, C.D.; Pawaskar, P.N. Dual-Functional Electrodeposited Vertically Grown $Ag-La_2O_3$ Nanoflakes for Non-Enzymatic Glucose Sensing and Energy Storage Application. *Surf. Interfaces* **2021**, *23*, 101018. [CrossRef]
17. Bulakhe, R.N.; Ryu, C.; Gunjakar, J.L.; In, J.B. Chemical route to the synthesis of novel ternary $CuCr_2S_4$ cathodes for asymmetric supercapacitors. *J. Energy Storage* **2022**, *56*, 106175.
18. Thangavel, R.; Ganesan, B.K.; Thangavel, V.; Yoon, W.-S.; Lee, Y.-S. Emerging Materials for Sodium-Ion Hybrid Capacitors: A Brief Review. *ACS Appl. Energy Mater.* **2021**, *4*, 13376–13394. [CrossRef]
19. Bhalerao, A.B.; Bulakhe, R.N.; Deshmukh, P.R.; Shim, J.-J.; Nandurkar, K.N.; Wagh, B.G.; Vattikuti, S.V.P.; Lokhande, C.D. Chemically synthesized 3D nanostructured polypyrrole electrode for high performance supercapacitor applications. *J. Mater. Sci. Mater. Electron.* **2018**, *29*, 15699–15707. [CrossRef]
20. Iqbal, M.Z.; Khan, M.W.; Shaheen, M.; Siddique, S.; Aftab, S.; Alzaid, M.; Iqbal, M.J. Evaluation of d-block metal sulfides as electrode materials for battery-supercapacitor energy storage devices. *J. Energy Storage* **2022**, *55*, 105418. [CrossRef]
21. Khan Abdul, S.; Anuj, K.; Amjad, F.; Mohammad, T.; Muhammad, A.; Muhammad, U.; Akmal, A.; Saira, A.; Lujun, P.; Ghulam, Y. Benchmarking the charge storage mechanism in nickel cobalt sulfide nanosheets anchored on carbon nanocoils/carbon nanotubes nano-hybrid for high performance supercapacitor electrode. *J. Energy Storage* **2022**, *56*, 106041.
22. Xu, H.; Chen, P.; Zhu, Y.; Bao, Y.; Ma, J.; Zhao, X.; Chen, Y. Self-assembly and controllable synthesis of high-rate porous $NiCo_2S_4$ electrode materials for asymmetric supercapacitors. *J. Electroanal. Chem.* **2022**, *921*, 116688. [CrossRef]
23. Lv, X.; Min, X.; Feng, L.; Lin, X.; Ni, Y. A novel $NiMn_2O_4@NiMn_2S_4$ core-shell nanoflower@ nanosheet as a high-performance electrode material for battery-type capacitors. *Electrochim. Acta* **2022**, *415*, 140254.
24. Du, D.; Lan, R.; Humphreys, J.; Xu, W.; Xie, K.; Wang, H.; Tao, S. Synthesis of $NiMoS_4$ for high-performance hybrid supercapacitors. *J. Electrochem. Soc.* **2017**, *164*, A2881. [CrossRef]
25. Du, D.; Lan, R.; Humphreys, J.; Amari, H.; Tao, S. Preparation of nanoporous nickelcopper sulfide on carbon cloth for high-performance hybrid supercapacitors. *Electrochim. Acta* **2018**, *273*, 170–180. [CrossRef]

26. Hai, Y.; Tao, K.; Dan, H.; Liu, L.; Gong, Y. Cr-doped (Co, Ni)$_3$S$_4$/Co$_9$S$_8$/Ni$_3$S$_2$ nanowires/nanoparticles grown on Ni foam for hybrid supercapacitor. *J. Alloy Compd.* **2020**, *835*, 155254.
27. Zhao, C.S.; Gao, H.P.; Chen, C.M.; Wu, H. Reduction of graphene oxide in Li-ion batteries. *J. Mater. Chem. A* **2015**, *3*, 18360–18364. [CrossRef]
28. Kim, S.; Park, G.; Sennu, P.; Lee, S.; Choi, K.; Oh, J.; Lee, Y.S.; Park, S. Effect of degree of reduction on the anode performance of reduced graphene oxide in Li-ion batteries. *Rsc Adv.* **2015**, *5*, 86237–86241. [CrossRef]
29. Tran, C.V.; Khandelwal, M.; Lee, J.; Nguyen, A.P.; In, J.B. Controllable self-propagating reduction of graphene oxide films for energy-efficient fabrication. *Int. J. Energ. Res.* **2022**, *46*, 6876–6888. [CrossRef]
30. Ruidíaz-Martínez, M.; Álvarez, M.A.; López-Ramón, M.V.; Cruz-Quesada, G.; Rivera-Utrilla, J.; Sánchez-Polo, M. Hydrothermal Synthesis of rGO-TiO$_2$ Composites as High-Performance UV Photocatalysts for Ethylparaben Degradation. *Catalysts* **2020**, *10*, 520. [CrossRef]
31. Chen, Y.; An, D.; Sun, S.; Gao, J.; Qian, L. Reduction and Removal of Chromium VI in Water by Powdered Activated Carbon. *Materials* **2018**, *11*, 269. [CrossRef] [PubMed]
32. Sun, F.; Zhu, Y.; Liu, X.; Chi, Z. Highly efficient removal of Se(IV) using reduced graphene oxide-supported nanoscale zero-valent iron (nZVI/rGO): Selenium removal mechanism. *Environ. Sci. Pollut. Res.* **2023**, *30*, 27560–27569. [CrossRef] [PubMed]
33. Zhuang, W.; Li, Z.; Song, M.; Zhu, W.; Tian, L. Synergistic improvement in electron transport and active sites exposure over RGO supported NiP/Fe$_4$P for oxygen evolution reaction. *Ionics* **2022**, *28*, 1359–1366. [CrossRef]
34. Ling, L.; Zhang, C.; Lai, D.; Su, M.; Gao, F.; Lu, Q. Simultaneous phosphorization and sulfuration to Synergistically promote the supercapacitor performance of heterogeneous (Co$_x$Ni$_{1-x}$)$_2$P/Co$_x$Ni$_{1-x}$S hydrangea-like microspheres. *J. Power Sources* **2023**, *581*, 233487. [CrossRef]
35. Chen, W.; Yuan, P.; Guo, S.; Gao, S.; Wang, J.; Li, M.; Liu, F.; Wang, J.; Cheng, J.P. Formation of mixed metal sulfides of Ni$_x$Cu$_{1-x}$Co$_2$S$_4$ for high-performance supercapacitors. *J. Electroanal. Chem.* **2019**, *836*, 134–142. [CrossRef]
36. Fu, Z.; Hu, J.; Hu, W.; Yang, S.; Luo, Y. Quantitative analysis of Ni^{2+}/Ni^{3+} in Li[Ni$_x$Mn$_y$Co$_z$]O$_2$ cathode materials: Non-linear least-squares fitting of XPS spectra. *Appl. Surf. Sci.* **2018**, *441*, 1048–1056. [CrossRef]
37. Liang, H.; Ma, K.; Zhao, X.; Geng, Z.; She, D.; Hu, H. Enhancement of Cr(VI) adsorption on lignin-based carbon materials by a two-step hydrothermal strategy: Performance and mechanism. *Int. J. Biol. Macromol.* **2023**, *252*, 126432. [CrossRef]
38. Bulakhe, R.N.; Arote, S.A.; Kwon, B.; Park, S.; In, I. Facile synthesis of nickel cobalt sulfide nano flowers for high performance supercapacitor applications. *Mater. Today Chem.* **2020**, *15*, 100210. [CrossRef]
39. Yang, X.; Luo, Z.; Wang, D.; Deng, C.; Zhao, Y.; Tang, F. Simple hydrothermal preparation of sulfur fluoride-doped g-C$_3$N$_4$ and its photocatalytic degradation of methyl orange. *Mater. Sci. Eng. B* **2023**, *288*, 116216. [CrossRef]
40. Zhang, Z.; Yang, J.; Liu, J.; Gu, Z.-G.; Yan, X. Sulfur-doped NiCo carbonate hydroxide with surface sulfate groups for highly enhanced electro-oxidation of urea. *Electrochim. Acta* **2022**, *426*, 140792. [CrossRef]
41. Xue, Y.; Fang, X.; Jiang, H.; Wu, J.; Liu, H.; Li, X.; He, P.; Li, F.; Qi, Y.; Gao, Q.; et al. Hierarchical microsphere Flower-like SnIn4S8 with active sulfur sites for adsorption and removal of mercury from coal-fired flue gas. *Chem. Eng. J.* **2023**, *472*, 145105. [CrossRef]
42. Ding, J.; Yue, R.; Zhu, X.; Liu, W.; Pei, H.; He, S.; Mo, Z. Flower-like Co3Ni1B nanosheets based on reduced graphene oxide (rGO) as an efficient electrocatalyst for the oxygen evolution reaction. *New J. Chem.* **2022**, *46*, 13524–13532. [CrossRef]
43. Bulakhe, R.N.; Shin, S.C.; In, J.B.; In, I. Chemically synthesized mesoporous nickel cobaltite electrodes of different morphologies for high-performance asymmetric supercapacitors. *J. Energy Storage* **2022**, *55*, 105730.
44. Condon, J.B. *Surface Area and Porosity Determinations by Physisorption: Measurement, Classical Theories and Quantum Theory*; Elsevier: Amsterdam, The Netherlands, 2019.
45. Palem, R.R.; Shimoga, G.; Rabani, I.; Bathula, C.; Seo, Y.-S.; Kim, H.-S.; Kim, S.-Y.; Lee, S.-H. Ball-milling route to design hierarchical nanohybrid cobalt oxide structures with cellulose nanocrystals interface for supercapacitors. *Int. J. Energ. Res.* **2022**, *46*, 8398–8412. [CrossRef]
46. Liu, J.; Wang, J.; Xu, C.; Jiang, H.; Li, C.; Zhang, L.; Lin, J.; Shen, Z.X. Advanced Energy Storage Devices: Basic Principles, Analytical Methods, and Rational Materials Design. *Adv. Sci.* **2018**, *5*, 1700322. [CrossRef]
47. Ertas, M.; Walczak, R.M.; Das, R.K.; Rinzler, A.G.; Reynolds, J.R. Supercapacitors based on polymeric dioxypyrroles and single walled carbon nanotubes. *Chem. Mater.* **2012**, *24*, 433–443. [CrossRef]
48. Kumar, A.; Das, D.; Sarkar, D.; Nanda, K.K.; Patil, S.; Shukla, A. Asymmetric Supercapacitors with Nanostructured RuS2. *Energy Fuels* **2021**, *35*, 12671–12679. [CrossRef]
49. Pazhamalai, P.; Krishnamoorthy, K.; Sahoo, S.; Mariappan, V.K.; Kim, S.J. Copper tungsten sulfide anchored on Ni-foam as a high-performance binder free negative electrode for asymmetric supercapacitor. *Chem. Eng. J.* **2019**, *359*, 409–418. [CrossRef]
50. Wang, Z.; Zhu, Z.; Zhang, Q.; Zhai, M.; Gao, J.; Chen, C.; Yang, B. Fabrication of N-doped carbon coated spinel copper cobalt sulfide hollow spheres to realize the improvement of electrochemical performance for supercapacitors. *Ceram. Int.* **2019**, *45*, 21286–21292. [CrossRef]
51. Kandhasamy, N.; Preethi, L.K.; Mani, D.; Walczak, L.; Mathews, T.; Venkatachalam, R. RGO nanosheet wrapped β-phase NiCu$_2$S nanorods for advanced supercapacitor applications. *Environ. Sci. Pollut. Res.* **2023**, *30*, 18546–18562.

52. Sathish, S.; Navamathavan, R. Electrochemical Investigation of Ni-Co-Zn-S/AC Nano Composite for High-Performance Energy Storage Applications. *ECS J. Solid State Sci. Technol.* **2022**, *11*, 101010.
53. Mathis, T.S.; Kurra, N.; Wang, X.; Pinto, D.; Simon, P.; Gogotsi, Y. Energy Storage Data Reporting in Perspective—Guidelines for Interpreting the Performance of Electrochemical Energy Storage Systems. *Adv. Energy Mater.* **2019**, *9*, 1902007. [CrossRef]

Disclaimer/Publisher's Note: The statements, opinions and data contained in all publications are solely those of the individual author(s) and contributor(s) and not of MDPI and/or the editor(s). MDPI and/or the editor(s) disclaim responsibility for any injury to people or property resulting from any ideas, methods, instructions or products referred to in the content.

Article

Anodizing Tungsten Foil with Ionic Liquids for Enhanced Photoelectrochemical Applications

Elianny Da Silva, Ginebra Sánchez-García, Alberto Pérez-Calvo, Ramón M. Fernández-Domene, Benjamin Solsona * and Rita Sánchez-Tovar *

IQCATAL Group (Heterogeneous Catalysis), Chemical Engineering Department (ETSE), Universitat de València, Av. Universitat s/n, 46100 Burjassot-Valencia, Spain; elianny.silva@uv.es (E.D.S.); ginebra.sanchez@uv.es (G.S.-G.); alberto.perez-calvo@uv.es (A.P.-C.); ramon.fernandez@uv.es (R.M.F.-D.)
* Correspondence: benjamin.solsona@uv.es (B.S.); rita.sanchez@uv.es (R.S.-T.)

Abstract: This research examines the influence of adding a commercial ionic liquid to the electrolyte during the electrochemical anodization of tungsten for the fabrication of WO_3 nanostructures for photoelectrochemical applications. An aqueous electrolyte composed of 1.5 M methanesulfonic acid and 5% v/v [BMIM][BF_4] or [EMIM][BF_4] was used. A nanostructure synthesized in an ionic-liquid-free electrolyte was taken as a reference. Morphological and structural studies of the nanostructures were performed via field emission scanning electron microscopy and X-ray diffraction analyses. Electrochemical characterization was carried out using electrochemical impedance spectroscopy and a Mott–Schottky analysis. From the results, it is highlighted that, by adding either of the two ionic liquids to the electrolyte, well-defined WO_3 nanoplates with improved morphological, structural, and electrochemical properties are obtained compared to samples synthesized without ionic liquid. In order to evaluate their photoelectrocatalytic performance, the samples were used as photocatalysts to generate hydrogen by splitting water molecules and in the photoelectrochemical degradation of methyl red dye. In both applications, the nanostructures synthesized with the addition of either of the ionic liquids showed a better performance. These findings confirm the suitability of ionic liquids, such as [BMIM][BF_4] and [EMIM][BF_4], for the synthesis of highly efficient photoelectrocatalysts via electrochemical anodization.

Keywords: ionic liquid; organic dye degradation; tungsten oxide; photoelectrocatalysis; water splitting

Citation: Da Silva, E.; Sánchez-García, G.; Pérez-Calvo, A.; Fernández-Domene, R.M.; Solsona, B.; Sánchez-Tovar, R. Anodizing Tungsten Foil with Ionic Liquids for Enhanced Photoelectrochemical Applications. *Materials* **2024**, *17*, 1243. https://doi.org/10.3390/ma17061243

Academic Editors: Jinsheng Zhao and Zhenyu Yang

Received: 6 February 2024
Revised: 29 February 2024
Accepted: 6 March 2024
Published: 8 March 2024

Copyright: © 2024 by the authors. Licensee MDPI, Basel, Switzerland. This article is an open access article distributed under the terms and conditions of the Creative Commons Attribution (CC BY) license (https://creativecommons.org/licenses/by/4.0/).

1. Introduction

In the last two decades, due to the advantages of nanostructured materials, such as their large surface area in relation to their volume and their high activity, they have been incorporated into a wide variety of energy applications, such as the production of hydrogen from water molecules, solar panels, etc., as well as environmental applications such as wastewater treatment, treatment of air effluents, and so on [1–3]. Recently, the use of ionic liquids (ILs) in the synthesis of nanostructured catalysts has become widespread, since they offer very attractive alternatives to traditional liquid organic solvents and solid salts. This is a result of the suitable physicochemical characteristics of these ILs, which include a greater conductivity, a low vapor pressure, nonflammability, and good thermal and chemical stability [4–6], in addition to their selective combination of cations or anions depending on the subsequent application and their treatability and reusability, greatly improving their environmental impacts and contributing to sustainability [5,6].

Electrochemical anodization is one of the many methods for synthesizing nanostructures from semiconductor metal oxides. It offers sufficient scalability and comparatively easy operation, giving excellent control over the morphology of the nanostructures and their physicochemical properties [7]. Anodization is carried out at moderate temperatures and atmospheric pressure, and the morphology and dimensions of the nanostructures can be altered by varying the number of operation parameters, including the temperature, electrolyte composition, and applied cell potential. This makes anodization a very suitable technique with a minimal environmental impact [8]. Some other methods have been studied to create WO_3 nanostructures with a wide variety of morphologies (nanowires, nanopores, nanoflakes, nanoplates, etc.) such as hydrothermal methods [9], solvothermal methods [10], deposition processes (laser deposition [11], sol–gel [12], electrodeposition [13,14], chemical vapor deposition [15], RF sputtering [16,17], spin-coating [18]), and combustion. However, they use corrosive reactants and large amounts of energy and they are more complex to operate in comparison with anodization techniques.

Numerous semiconductor metal oxides have been studied by scientists for a variety of environmental and energy applications due to their photocatalytic activity [19,20]. By shining light on a nanostructured photocatalyst and applying a potential at the same time, matter such as dyes can be degraded. In this process, a positively charged hole is simultaneously created in the valence band and an excited electron in the valence band moves into the conduction band. They migrate to the surface of the photocatalyst, where they participate in the redox reaction of water and produce OH^- radicals, which are responsible for oxidizing the organic species present [21,22]. In the same way, during photoelectrochemical water splitting, electron–hole pairs are photogenerated in the conduction and valence layers, respectively. Then, they separate, and holes oxidize the water molecules to create O_2 and H^+ on the surface of the semiconductor, and electrons in the counter electrode reduce H^+ into gaseous hydrogen [23–25].

Most of the many investigations that have been conducted with commercial ILs as solvents in the creation of nanostructured catalysts are centered on the fabrication of TiO_2 nanostructures. The first investigation into ionic-liquid-based titanium anodization was reported by Schmuki et al. They showed that self-organized layers of TiO_2 NTs may be produced directly on a titanium surface using anodization in 1-n-butyl-3-methyl-imidazolium tetrafluoroborate, [BMIN][BF_4] [26]. Wender and colleagues subsequently synthesized titanium dioxide nanostructures with the same IL to evaluate them as photocatalysts for methyl orange photodegradation and hydrogen evolution from water/methanol mixtures [27]. Other studies have also shown improvements in the behavior of anodized titanium foils using this ionic liquid and some others like 1-ethyl-3 methylimidazolium tetrafluoroborate, [EMIN][BF_4]; 1-butyl-3-methyl-imidazolium chloride, [BMIM][Cl]; 3-methyl-1-octyimidazolium tetrafluoroborate, [OMIM][BF_4]; and 1-butylpiridinium chloride, [BPy][Cl] [28–30].

However, these different compounds have been rarely employed as solvents during electrochemical anodization on other types of semiconductor metal oxides and they could greatly contribute to enhancing their photoelectrochemical performance. Considering this, tungsten oxide (WO_3) is drawing a lot of attention because of its natural abundance, sufficient conductivity, resistance to photocorrosion, low band gap value (between 2.5 and 3.0 eV [31–33]), and large hole diffusion (150 nm) [34]. Furthermore, the chemistry of tungsten and the ability of various compounds (ligands) to generate tungsten complexes and alter the electrolyte composition during anodization provide the opportunity to optimize the size and morphology of nanostructures [35]. In this regard, ILs may be employed as complexing agents for tungsten during the anodization process to produce WO_3 nanostructures with different morphologies and sizes. The ionic liquids [BMIN][BF_4] and [EMIM][BF_4] contain fluoride species [F^-] that act as monodentate ligands which form stronger bonds (higher stability) with tungsten, increasing the dissolution rate of the WO_3 layer, which will then precipitate on the surface. They also delay the precipitation of tungstic acids on the electrode surface, forming nanostructures with better behavior [36]. In addition, the organic part of these molecules is short, which could facilitate interactions between the inorganic part [BF_4]$^-$ and the electrolyte and substrate.

However, not many studies report the use of these materials in the synthesis of tungsten oxide nanostructures. N-methyl-pyrrolidinium tetrafluoroborate, a protic ionic liquid (PIL), was used by Go et al. to synthesize WO_3 nanoplates, and their findings demonstrated an improvement in all electrochromic parameters when compared to an acid medium electrolyte without ILs [18].

Likewise, within the different applications mentioned, and in order to evaluate the photoelectrocatalytic performance of the WO_3 nanostructures synthesized with the commercial ionic liquids [BMIN][BF_4] and [EMIN][BF_4], they will be used to carry out the splitting of water molecules for hydrogen production (with implications for energy applications) and the decomposition of the methyl red dye (with implications for the environment). Methyl red (MR) is an anionic organic dye that is produced as waste in the course of many different industries operations, including paper, pulp, plastic, leather, and textile industries, and so forth [37]. Various physicochemical and biological approaches, including chemical oxidation, ion exchange, and biodegradation, have been employed over the years to remove this pollutant from wastewater because of its high toxicity and health concerns to humans. Nevertheless, all of them have certain drawbacks, some of which include producing a lot of extra waste or their very high costs [2,37]. However, photoelectrocatalytic decomposition is promoted as a quick and easy method for removing this kind of organic contaminant.

The primary goal of this work is to conduct a comprehensive analysis of the physicochemical and electrochemical properties of WO_3 nanostructures anodized with the commercial ILs [BMIN][BF_4] and [EMIN][BF_4], ionic liquids which have not been traditionally used for this purpose. Furthermore, this investigation uses the nanostructures with the promising properties provided by ILs in applications in which they have not been tested before, such as the photoelectrochemical splitting of water and in the photodegradation of methyl red.

2. Materials and Methods

2.1. Synthesis by Electrochemical Anodization

Tungsten oxide nanostructures were synthesized by electrochemical anodization at 50 °C. A tungsten foil (with 1.32 cm^2 exposed to the electrolyte) was used, and a potential difference between the tungsten sheet and the cathode (platinum foil) of 20 V for 4 h was applied. The anodization system was composed of a vertical cell consisting of a single compartment where the area exposed to the electrolyte was controlled by an O-ring. The synthesis of the nanostructures was carried out in an aqueous electrolyte with 1.5 M methanesulfonic acid and 5% v/v of [BMIM][BF_4] or [EMIM][BF_4] (labeled as BMIM and EMIM, respectively). To compare the obtained results, an ionic-liquid-free electrolyte was utilized (identified as Blank). After, the samples were rinsed with water, dried with a nitrogen stream, and annealed at 600 °C for 4 h.

2.2. Morphological and Structural Characterization of the Nanostructures (FESEM and XRD)

Morphological characterization of the nanostructures was performed using a field emission scanning electron microscope (FESEM) Hitachi S4800, at an accelerating potential of 5 kV. Using this technique, images were taken of the top of the nanostructures and of the cross-section (scratching the surface of the samples).

Additionally, the samples were subjected to X-ray diffraction (XRD) analysis, making use of a Bruker D8AVANCE diffractometer (Bruker, Billerica, MA, USA) fitted with a Cu Kα1 monochromatic source.

2.3. Electrochemical Characterization

Electrochemical characterization was performed via electrochemical impedance spectroscopy (EIS). A three-electrode single compartment cell was used: the nanostructure was the working electrode (with an exposed area of 0.5 cm^2), a platinum tip was the counter electrode and an Ag/AgCl electrode was the reference electrode. This test was carried out in 0.1 M H_2SO_4 as the electrolyte. Using a potentiostat (PalmSens4, PalmSens, Houten, The Netherlands) at a constant potential of 1 $V_{Ag/AgCl}$, a frequency scan was applied from 10 kHz to 10 mHz with an amplitude of 10 mV. Furthermore, in a similar experimental setup, a Mott–Schottky analysis was performed, scanning the potential from 1 to 0 $V_{Ag/AgCl}$ with a frequency of 5000 Hz.

2.4. Application of WO$_3$ Nanostructures

2.4.1. Photoelectrochemical Production of Hydrogen

The tungsten oxide nanostructures were used to produce hydrogen from the photoelectrochemical splitting of water. These tests were carried out in a three-electrode single compartment cell using the samples as a working electrode, a platinum tip as a counter electrode, and an Ag/AgCl reference electrode. The samples were masked with an O-ring in the cell, exposing 0.5 cm^2 of each one to the 0.1 M H$_2$SO$_4$ electrolyte. During this analysis, the surface of the WO$_3$ nanostructures was illuminated with a UV light (λ = 365 nm, 100 mW·cm^{-2}) while scanning the potential from 0 to 1.04 $V_{Ag/AgCl}$ at a rate of 0.005 V/s (using a PalmSens4 potentiostat). Then, the same nanostructure was scanned in the same potential range, but this time in the dark.

2.4.2. Photoelectrodegradation of Methyl Red

The application of the samples as photoelectrocatalysts in the photoelectrochemical degradation of methyl red dye in sulfuric acid 0.1 M was tested. A glass cell with a quartz window and three electrodes immersed in the dye solution was used for this analysis. The nanostructure was used as a working electrode (with an exposed area of 1.32 cm^2), a platinum tip was used as a counter electrode and an Ag/AgCl electrode was used as a reference. The nanostructure was illuminated with a UV lamp (λ = 365 nm, 100 mW·cm^{-2}), and with a potentiostat (PalmSens4), a potential of 1 V was applied.

Dye degradation was monitored every 10 min for 1 h using a quartz cell in a spectrophotometer (Cecil CE 1011, Cecil Instruments Limited, Cambridge, UK) at a wavelength of 517 nm. Figure 1 displays the UV absorbance spectra of the methyl red solution; the maximum absorption peak is in accordance with the literature [38]. The calibration of absorbance vs. concentration obtained at 517 nm is shown in the inset in Figure 1.

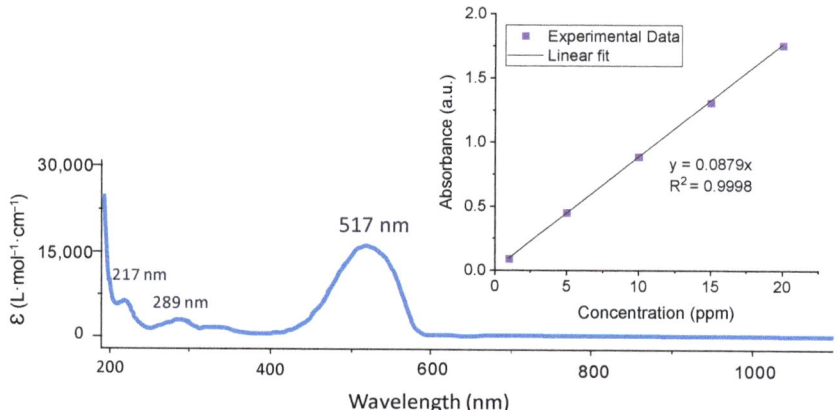

Figure 1. UV absorbance spectra of methyl red (inset: relationship obtained between absorbance "y" and methyl red concentration "x" at a wavelength of 517 nm).

To check the accuracy of the method, the % recovery of three series of three samples with different concentrations (1, 5, 10 ppm) of methyl red has been calculated, showing that the method has a recovery percentage between 97.84 and 99.85. The precision of the method was also determined using three samples with different concentrations (1, 5, 10 ppm) of methyl red and analyzing each one three times and calculating the relative standard deviation (RSD). The results show that the method precisely obtains values below 1.8% RSD.

3. Results and Discussion

3.1. Synthesis by Electrochemical Anodization

It is important to mention that different concentrations (5%, 10%, 30%) of [BMIM][BF$_4$] and [EMIM][BF$_4$] were used to anodize tungsten. However, the best photoelectrocatalytic performance (Figure S1) was achieved by low-IL-concentration nanostructures (5%). Figure 2 displays the current density–time data registered during anodization in the blank electrolyte (without ILs) and in both ionic liquids (EMIM and BMIM). A magnified inset plot is shown to enable a proper view the current density transients of each sample. In this figure, the current density values reached during the synthesis process and their shapes can be examined.

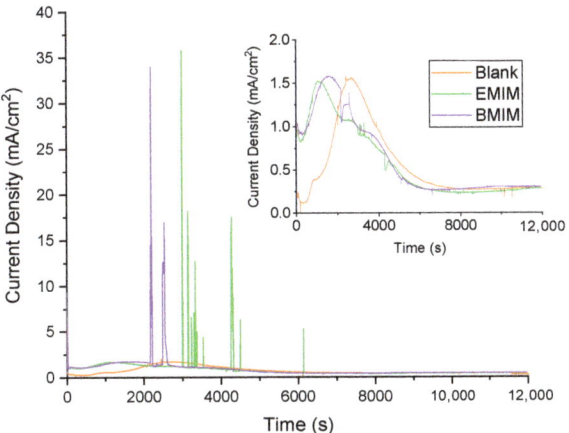

Figure 2. Current density curves recorded during electrochemical anodization at 20 V and 50 °C in different electrolytes (with and without IL).

The blank electrolyte sample exhibits the shape of a typical dissolution–precipitation formation mechanism [39,40]. First, the current density steadily drops due to the growth of a compact WO$_3$ layer; then, the high concentration of protons in the solution causes the dissolution of this layer and a resulting current density increase (second stage). Finally, the current density drops once more when the soluble tungsten species (created immediately before) reach supersaturation levels and precipitate as tungstic acid (hydrated and amorphous WO$_3$·2H$_2$O) in the form of nanostructures on the sample. Equations (1)–(3) represent these three processes that take place during anodization:

$$W + 3H_2O \rightarrow WO_3 + 6H^+ + 6e^- \quad (1)$$

$$WO_3 + 2H^+ \rightarrow WO_2^{2+} + H_2O \quad (2)$$

$$WO_2^{2+} + 3H_2O \rightarrow WO_3 \cdot 2H_2O + 2H^+ \quad (3)$$

Some differences can be elucidated when 5% of either ionic liquid was added to the electrolyte, for instance, in the moment the current density reaches its maximum. This maximum is reached sooner in the presence of 5% IL due to the lower dielectric constants of the ionic liquids and, therefore, of the anodizing electrolyte containing them. A lower dielectric constant facilitates the precipitation of the soluble tungsten species generated in stage two of the anodization process. Therefore, it is reasonable that the current density peak belonging to this part of the synthesis is reached earlier in electrolytes with lower dielectric constants (EMIM or BMIM). Moreover, this behavior is reflected in the total charge of the anodization process of the samples (Table 1). The addition of either of the ionic liquids to the electrolyte accelerates the dissolution–precipitation mechanism associated with the formation of WO$_3$ nanostructures and, therefore, the total charge density is higher.

Table 1. Total charge during electrochemical anodization of tungsten.

Sample	Total Charge (C)
Blank	9.5
EMIM	10.2
BMIM	10.8

On the other hand, in the current transients recorded during anodization for the samples synthesized with EMIM or BMIM, diverse peaks can be appreciated, which correspond to the formation of pitting or the localized dissolution of tungsten due to the presence of BF^- ions that tend to form soluble tungsten complexes [41,42]. This result has been observed before when using other fluoride-containing electrolytes. As has been outlined before, with NaF, fluoride ions formed strong bonds with tungsten since they act as monodentate ligands. Then, due to the acidic electrolyte, fluoride ions encouraged localized dissolution of the formed WO_3 layer and developed soluble fluoride complexes [36]. Specifically, methanesulfonic acid can excellently solubilize metal salts and it is an environmentally friendly electrolyte [43].

Furthermore, comparing both IL curves, it can be noted how stage two occurs earlier for the EMIM electrolyte than for the BMIM electrolyte. This can be attributed to the organic contribution, which is the result of larger organic molecules in the electrolyte. This leads to steric hindrance in the interaction between the oxide layer and fluoride ions.

3.2. Morphological and Structural Characterization of the Nanostructures (FESEM and XRD)

The FESEM images shown in Figure 3 were taken to study the influence of ionic liquids on the WO_3 nanostructures synthesized via electrochemical anodization. The images show the morphology of the samples synthesized without ionic liquid and of samples for which 5% v/v EMIM or BMIM was added to the anodization electrolyte. In the three studied cases, the formation of nanoplates can be observed, which increase in length and number with the addition of ionic liquid. This behavior is in agreement with what was expected according to the anodization curves studied in the previous section, since the nanoplates synthesized with EMIM or BMIM begin to precipitate before those formed with the free electrolyte. According to Figure 2, the second stage (precipitation stage) starts at 1100 or 1700 s for the samples synthesized with EMIM or BMIM, respectively, while, for the blank electrolyte, it starts at ~2000 s. Therefore, the nanoplates synthesized with ionic liquid have more time to form and grow. All of this leads to a higher surface area of the nanostructures. Moreover, the larger number of nanoplates in the EMIM sample can also be explained by the precipitation stage beginning earlier, as the peak is reached before for this nanostructure (Figure 2).

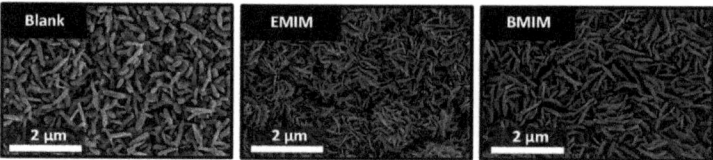

Figure 3. FESEM images of tungsten oxide nanostructures synthesized via electrochemical anodization with different electrolyte solutions (with and without IL).

Table 2 shows three parameters related to WO_3 nanostructures: nanoplate length, nanoplate thickness and WO_3 layer thickness. These measurements were determined from the top view and cross-sectional images of the nanostructures taken via FESEM (see Supporting Information, Figure S2). The results indicate that the nanoplates obtained by adding EMIM or BMIM to the anodization electrolyte are longer and thinner. In addition, the thickness of the WO_3 layer increases with the presence of ionic liquids. This behavior is consistent with what has been previously mentioned regarding the earlier precipitation of tungsten species during anodization. The BF$^-$ ions from ionic liquids tend to form soluble tungsten complexes and, as the anodization process continues, the supersaturation condition is reached faster, so that soluble species start to precipitate on the sample surface and favor the formation of thicker layers.

Table 2. Morphological parameters of WO$_3$ nanostructures.

Sample	Nanoplate Length (μm)	Nanoplate Thickness (μm)	WO$_3$ Layer Thickness (μm)
Blank	0.48 ± 0.09	0.09 ± 0.02	0.7 ± 0.1
EMIM	0.60 ± 0.05	0.06 ± 0.01	1.5 ± 0.2
BMIM	0.64 ± 0.06	0.07 ± 0.01	1.2 ± 0.1

Figure 4 shows the X-ray diffractogram of WO$_3$ nanostructures synthesized by electrochemical anodization after being calcined at 600 °C for 4 h. At this annealing temperature (600 °C), the WO$_3$ nanostructures show a high crystalline structure, a lower resistance to charge transfer, and a higher dopant density, leading to a better photoelectrochemical performance [44].

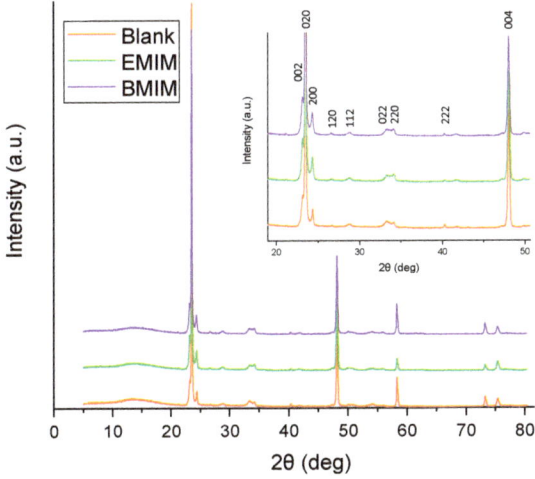

Figure 4. X-ray diffraction patterns for WO$_3$ nanostructures after anodization in different electrolytes (with and without IL) and magnification of the different monoclinic phase peaks.

As can be seen, the three samples present the characteristic peaks of monoclinic tungsten (JCPDS no. 43-1035). This result agrees with previous studies, as other researchers have reported that after annealing, amorphous WO$_3$ turned into a crystalline monoclinic phase [45–47]. Therefore, it is clear that the addition of ionic liquid in the electrolyte does not affect the crystalline properties of the synthesized nanostructures. Using Scherrer's Equation (4), the size of the crystallites has been determined (Table 3).

$$D = \frac{\kappa \lambda}{FWHM \cdot cos(\theta)} \quad (4)$$

where D is the crystallite size (nm), κ is the shape factor (0.9), λ is the X-ray wavelength (0.1542 nm), $FWHM$ is the width at half maximum intensity (rad), and θ is the Bragg angle [48]. In this case, the triplet of peaks appearing at 23.15, 23.48, and 24.25°, corresponding to the crystallographic planes (002), (020), and (200) and which can be assigned to the main monoclinic WO$_3$ peaks (JCPDS no. 43-1035), have been used. The rest of the peaks also belong to the monoclinic WO_3 phase, which has been described as the most stable one [49,50].

Table 3. Crystallite size determined via the Scherrer equation.

Sample	Crystallite Size (nm)
Blank	53.5
EMIM	49.7
BMIM	49.7

The obtained results indicate that the addition of ionic liquid in the electrolyte causes a decrease in the crystallite size. This behavior is due to the presence of $[BF_4]^-$ ions in the electrolyte since they favor the formation of tungsten complexes and influence the dissolution rate of WO_3. As a consequence, the surface area of the nanostructures increases, leading to the decrease in the crystallite size of the samples. This result is favorable for the photocatalytic performance of the nanostructures, since for samples with smaller crystallite sizes, the transfer of photogenerated charge carrier pairs is enhanced and their recombination probability is reduced [51,52].

3.3. Electrochemical Characterization

Electrochemical impedance spectroscopy tests were performed to learn about the resistance to charge transfer processes in each photoelectrode. Figure 5 shows the Nyquist (A) and Bode module (B) data for the WO_3 nanostructures in the frequency range of 100 kHz to 1 mHz obtained at a potential of 1 V versus Ag/AgCl 3 M KCl (Bode phase plots in Figure S3).

Regardless of the sample, every Nyquist plot presents two semicircles, each one corresponding to one of the regions of the Bode module plots with different slopes. First, the charge transfer response of the oxide/electrolyte interface can be linked to the semicircle obtained at high and intermediate frequencies (see inset of Figure 5A), which can reveal information about the active surface area of the catalysts. The one at low frequencies is usually related to the compact layer of oxide formed under the nanostructure [53,54].

Generally, the semicircle amplitude is proportional to the impedance of the related electrochemical process. Observing Figure 5, the Blank sample presents the largest semicircle. This fact is confirmed in Figure 5B, where the values registered at the lowest frequency in the plot represent the total resistance offered by each nanostructure (R_T). These results are presented in Table 4. As observed, the Blank sample presents a higher R_T. It is confirmed that the addition of ionic liquid to the anodizing electrolyte decreases the total resistance to charge transfer of the tungsten nanostructures. It is important to highlight that the total resistance (R_T) is lower for the WO_3 nanostructure synthesized with EMIM than for the one synthesized with BMIM. This result is due to the shorter structural chain of EMIM which, therefore, makes it a slightly better complexing agent than BMIM. Additionally, EMIM nanostructures with larger surface areas have been obtained and this decreases the resistance they offer and increases their donor density [55].

Table 4. Total (Bode module plots) and active part (equivalent circuit fitting results) resistances of the nanocatalysts.

Nanostructure	R_T (kOhm·cm^2)	R_1 (kOhm·cm^2)
Blank	163.15	8.19
EMIM	53.95	1.38
BMIM	116.05	1.98

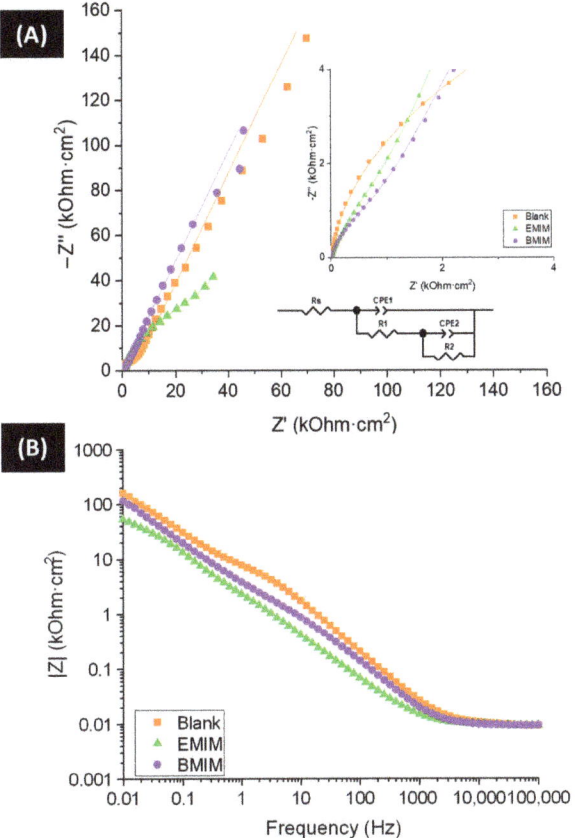

Figure 5. (**A**) Nyquist plots of WO_3 nanostructures formed in electrolytes with and without IL (continuous line represents the fitting to the equivalent circuit), (**B**) Bode module plot of WO_3 nanostructures formed in electrolytes with and without IL.

To quantitatively analyze the EIS results, an electrical equivalent circuit with two parallel R-C time constants was used, as illustrated in Figure 5A (depicted in more detail in Figure S4). In this circuit, the non-ideality of the system has been considered by using constant phase elements (CPEs) as a substitute for pure capacitors [56,57]. An impedance fitting analysis was performed with the software ZView4, following the equation shown below (5):

$$Z = R_S + \frac{R_1 + R_2 + R_1 R_2 Y_2 (j\omega)^{\alpha_2}}{1 + R_1 Y_1 (j\omega)^{\alpha_1} + R_2 Y_2 (j\omega)^{\alpha_2} + R_2 Y_1 (j\omega)^{\alpha_1} + R_1 R_2 Y_1 Y_2 (j\omega)^{\alpha_1} (j\omega)^{\alpha_2}} \quad (5)$$

From this fitting, the charge transfer resistance in the active parts of the nanostructure/electrolyte interface (R_1) can be quantified. The results are displayed in Table 4, taking into consideration that, for all cases, the chi-squared values were lower than 10^{-3}, validating the circuit used. The rest of the parameters in Equation (5) appear in Table S1. As expected, R_2 values (bulk section) are extremely high as they belong to the compact oxide layer beneath the active part of the nanocatalysts. Therefore, these values are not considered at any point, since this part of the samples does not take part in the studied reactions (only the nanostructured top layers take part).

As expected, charge transfer in the active part of the WO$_3$ nanostructures is also improved by the addition of ionic liquid during the anodization process, which can be related to the higher surface area of the IL samples, as described before. Generally, this effect would be beneficial for the photoelectrochemical performance of these catalysts since a low resistance improves electron transfer and hole diffusion, leading to a better photoresponse [58–60].

These findings are in agreement with what the Mott–Schottky plots show in Figure 6A. The synthesized tungsten oxide nanostructures are n-type semiconductors, as the positive slopes of the plots reveal. Therefore, electrons are the dominant charge carriers, and with the use of the Mott–Schottky Equation (6), it is possible to calculate the donor density (N_D) of each sample:

$$\frac{1}{C^2} = \frac{1}{C_H^2} + \frac{2}{e \cdot \varepsilon_T \cdot \varepsilon_0 \cdot N_D} \cdot \left(E - E_{FB} - \frac{k \cdot T}{e} \right) \quad (6)$$

where C_{SC} is the space charge layer capacitance, C_H is the Helmholtz layer capacitance, e the electron charge (1.60×10^{-19} C), ε_0 is the vacuum permittivity (8.85×10^{-14} F/cm), ε is the dielectric constant of WO$_3$ (50 [61–63]), E is the applied potential, k is the Boltzmann constant (1.38×10^{-23} J/K), and T is the absolute temperature [64]. The values of N_D for each nanostructure are shown in Figure 6B.

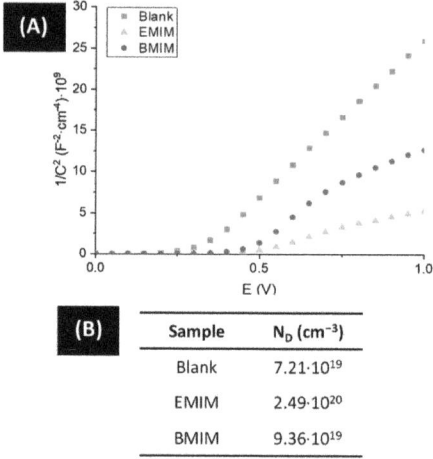

Figure 6. (**A**) Mott–Schottky plots obtained at a frequency of 5 kHz for WO$_3$ nanostructures anodized with varied ILs. (**B**) Donor density (N_D) calculated from MS plots for the nanostructures synthesized with different ILs.

For tungsten oxide, the donor density is linked to oxygen vacancies, since these are the dominant defects in these nanostructures. From Figure 6B, it can be inferred that the addition of ionic liquid in the anodization process increases the donor density of the samples. A higher number of oxygen vacancies has a positive impact on WO$_3$ nanostructures when used as photoanodes, since the electrical conductivity increases and, therefore, their photoelectrochemical performance improves. However, oxygen vacancies can also act as electron traps, facilitating the recombination of photogenerated electron/hole pairs [65,66], which could lead to worse photoelectrochemical behavior of the nanostructures.

3.4. Application of WO$_3$ Nanostructures

3.4.1. Photoelectrochemical Production of Hydrogen

Figure 7A shows the results obtained in the photoelectrochemical separation of water molecules after exposing the samples to illumination with a UV lamp and making a potential scan in the positive direction. According to what can be seen in the graph, the three samples present a good photoresponse, since an increase in the current density was recorded when illuminating the surface of the nanostructure. It is worth stating that when nanostructures are in the dark, the current density recorded is very close to 0 in all cases. Figure 7B exhibits the theoretical molar flow of hydrogen produced with the nanostructures at a potential of 1 V, calculated with Faraday's law. According to these results, the nanostructures synthesized with ionic liquid exhibit a hydrogen production that is about 130% higher than that obtained with the blank sample. Furthermore, compared to what has been reported in the literature, these nanostructures exhibit a much higher hydrogen production than others. For example, other IL-synthesized nanostructures generate 81–82 µmol during water splitting, while CuCl/WO$_3$ samples only generate 5 µmol after 30 min of exposure to UV light [67]. These results suggest that the proposed method of synthesis allows for obtaining efficient WO$_3$ nanostructures with a better photoresponse, since their production of hydrogen is 16 times higher. In other study [68], tungsten oxide nanostructures were illuminated with UV light for water splitting applications, and the current density reached maximum values of 0.5 mA/cm^2 when applying $1V_{Ag/AgCl}$ using a light intensity of 50 mW/cm^2. The current density values were more than 15 times lower than the ones achieved in this work at the same applied potential with 100 mW/cm^2.

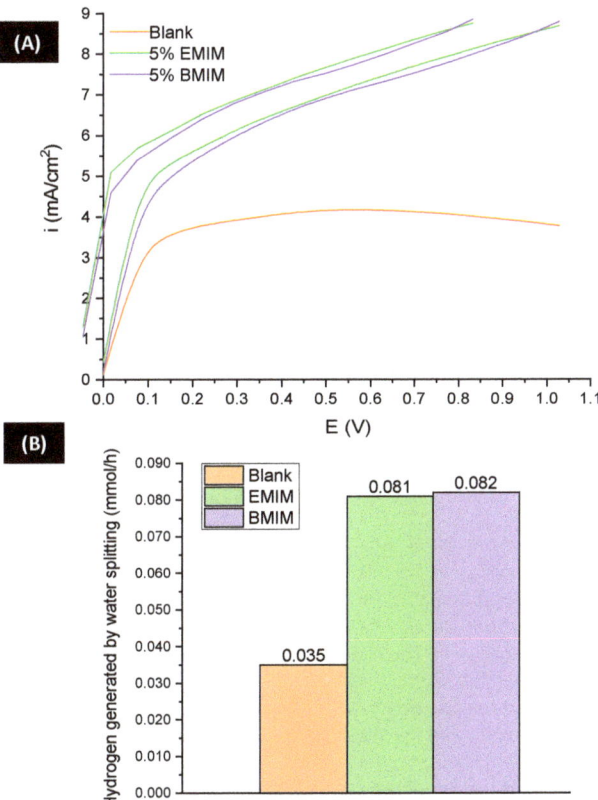

Figure 7. (**A**) Photocurrent transient vs. potential of tungsten oxide nanostructures synthesized by electrochemical anodization in different electrolyte solutions (with and without IL). (**B**) Number of moles of hydrogen generated during the splitting of water molecules.

According to the morphological properties of the samples, it can be seen that the nanostructures synthesized with ionic liquid had a thicker nanostructured layer, an observation that favors the photoelectrocatalytic performance by increasing the surface area of the nanostructures [36,69].

Additionally, the three samples had a good crystalline structure; however, those synthesized with EMIM and BMIM presented a smaller crystallite size. Therefore, the probability of recombination of the electron/hole pair is lower than for the sample synthesized with the blank electrolyte [70].

Electrochemical characterization of the samples indicated that those that were anodized under the presence of ionic liquid had a considerably lower charge transfer resistance than the one synthesized with an ionic-liquid-free electrolyte. Therefore, it is implied that these nanostructures have better photoelectrocatalytic behavior, since electron transfer and hole diffusion are improved. A similar effect is obtained as a result of the number of defects present in the nanostructures, since other studies have shown that, with a higher number of vacancies, charge transfer processes are improved due to the better mobility of electrons [71,72].

3.4.2. Photoelectrodegradation of Methyl Red

Tungsten oxide nanostructures were tested as photoelectrocatalysts in the degradation of methyl red dye in a 0.1 M sulfuric acid solution. The evolution of dye degradation over time can be seen in Figure 8.

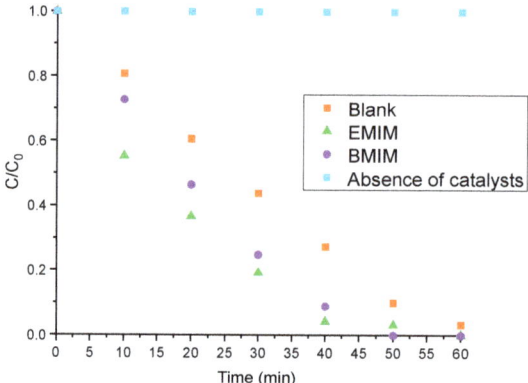

Figure 8. C/C_0 of methyl red as a function of time during its photoelectrochemical degradation using a WO_3 nanostructure synthesized with ionic liquid.

In this analysis, a starting dye concentration of 20 ppm was used and, measuring the absorbance at ~517 nm using a spectrophotometer, the elimination of the dye was evaluated every 10 min. According to the obtained results, the concentration of methyl red without catalysts remained constant during the experiment, showing that there is no degradation of the dye. On the other hand, the nanostructures synthesized with ionic liquid exhibit a better performance than the sample synthesized with a blank electrolyte, since complete degradation of the dye was achieved after 60 or 50 min when using the nanostructures synthesized with EMIM or BMIN, respectively, while, for the reference sample, 60 min was not enough to degrade the whole dye. It is important to note that the sample anodized without ionic liquid always had a lower percentage of elimination than the anodized samples with ionic liquids. Taking a closer look at the degradation at 40 min, both nanostructures anodized with EMIM and BMIM achieve more than 90% MR degradation, while for the blank this value was close to 70%, which means that it is possible to reduce the energy consumption necessary to degrade a high percentage of methyl red dye using ILs as electrolytes. At 60 min, the sample without ILs managed to degrade most of the dye because it is also a nanostructured area with photoelectrocatalytic activity and, with longer durations, it has the capacity to completely degrade methyl red.

In Figure S5, the degradation efficiency of each nanostructure is calculated. It can be appreciated how the EMIM nanostructure presents a better degradation percentage during most of the test, although the EMIM nanostructure requires 10 min more than the BMIM sample to achieve 100% degradation efficiency. These results are related to what has been mentioned above; that is, EMIM has a better complexing activity than BMIM, which leads to a greater dissolution of the oxide layer during the anodization and subsequently a greater precipitation with nanoplates of smaller sizes and with greater surface areas that benefit the photoelectrocalytic activity. Therefore, the use of EMIM seems more promising due to the obtained results. Despite this, Figure S5 reinforces the fact that both ionic liquids improve the photoelectrochemical behavior of the tungsten oxide nanostructures and both have a similar photoelectrocatalytic performance.

Considering the above-mentioned observations and the results of water splitting, the samples synthesized with EMIM or BMIM present a higher photoelectrocatalytic performance, since, as previously determined, their morphological, structural, and electrochemical properties were improved in relation to the sample synthesized with an ionic-liquid-free electrolyte [73]. In addition, the degradation efficiency obtained in this study is better than the 97% degradation of methyl red reached in 160 min with WO_3 nanostructures under visible light illumination, but more favorable conditions were used in the latter example, such as a lamp of 500 W, an active surface area of 64 cm^2, and a greater external potential of 1.5 V [74]. In similar degradation conditions but with another organic dye, methyl orange (similar structure), our nanostructure also provided better results [75]. Compared with other metal oxides such as TiO_2 doped with Au, the WO_3 nanostructures fabricated in the presence of EMIM and BMIM exhibit better degradation yields in less time, although in that study, two ultraviolet lamps were used during the degradation tests [76].

4. Conclusions

In this study, we proved that with the incorporation of commercial ionic liquids (EMIM or BMIM) to the electrolyte during electrochemical anodization of tungsten, it is possible to synthesize more efficient nanostructures for photoelectrochemical applications, such as photoelectrochemical hydrogen production and the degradation of methyl red dye.

To be more specific, with 5% v/v of either IL, the anodization process was faster, and this led to higher-surface-area nanostructures, since the nanoplate length increased from 0.48 µm (blank) to 0.60 µm (EMIM) and 0.64 µm (BMIM) and the layer thickness increased from 0.7 µm using the blank electrolyte to 1.5 µm and 1.2 µm when using EMIM and BMIM, respectively.

Furthermore, the formation of a WO_3 monoclinic crystalline phase in the samples was improved by the addition of EMIM/BMIM, since the crystallite size decreased from 53.5 nm for the sample anodized in the blank electrolyte to 49.7 nm for the IL-anodized sample.

Additionally, the use of ILs during electrochemical anodization of tungsten resulted in nanostructures with a higher number of oxygen vacancies, as the number rose from 7.21×10^{19} cm^{-3} to 2.46×10^{20} cm^{-3} and 9.36×10^{19} cm^{-3} for EMIM and BMIM samples, respectively. This led to a higher electrical conductivity that improved the photoelectrochemical performance.

In particular, during the splitting of water using EMIM and BMIM samples, the generation of hydrogen was increased by 132% and 135%, respectively, when compared to the hydrogen production of the blank sample. Moreover, with the application of IL samples as photoelectrocatalysts in the photoelectrodegradation of methyl red, a 100% degradation efficiency was achieved after 50/60 min. To sum up, these results exhibit how IL-anodized nanostructures can be used for efficient and fast methyl red degradation involving a lower energy consumption, since less time is needed for a higher percentage of elimination.

Supplementary Materials: The following supporting information can be downloaded at: https://www.mdpi.com/article/10.3390/ma17061243/s1, Figure S1: Photocurrent transient vs. potential of WO_3 nanostructures synthesized by electrochemical anodization in different electrolytes (with and without IL) and with different concentrations of BMIM and EMIM (5, 10 and 30%); Figure S2:

FESEM images of the WO_3 nanostructure synthesized by electrochemical anodization with blank electrolyte. Top view (left) and cross section (right); Figure S3: Bode-Module and Bode-phase plots WO_3 nanostructures formed in electrolytes with and without IL; Figure S4: Equivalent circuit used for EIS fitting; Table S1: Results of EIS fitting analysis; Figure S5: Methyl red degradation efficiency of the different nanostructures.

Author Contributions: The manuscript was written through contributions from all authors. Methodology, E.D.S. and G.S.-G.; Formal analysis, E.D.S. and G.S.-G.; Resources, B.S. and R.S.-T.; Writing—original draft, E.D.S., G.S.-G. and A.P.-C.; Writing—review & editing, E.D.S., G.S.-G., A.P.-C., R.M.F.-D., B.S. and R.S.-T.; Supervision, R.M.F.-D., B.S. and R.S.-T. All authors have read and agreed to the published version of the manuscript.

Funding: This research was funded by Generalitat Valenciana with CIAICO/2021/094, CIGRIS/2022/198, CIACIF/2021/010 and CIGE/2022/166, by Ministerio de Ciencia e Innovación-Agencia Estatal de Investigación through the projects PID2021-126235OB C33 and TED2021-129555B-I00 and by European Union NextGenerationEU/PRTR.

Data Availability Statement: Data are contained within the article and supplementary materials.

Acknowledgments: Authors would like to acknowledge SCSIE-UV for assistance in characterization of materials.

Conflicts of Interest: The authors declare no conflicts of interest.

References

1. Al-Aisaee, N.; Alhabradi, M.; Yang, X.; Alruwaili, M.; Rasul, S.; Tahir, A.A. Fabrication of WO_3/Fe_2O_3 Heterostructure Photoanode by PVD for Photoelectrochemical Applications. *Sol. Energy Mater. Sol. Cells* **2023**, *263*, 112561. [CrossRef]
2. Ebrahimi, H.R.; Modrek, M. Photocatalytic Decomposition of Methyl Red Dye by Using Nanosized Zinc Oxide Deposited on Glass Beads in Various PH and Various Atmosphere. *J. Chem.* **2013**, *2013*, 151034. [CrossRef]
3. Conrad, C.L.; Elias, W.C.; Garcia-Segura, S.; Reynolds, M.A.; Wong, M.S. A Simple and Rapid Method of Forming Double-Sided TiO2 Nanotube Arrays. *ChemElectroChem* **2022**, *9*, e202200081. [CrossRef]
4. Sánchez-García, G.; Da Silva, E.; Fernández-Domene, R.M.; Cháfer, A.; González-Alfaro, V.; Solsona, B.; Sánchez-Tovar, R. TiO_2 Nanostructures Synthesized by Electrochemical Anodization in Green Protic Ionic Liquids for Photoelectrochemical Applications. *Ceram. Int.* **2023**, *49*, 26900–26909. [CrossRef]
5. Hussain, S.M.S.; Adewunmi, A.A.; Alade, O.S.; Murtaza, M.; Mahboob, A.; Khan, H.J.; Mahmoud, M.; Kamal, M.S. A Review of Ionic Liquids: Recent Synthetic Advances and Oilfield Applications. *J. Taiwan Inst. Chem. Eng.* **2023**, *153*, 105195. [CrossRef]
6. Sharma, P.; Sharma, S.; Kumar, H. Introduction to Ionic Liquids, Applications and Micellization Behaviour in Presence of Different Additives. *J. Mol. Liq.* **2024**, *393*, 123447. [CrossRef]
7. Wei, W.; Liu, Y.; Yao, X.; Hang, R. Na-Ti-O Nanostructured Film Anodically Grown on Titanium Surface Have the Potential to Improve Osteogenesis. *Surf. Coat. Technol.* **2020**, *397*, 125907. [CrossRef]
8. Tacca, A.; Meda, L.; Marra, G.; Savoini, A.; Caramori, S.; Cristino, V.; Bignozzi, C.A.; Pedro, V.G.; Boix, P.P.; Gimenez, S.; et al. Photoanodes Based on Nanostructured WO_3 for Water Splitting. *ChemPhysChem* **2012**, *13*, 3025–3034. [CrossRef]
9. Amano, F.; Tian, M.; Ohtani, B.; Chen, A. Photoelectrochemical Properties of Tungsten Trioxide Thin Film Electrodes Prepared from Facet-Controlled Rectangular Platelets. *J. Solid State Electrochem.* **2012**, *16*, 1965–1973. [CrossRef]
10. Rodríguez-Pérez, M.; Chacón, C.; Palacios-González, E.; Rodríguez-Gattorno, G.; Oskam, G. Photoelectrochemical Water Oxidation at Electrophoretically Deposited WO_3 Films as a Function of Crystal Structure and Morphology. *Electrochim. Acta* **2014**, *140*, 320–331. [CrossRef]
11. Rougier, A.; Portemer, F.; Quede, A.; El Marssi, M.; Francè, F.F. Characterization of Pulsed Laser Deposited WO_3 Thin Films for Electrochromic Devices. *Appl. Surf. Sci.* **1999**, *153*, 1–9. [CrossRef]
12. Aliannezhadi, M.; Abbaspoor, M.; Shariatmadar Tehrani, F.; Jamali, M. High Photocatalytic WO_3 Nanoparticles Synthesized Using Sol-Gel Method at Different Stirring Times. *Opt. Quantum Electron.* **2023**, *55*, 250. [CrossRef]
13. Poongodi, S.; Kumar, P.S.; Mangalaraj, D.; Ponpandian, N.; Meena, P.; Masuda, Y.; Lee, C. Electrodeposition of WO_3 Nanostructured Thin Films for Electrochromic and H_2S Gas Sensor Applications. *J. Alloys Compd.* **2017**, *719*, 71–81. [CrossRef]
14. Pligovka, A. Reflectant Photonic Crystals Produced via Porous-Alumina-Assisted-Anodizing of Al/Nb and Al/Ta Systems. *Surf. Rev. Lett.* **2021**, *28*, 1–7. [CrossRef]
15. Deshpande, R.; Lee, S.H.; Mahan, A.H.; Parilla, P.A.; Jones, K.M.; Norman, A.G.; To, B.; Blackburn, J.L.; Mitra, S.; Dillon, A.C. Optimization of Crystalline Tungsten Oxide Nanoparticles for Improved Electrochromic Applications. *Solid. State Ion.* **2007**, *178*, 895–900. [CrossRef]
16. Pan, L.; Han, Q.; Dong, Z.; Wan, M.; Zhu, H.; Li, Y.; Mai, Y. Reactively Sputtered WO3 Thin Films for the Application in All Thin Film Electrochromic Devices. *Electrochim. Acta* **2019**, *328*, 135107. [CrossRef]

17. Pligovka, A.; Lazavenka, A.; Zakhlebayeva, A. Electro-Physical Properties of Niobia Columnlike Nanostructures via the Anodizing of Al/Nb Layers. In Proceedings of the IEEE Conference on Nanotechnology, 2018 IEEE 18th International Conference on Nanotechnology, Cork, Ireland, 23–26 July 2018; IEEE Computer Society: Washington, DC, USA, 2018; Volume 2018.
18. Go, G.H.; Shinde, P.S.; Doh, C.H.; Lee, W.J. PVP-Assisted Synthesis of Nanostructured Transparent WO_3 Thin Films for Photoelectrochemical Water Splitting. *Mater. Des.* **2016**, *90*, 1005–1009. [CrossRef]
19. Shimosako, N.; Sakama, H. Influence of Vacuum Environment on Photocatalytic Degradation of Methyl Red by TiO_2 Thin Film. *Acta Astronaut.* **2021**, *178*, 693–699. [CrossRef]
20. Davi, M.; Ogutu, G.; Schrader, F.; Rokicinska, A.; Kustrowski, P.; Slabon, A. Enhancing Photoelectrochemical Water Oxidation Efficiency of $WO_3/\alpha\text{-}Fe_2O_3$ Heterojunction Photoanodes by Surface Functionalization with CoPd Nanocrystals. *Eur. J. Inorg. Chem.* **2017**, *2017*, 4267–4274. [CrossRef]
21. Maldonado, M.I.; Passarinho, P.C.; Oller, I.; Gernjak, W.; Fernández, P.; Blanco, J.; Malato, S. Photocatalytic Degradation of EU Priority Substances: A Comparison between TiO_2 and Fenton plus Photo-Fenton in a Solar Pilot Plant. *J. Photochem. Photobiol. A Chem.* **2007**, *185*, 354–363. [CrossRef]
22. Luo, J.; Hepel, M. Photoelectrochemical Degradation of Naphthol Blue Black Diazo Dye on WO_3 Film Electrode. *Electrochim. Acta* **2001**, *46*, 2913–2922. [CrossRef]
23. Sreekantan, S.; Saharudin, K.A.; Basiron, N.; Wei, L.C. New-Generation Titania-Based Catalysts for Photocatalytic Hydrogen Generation. In *Nanostructured, Functional, and Flexible Materials for Energy Conversion and Storage Systems*; Elsevier: Amsterdam, The Netherlands, 2020; pp. 257–292. ISBN 9780128195529.
24. Kumar, M.; Meena, B.; Subramanyam, P.; Suryakala, D.; Subrahmanyam, C. Recent Trends in Photoelectrochemical Water Splitting: The Role of Cocatalysts. *NPG Asia Mater.* **2022**, *14*, 88. [CrossRef]
25. Chen, X.; Yang, J.; Cao, Y.; Kong, L.; Huang, J. Design Principles for Tungsten Oxide Electrocatalysts for Water Splitting. *ChemElectroChem* **2021**, *8*, 4427–4440. [CrossRef]
26. Paramasivam, I.; Macak, J.M.; Selvam, T.; Schmuki, P. Electrochemical Synthesis of Self-Organized TiO_2 Nanotubular Structures Using an Ionic Liquid (BMIM-BF4). *Electrochim. Acta* **2008**, *54*, 643–648. [CrossRef]
27. Wender, H.; Feil, A.F.; Diaz, L.B.; Ribeiro, C.S.; Machado, G.J.; Migowski, P.; Weibel, D.E.; Dupont, J.; Teixeira, S.R. Self-Organized TiO_2 Nanotube Arrays: Synthesis by Anodization in an Ionic Liquid and Assessment of Photocatalytic Properties. *ACS Appl. Mater. Interfaces* **2011**, *3*, 1359–1365. [CrossRef]
28. Li, H.; Qu, J.; Cui, Q.; Xu, H.; Luo, H.; Chi, M.; Meisner, R.A.; Wang, W.; Dai, S. TiO_2 Nanotube Arrays Grown in Ionic Liquids: High-Efficiency in Photocatalysis and Pore-Widening. *J. Mater. Chem.* **2011**, *21*, 9487–9490. [CrossRef]
29. Pancielejko, A.; Mazierski, P.; Lisowski, W.; Zaleska-Medynska, A.; Łuczak, J. Ordered TiO_2 Nanotubes with Improved Photoactivity through Self-Organizing Anodization with the Addition of an Ionic Liquid: Effects of the Preparation Conditions. *ACS Sustain. Chem. Eng.* **2019**, *7*, 15585–15596. [CrossRef]
30. Heydari Dokoohaki, M.; Mohammadpour, F.; Zolghadr, A.R. New Insight into Electrosynthesis of Ordered TiO_2 Nanotubes in EG-Based Electrolyte Solutions: Combined Experimental and Computational Assessment. *Phys. Chem. Chem. Phys.* **2020**, *22*, 22719–22727. [CrossRef]
31. Samuel, O.; Othman, M.H.D.; Kamaludin, R.; Sinsamphanh, O.; Abdullah, H.; Puteh, M.H.; Kurniawan, T.A. WO_3–Based Photocatalysts: A Review on Synthesis, Performance Enhancement and Photocatalytic Memory for Environmental Applications. *Ceram. Int.* **2022**, *48*, 5845–5875. [CrossRef]
32. Dong, P.; Hou, G.; Xi, X.; Shao, R.; Dong, F. WO_3-Based Photocatalysts: Morphology Control, Activity Enhancement and Multifunctional Applications. *Env. Sci. Nano* **2017**, *4*, 539–557. [CrossRef]
33. Kwong, W.L.; Savvides, N.; Sorrell, C.C. Electrodeposited Nanostructured WO_3 Thin Films for Photoelectrochemical Applications. *Electrochim. Acta* **2012**, *75*, 371–380. [CrossRef]
34. Kolaei, M.; Lee, B.K.; Masoumi, Z. Enhancing the Photoelectrochemical Activity and Stability of Plate-like WO_3 Photoanode in Neutral Electrolyte Solution Using Optimum Loading of $BiVO_4$ Layer and NiFe–LDH Electrodeposition. *J. Alloys Compd.* **2023**, *968*, 172133. [CrossRef]
35. Watcharenwong, A.; Chanmanee, W.; de Tacconi, N.R.; Chenthamarakshan, C.R.; Kajitvichyanukul, P.; Rajeshwar, K. Anodic Growth of Nanoporous WO3 Films: Morphology, Photoelectrochemical Response and Photocatalytic Activity for Methylene Blue and Hexavalent Chrome Conversion. *J. Electroanal. Chem.* **2008**, *612*, 112–120. [CrossRef]
36. Fernández-Domene, R.M.; Sánchez-Tovar, R.; Lucas-Granados, B.; Roselló-Márquez, G.; García-Antón, J. A Simple Method to Fabricate High-Performance Nanostructured WO_3 Photocatalysts with Adjusted Morphology in the Presence of Complexing Agents. *Mater. Des.* **2017**, *116*, 160–170. [CrossRef]
37. Ullah, I.; Tariq, M.; Muhammad, M.; Khan, J.; Rahim, A.; Abdullah, A.Z. UV Photocatalytic Remediation of Methyl Red in Aqueous Medium by Sulfate ($SO_4\bullet-$) and Hydroxyl ($\bullet OH$) Radicals in the Presence of $Fe_2+$and $Co@TiO_2$ NPs Photocatalysts. *Colloids Surf. A Physicochem. Eng. Asp.* **2023**, *679*, 132614. [CrossRef]
38. Galenda, A.; Crociani, L.; Habra, N.E.; Favaro, M.; Natile, M.M.; Rossetto, G. Effect of Reaction Conditions on Methyl Red Degradation Mediated by Boron and Nitrogen Doped TiO_2. *Appl. Surf. Sci.* **2014**, *314*, 919–930. [CrossRef]
39. Lassner, E.; Schubert, W.-D. *Tungsten: Properties, Chemistry, Technology of the Element, Alloys, and Chemical Compounds*; Springer: Berlin/Heidelberg, Germany, 1999; Volume 422.

40. Paola, A.D.; Quarto, F.D.; Sunseri, C. Anodic Oxide Films on Tungsten—I. The Influence of Anodizing Parameters on Charging Curves and Film Composition. *Corros. Sci.* **1980**, *20*, 1067–1078. [CrossRef]
41. Fernández-Domene, R.M.; Sánchez-Tovar, R.; Segura-Sanchís, E.; García-Antón, J. Novel Tree-like WO_3 Nanoplatelets with Very High Surface Area Synthesized by Anodization under Controlled Hydrodynamic Conditions. *Chem. Eng. J.* **2016**, *286*, 59–67. [CrossRef]
42. Freitas, R.G.; Justo, S.G.; Pereira, E.C. The Influence of Self-Ordered TiO_2 Nanotubes Microstructure towards Li+ Intercalation. *J. Power Sources* **2013**, *243*, 569–572. [CrossRef]
43. Roselló-Márquez, G.; Fernández-Domene, R.M.; Sánchez-Tovar, R.; García-Antón, J. Photoelectrocatalyzed Degradation of Organophosphorus Pesticide Fenamiphos Using WO_3 Nanorods as Photoanode. *Chemosphere* **2020**, *246*, 125677. [CrossRef]
44. Roselló-Márquez, G.; Fernández-Domene, R.M.; Sánchez-Tovar, R.; García-Antón, J. Influence of Annealing Conditions on the Photoelectrocatalytic Performance of WO_3 Nanostructures. *Sep. Purif. Technol.* **2020**, *238*, 116417. [CrossRef]
45. Wei, W.; Shaw, S.; Lee, K.; Schmuki, P.; Wei, W.; Shaw, S.; Lee, K.; Schmuki, P. Rapid Anodic Formation of High Aspect Ratio WO_3 Layers with Self-Ordered Nanochannel Geometry and Use in Photocatalysis. *Chem. A Eur. J.* **2012**, *18*, 14622–14626. [CrossRef]
46. Liu, Q.; Chen, Q.; Bai, J.; Li, J.; Li, J.; Zhou, B. Enhanced Photoelectrocatalytic Performance of Nanoporous WO_3 Photoanode by Modification of Cobalt-Phosphate (Co-Pi) Catalyst. *J. Solid. State Electrochem.* **2014**, *18*, 157–161. [CrossRef]
47. Yousif, A.A.; Khudadad, A.I. Effects of Annealing Process on the WO_3 Thin Films Prepared by Pulsed Laser Deposition. *IOP Conf. Ser. Mater. Sci. Eng.* **2020**, *745*, 012064. [CrossRef]
48. Hatel, R.; Baitoul, M. Nanostructured Tungsten Trioxide (WO_3): Synthesis, Structural and Morphological Investigations. *J. Phys. Conf. Ser.* **2019**, *1292*, 012014. [CrossRef]
49. Shabdan, Y.; Markhabayeva, A.; Bakranov, N.; Nuraje, N. Photoactive Tungsten-Oxide Nanomaterials for Water-Splitting. *Nanomaterials* **2020**, *10*, 1871. [CrossRef]
50. Murillo-Sierra, J.C.; Hernández-Ramírez, A.; Hinojosa-Reyes, L.; Guzmán-Mar, J.L. A Review on the Development of Visible Light-Responsive WO_3-Based Photocatalysts for Environmental Applications. *Chem. Eng. J. Adv.* **2021**, *5*, 100070. [CrossRef]
51. Huang, Z.F.; Song, J.; Pan, L.; Zhang, X.; Wang, L.; Zou, J.J. Tungsten Oxides for Photocatalysis, Electrochemistry, and Phototherapy. *Adv. Mater.* **2015**, *27*, 5309–5327. [CrossRef]
52. Guo, Y.; Quan, X.; Lu, N.; Zhao, H.; Chen, S. High Photocatalytic Capability of Self-Assembled Nanoporous WO_3 with Preferential Orientation of (002) Planes. *Env. Sci. Technol.* **2007**, *41*, 4422–4427. [CrossRef]
53. Fernández-Domene, R.M.; Roselló-Márquez, G.; Sánchez-Tovar, R.; Lucas-Granados, B.; García-Antón, J. Photoelectrochemical Removal of Chlorfenvinphos by Using WO_3 Nanorods: Influence of Annealing Temperature and Operation PH. *Sep. Purif. Technol.* **2019**, *212*, 458–464. [CrossRef]
54. Palmas, S.; Polcaro, A.M.; Ruiz, J.R.; Da Pozzo, A.; Mascia, M.; Vacca, A. TiO_2 Photoanodes for Electrically Enhanced Water Splitting. *Int. J. Hydrog. Energy* **2010**, *35*, 6561–6570. [CrossRef]
55. Ilka, M.; Bera, S.; Kwon, S.H. Influence of Surface Defects and Size on Photochemical Properties of SnO_2 Nanoparticles. *Materials* **2018**, *11*, 904. [CrossRef] [PubMed]
56. Kumbhar, V.S.; Lee, J.; Choi, Y.; Lee, H.; Ryuichi, M.; Nakayama, M.; Lee, W.; Oh, H.; Lee, K. Electrochromic and Pseudocapacitive Behavior of Hydrothermally Grown WO_3 Nanostructures. *Thin Solid. Film.* **2020**, *709*, 138214. [CrossRef]
57. Amano, F.; Koga, S. Electrochemical Impedance Spectroscopy of WO_3 Photoanodes on Different Conductive Substrates: The Interfacial Charge Transport between Semiconductor Particles and Ti Surface. *J. Electroanal. Chem.* **2022**, *921*, 116685. [CrossRef]
58. Batista-Grau, P.; Fernández-Domene, R.M.; Sánchez-Tovar, R.; Blasco-Tamarit, E.; Solsona, B.; García-Antón, J. Indirect Charge Transfer of Holes via Surface States in ZnO Nanowires for Photoelectrocatalytic Applications. *Ceram. Int.* **2022**, *48*, 21856–21867. [CrossRef]
59. Dhandole, L.K.; Koh, T.S.; Anushkkaran, P.; Chung, H.S.; Chae, W.S.; Lee, H.H.; Choi, S.H.; Cho, M.; Jang, J.S. Enhanced Charge Transfer with Tuning Surface State in Hematite Photoanode Integrated by Niobium and Zirconium Co-Doping for Efficient Photoelectrochemical Water Splitting. *Appl. Catal. B* **2022**, *315*, 121538. [CrossRef]
60. Huang, M.C.; Wang, T.; Wu, B.J.; Lin, J.C.; Wu, C.C. Anodized ZnO Nanostructures for Photoelectrochemical Water Splitting. *Appl. Surf. Sci.* **2016**, *360*, 442–450. [CrossRef]
61. Faughnan, B.W.; Crandall, R.S.; Lampert, M.A. Model for the Bleaching of WO_3 Electrochromic Films by an Electric Field. *Appl. Phys. Lett.* **1975**, *27*, 275–277. [CrossRef]
62. Yagi, M.; Maruyama, S.; Sone, K.; Nagai, K.; Norimatsu, T. Preparation and Photoelectrocatalytic Activity of a Nano-Structured WO_3 Platelet Film. *J. Solid. State Chem.* **2008**, *181*, 175–182. [CrossRef]
63. Liu, Y.; Li, Y.; Li, W.; Han, S.; Liu, C. Photoelectrochemical Properties and Photocatalytic Activity of Nitrogen-Doped Nanoporous WO_3 Photoelectrodes under Visible Light. *Appl. Surf. Sci.* **2012**, *258*, 5038–5045. [CrossRef]
64. Bonham, D.B.; Orazem, M.E. A Mathematical Model for the Influence of Deep-Level Electronic States on Photoelectrochemical Impedance Spectroscopy: II. Assessment of Characterization Methods Based on Mott-Schottky Theory. *J. Electrochem. Soc.* **1992**, *139*, 127–131. [CrossRef]
65. Irie, H.; Watanabe, Y.; Hashimoto, K. Nitrogen-Concentration Dependence on Photocatalytic Activity of TiO_2-XNx Powders. *J. Phys. Chem. B* **2003**, *107*, 5483–5486. [CrossRef]

66. Wang, D.; Zhang, X.; Sun, P.; Lu, S.; Wang, L.; Wang, C.; Liu, Y. Photoelectrochemical Water Splitting with Rutile TiO$_2$ Nanowires Array: Synergistic Effect of Hydrogen Treatment and Surface Modification with Anatase Nanoparticles. *Electrochim. Acta* **2014**, *130*, 290–295. [CrossRef]
67. Takagi, M.; Kawaguchi, M.; Yamakata, A. Enhancement of UV-Responsive Photocatalysts Aided by Visible-Light Responsive Photocatalysts: Role of WO$_3$ for H$_2$ Evolution on CuCl. *Appl. Catal. B* **2020**, *263*, 118333. [CrossRef]
68. Levinas, R.; Tsyntsaru, N.; Murauskas, T.; Cesiulis, H. Improved Photocatalytic Water Splitting Activity of Highly Porous WO$_3$ Photoanodes by Electrochemical H+ Intercalation. *Front. Chem. Eng.* **2021**, *3*, 760700. [CrossRef]
69. Fernández-Domene, R.M.; Roselló-Márquez, G.; Sánchez-Tovar, R.; Cifre-Herrando, M.; García-Antón, J. Synthesis of WO$_3$ Nanorods through Anodization in the Presence of Citric Acid: Formation Mechanism, Properties and Photoelectrocatalytic Performance. *Surf. Coat. Technol.* **2021**, *422*, 127489. [CrossRef]
70. Devi, L.G.; Murthy, B.N.; Kumar, S.G. Photocatalytic Activity of TiO$_2$ Doped with Zn2+ and V5+ Transition Metal Ions: Influence of Crystallite Size and Dopant Electronic Configuration on Photocatalytic Activity. *Mater. Sci. Eng. B* **2010**, *166*, 1–6. [CrossRef]
71. Grushevskaya, S.; Belyanskaya, I.; Kozaderov, O. Approaches for Modifying Oxide-Semiconductor Materials to Increase the Efficiency of Photocatalytic Water Splitting. *Materials* **2022**, *15*, 4915. [CrossRef] [PubMed]
72. Sarkar, R.; Kar, M.; Habib, M.; Zhou, G.; Frauenheim, T.; Sarkar, P.; Pal, S.; Prezhdo, O.V. Common Defects Accelerate Charge Separation and Reduce Recombination in CNT/Molecule Composites: Atomistic Quantum Dynamics. *J. Am. Chem. Soc.* **2021**, *143*, 6649–6656. [CrossRef] [PubMed]
73. Zhu, T.; Chong, M.N.; Chan, E.S. Nanostructured Tungsten Trioxide Thin Films Synthesized for Photoelectrocatalytic Water Oxidation: A Review. *ChemSusChem* **2014**, *7*, 2974–2997. [CrossRef] [PubMed]
74. Hunge, Y.M.; Mohite, V.S.; Kumbhar, S.S.; Rajpure, K.Y.; Moholkar, A.V.; Bhosale, C.H. Photoelectrocatalytic Degradation of Methyl Red Using Sprayed WO3 Thin Films under Visible Light Irradiation. *J. Mater. Sci. Mater. Electron.* **2015**, *26*, 8404–8412. [CrossRef]
75. Mohite, S.V.; Ganbavle, V.V.; Rajpure, K.Y. Photoelectrocatalytic Activity of Immobilized Yb Doped WO$_3$ Photocatalyst for Degradation of Methyl Orange Dye. *J. Energy Chem.* **2017**, *26*, 440–447. [CrossRef]
76. Hernández, R.; Elizalde, E.A.; Domínguez, A.; Olvera-Rodríguez, I.; Esquivel, K.; Guzmán, C. Photoelectrocatalytic Degradation of Methyl Red Dye Using Au Doped TiO$_2$ Photocatalyst. In Proceedings of the 2016 12th Congreso Internacional de Ingenieria, CONIIN 2016, Santiago de Queretaro, Mexico, 1–6 May 2016; Institute of Electrical and Electronics Engineers Inc.: New York, NY, USA, 2016.

Disclaimer/Publisher's Note: The statements, opinions and data contained in all publications are solely those of the individual author(s) and contributor(s) and not of MDPI and/or the editor(s). MDPI and/or the editor(s) disclaim responsibility for any injury to people or property resulting from any ideas, methods, instructions or products referred to in the content.

Article

C_{60}- and CdS-Co-Modified Nano-Titanium Dioxide for Highly Efficient Photocatalysis and Hydrogen Production

Meifang Zhang [1], Xiangfei Liang [1,*], Yang Gao [1] and Yi Liu [2,*]

[1] Institute of Carbon Neutral New Energy Research, Yuzhang Normal University, Nanchang 330031, China; mfzhang@whu.edu.cn (M.Z.); ygao2023@126.com (Y.G.)
[2] School of Chemical and Environmental Engineering, Wuhan Polytechnic University, Wuhan 430023, China
* Correspondence: xfliang96@126.com (X.L.); yiliuchem@whu.edu.cn (Y.L.)

Abstract: The inherent properties of TiO_2, including a wide band gap and restricted spectral response range, hinder its commercial application and its ability to harness only 2–3% of solar energy. To address these challenges and unlock TiO_2's full potential in photocatalysis, C_{60}- and CdS-co-modified nano-titanium dioxide has been adopted in this work to reduce the band gap, extend the absorption wavelength, and control photogenerated carrier recombination, thereby enhancing TiO_2's light-energy-harnessing capabilities and hydrogen evolution capacity. Using the sol-gel method, we successfully synthesized CdS-C_{60}/TiO_2 composite nanomaterials, harnessing the unique strengths of CdS and C_{60}. The results showed a remarkable average yield of 34.025 μmol/h for TiO_2 co-modified with CdS and C_{60}, representing a substantial 17-fold increase compared to pure CdS. Simultaneously, the average hydrogen generation of C_{60}-modified CdS surged to 5.648 μmol/h, a notable two-fold improvement over pure CdS. This work opens up a new avenue for the substantial improvement of both the photocatalytic degradation efficiency and hydrogen evolution capacity, offering promise of a brighter future in photocatalysis research.

Keywords: co-modified; CdS-C_{60}/TiO_2; photocatalytic activity; hydrogen evolution

Citation: Zhang, M.; Liang, X.; Gao, Y.; Liu, Y. C_{60}- and CdS-Co-Modified Nano-Titanium Dioxide for Highly Efficient Photocatalysis and Hydrogen Production. *Materials* **2024**, *17*, 1206. https://doi.org/10.3390/ma17051206

Academic Editor: Ovidiu Oprea

Received: 18 December 2023
Revised: 6 February 2024
Accepted: 8 February 2024
Published: 5 March 2024

Copyright: © 2024 by the authors. Licensee MDPI, Basel, Switzerland. This article is an open access article distributed under the terms and conditions of the Creative Commons Attribution (CC BY) license (https:// creativecommons.org/licenses/by/ 4.0/).

1. Introduction

In recent years, photocatalytic technology has garnered significant attention from researchers owing to its cost-effectiveness, simplicity, and minimal environmental impact. This versatile technology leverages a range of photocatalysts to disintegrate a wide array of both organic and inorganic pollutants, offering the promise of converting CO_2 into valuable renewable chemicals and facilitating water decomposition [1–3]. Among these photocatalysts, titanium dioxide (TiO_2) stands out for its relatively high activity, rendering it highly effective in the purification of air and water, even in biomedical applications [4,5]. Consequently, its applications in the field of photocatalysis hold considerable promise [6,7].

Nevertheless, the inherent limitations of TiO_2 are its wide band gap and restricted spectral response range, which result in its ability to utilize only approximately 2–3% of solar energy and hinder its commercial application [8]. To address these challenges and unlock the full potential of TiO_2 in photocatalysis, modification techniques have emerged as a pivotal approach to enhancing its photocatalytic performance. These techniques aim to reduce the band gap, extend the absorption wavelength, and control photogenerated carrier recombination, thereby enhancing TiO_2's light-energy-harnessing capabilities [9]. Notably, the choice of dopants significantly influences the photocatalytic activity, affecting the recombination of photogenerated electrons (e^-) and holes (h^+) on the material's surface.

In the field of photoelectrochemical water decomposition, nano-TiO_2 coupling systems have been extensively researched [10,11]. Cadmium sulfide (CdS), with its narrow energy gap of 2.4 eV, has proved effective as a coupling agent for TiO_2, promoting the photocatalytic decomposition of water [12,13]. The strong coupling between TiO_2 and CdS allows for the direct combination of TiO_2 particles with dispersed CdS particles. Although CdS exhibits photoetching instability when used independently, the formation of composite nanomaterials with TiO_2, characterized by a wider band gap, effectively reduces the likelihood of photogenerated electron–hole recombination. Thus, the strategic coupling of TiO_2 with CdS mitigates the photoetching instability associated with CdS alone and expands the spectral response range [14–16].

Fullerene (C_{60}), renowned for its exceptional electron (e^-) transport capabilities, has emerged as an intriguing option for mitigating charge recombination. This has fueled increasing interest in TiO_2 catalysts supported by C_{60} [17,18]. C_{60} boasts an extensive three-dimensional π electron system, featuring a closed core–shell structure comprising 60 delocalized π electrons and 30 bonded molecular orbitals. Even minor structural variations or changes in the solvent properties can influence the electron transfer within the system, making the incorporation of C_{60} an attractive strategy for enhancing effective electron transfer [19,20]. With a band gap ranging from 1.6 to 1.9 eV, C_{60} holds considerable promise for practical applications, and catalysts modified with C_{60} are poised to achieve a remarkable photocatalytic performance.

Recognizing the complementary advantages of CdS and C_{60}, we have adopted a comprehensive approach to co-modifying TiO_2. Utilizing the sol-gel method, we have successfully synthesized CdS-C_{60}/TiO_2 composite nanomaterials with highly effective photocatalytic properties. Remarkably, the hydrogen production of CdS modified with C_{60} exhibited a two-fold increase compared to pure CdS, reaching an average of 5.648 μmol/h. Furthermore, TiO_2 co-modified with CdS and C_{60} demonstrated an impressive average yield of 34.025 μmol/h, representing a staggering 17-fold increase over pure CdS. This study underscores the synergistic benefits of combining CdS and C_{60} in enhancing the photocatalytic and hydrogen production performance of TiO_2, opening up new avenues for sustainable and efficient environmental remediation processes.

2. Materials and Methods

Materials: The cadmium chloride ($CdCl_2$, 99.0%) and sodium sulfide ($Na_2S \cdot 9H_2O$) were purchased from Tianjin Kemiou Chemical Reagent Co., Ltd. (Tianjin, China). The Malachite Green (MG), P25 (Degussa, Hamburg, Germany), and butyl titanate ($Ti(OBu)_4$) were all purchased from China National Pharmaceutical Group Chemical Reagent Co., Ltd. (Beijing, China). The C_{60} was purchased from Shanghai Macklin Biochemical Co., Ltd. (Shanghai, China). All the other reagents were analytical-grade. The pH of the solution was adjusted using nitric acid and sodium hydroxide solutions. All experiments used secondary water.

Synthesis of CdS: Dissolve 3.1148 g of $CdCl_2 \cdot 2.5H_2O$ in 20 mL of ethanol in the first bottle, and dissolve 2.4004 g of $Na_2S \cdot 9H_2O$ in 20 mL of ethanol in the second bottle. After stirring for 10 min, the second bottle is added to the first bottle slowly, drop by drop. Gel A is stirred for two hours at room temperature before being transferred into an alumina crucible. Place the sample into a muffle furnace and set the conditions to increase from room temperature to 450 °C, with a heating rate of 2 °C/min in an air atmosphere, and maintain it there for 2 h after aging at room temperature for 24 h. The calcination conditions are the same for each sample below. Finally, create a sample of CdS powder by grinding the acquired sample in an agate mortar [21–23].

Synthesis of CdS-TiO_2: Make gel A as directed by the conditions listed above. In a flask with a circular bottom and labeled "gel B", combine 20 mL of butyl titanate (TBT) with 30 mL of ethanol solution. The combination added dropwise is stirred at room temperature for 2 h. Prepare another combination using 20 mL of ethanol, 4 mL of distilled water, 2 mL of nitric acid, and 20 mL of gel A. They should be added dropwise to the room-temperature gel B solution, stirred as they are added, and left to react for 24 h. Age for 24 h in an alumina crucible. The sample is burned under the same conditions in a muffle furnace and then ground to produce a CdS-TiO_2 sample.

Synthesis of CdS-C$_{60}$: In order to create CdS, 30 mg of Ox-C$_{60}$ (also known as activated C$_{60}$) is added to a combination of distilled water and agitated for 2 h. An alumina crucible is used to age gel C for 24 h. Then, under the same conditions, it is calcined in a muffle furnace. The CdS-C$_{60}$ sample is obtained by grinding it after cooling to room temperature.

Synthesis of CdS-C$_{60}$-TiO$_2$: Synthesize gel B and gel C separately, measure 20 mL of ethanol and 4 mL of distilled water, and combine 2 mL of nitric acid and gel C to put together another mixture. The combination dropwise is added to the gel B solution at room temperature, stir whilst adding, and react for 24 h. Age it in an alumina crucible for 24 h and burn it in a muffle furnace underneath the identical conditions. Grind to achieve the CdS-C$_{60}$-TiO$_2$ sample. The experimental characterization is detailed in the Supporting Information (pages 2–4).

3. Results and Discussion

To explore the changes in the TiO$_2$ crystal morphology after the modification, we conducted X-ray diffraction (XRD) measurements for the pure TiO$_2$ and the modified TiO$_2$ materials in Figure 1a. As can be seen, the diffraction peaks of P25 (commercial TiO$_2$) appear at $2\theta = 25.08°$, $37.94°$, $48.14°$, $54.26°$, $55.26°$, $62.90°$, $68.98°$, $70.02°$, and $75.24°$, respectively, corresponding to the diffraction peaks of anatase-phase TiO$_2$ (101), (004), (200), (105), (211), (204), (116), (200), and (215), respectively, where $2\theta = 25.08°$ was the strongest diffraction peak. The diffraction peaks of the rutile phase (110), (101), and (200) crystal faces appear at $2\theta = 27.7°$, $36.2°$, and $42.82°$, indicating that TiO$_2$ is a mixed crystal structure dominated by anatase phases and rutile phases [24]. Comparing P25, CdS-TiO$_2$, and CdS-C$_{60}$/TiO$_2$, the rutile diffraction peak weakened in turn, and the anatase-phase diffraction peak increased in turn, indicating that the introduction of CdS into the samples can inhibit the growth of their rutile phase. As we all know, rutile TiO$_2$, as the most stable crystalline structure form, has a good crystallization state and fewer defects, resulting in electrons and holes that are easy to compound, with almost no photocatalytic activity. Therefore, the CdS-C$_{60}$/TiO$_2$ composite with less rutile TiO$_2$ will have a better catalytic performance. We also tested the XPS patterns of P25, CdS-TiO$_2$, and CdS-C$_{60}$/TiO$_2$, as shown in Figure S1 (Supporting Information). It was proven that there is no Ti2p separation between the C$_{60}$, CdS, and TiO$_2$ phases after preparation, but the shift in the oxygen atom toward a high binding energy means that the chemical bonds between the oxygen atom and other elements become tighter and more stable, which indicates that CdS-C$_{60}$ has a significant effect on the modification of TiO$_2$, which promotes the transfer of electrons between oxygen atoms and other elements, thereby increasing the activity of the catalyst.

Not only that but most UV radiation with a wavelength of less than 400 nm is absorbed by P25. The absorption bands of CdS, CdS-C$_{60}$, CdS-TiO$_2$, and CdS-C$_{60}$/TiO$_2$ all entered the visible light region. This is mainly because the band gap of CdS is 2.4 eV, the response characteristics of visible light are better, and the crystal structure of the original substance is changed after being modified with CdS. Compared with pure CdS and pure TiO$_2$, the absorbance of CdS-C$_{60}$ and CdS-C$_{60}$/TiO$_2$ increased, showing that the band gaps of various semiconductor catalysts overlapped in the ultraviolet and visible regions, respectively, which is conducive to improving their photocatalytic performance. The C$_{60}$-modified sample's absorption band edge shows some redshift, indicating that C$_{60}$ can narrow the catalyst's band gap, which is due to the interaction between the components, causing surface lattice oxygen vacancy and other defects [25–27]. The results above indicate that the crystal structure of TiO$_2$ is alterable by CdS and C$_{60}$, leading to the creation of defects on the surface lattice. The cause of the edge absorption is the defect absorption rather than the actual absorption of the lattice [28]. Additionally, the band gap energy of TiO$_2$ is decreased, enabling it to be activated by visible light and to exhibit photocatalytic activity. These changes not only enhance the utilization of light energy but also improve the overall efficiency of the photocatalysis process. Meanwhile, EDX fluorescence spectroscopy analyses on each sample were performed to detect the type and content of the characteristic elements contained in each sample, as shown in Figure 1c–e.

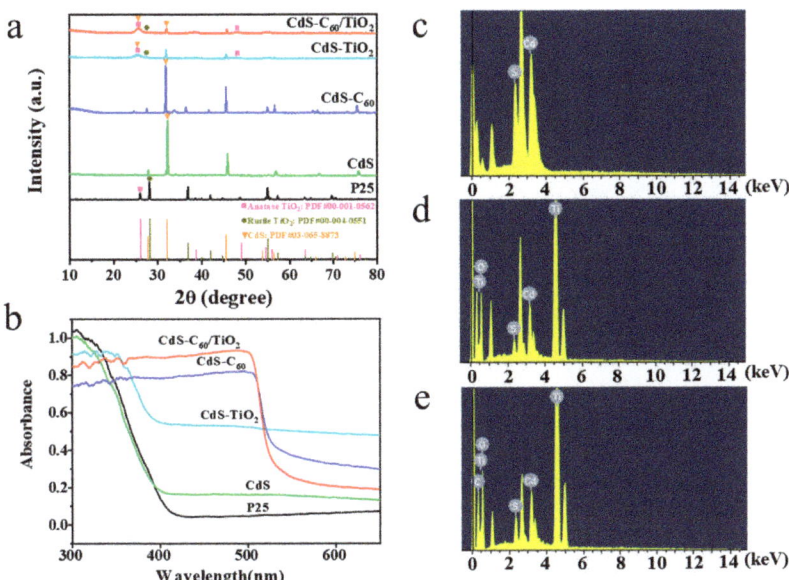

Figure 1. (**a**) XRD patterns and (**b**) UV–Vis diffuse reflectance spectra of P25, CdS, CdS-C_{60}, CdS-TiO_2, and CdS-C_{60}/TiO_2. Energy-Dispersive X-ray (EDX) elemental microanalysis of (**c**) CdS, (**d**) CdS-TiO_2, (**e**) CdS-C_{60}/TiO_2.

As the FT-IR spectra show in Figure 2a,b, all the catalyst samples have strong absorption peaks at 3420–3510 cm^{-1} and 1620–1640 cm^{-1}, corresponding to the bending vibration peaks of the H-O-H bonds, which indicates the bending vibration of the presence of numerous water molecules adsorbed onto the catalyst surface. These peaks confirm that water molecules adhere to the catalyst surface in a physical manner. Consequently, it can be concluded that these nanomaterials possess the ability to exhibit a significant photocatalytic effect, which may be the presence of the majority of hydroxyl groups on the surface, and under photoexcitation, the e$^-$-h$^+$ pair generated by the catalyst sample acts on the hydroxyl group to generate highly oxidizing ·OH. The abundant and diverse hydroxyl groups present on the surface also promote the creation of sites where O_2, CO_2, and CO molecules can be adsorbed [29]. The absorption peak at 1380 cm^{-1} of the spectrum indicates that O_2 molecules are adsorbed on the surface. Whether it is the formation of highly oxidizing free radicals or the provision of adsorption sites for small molecules, this facilitates the oxidation of organic material into a variety of small molecule compounds via photocatalysis. In comparison to P25, the absorption peaks of CdS-TiO_2 and CdS-C_{60}/TiO_2 shift toward higher frequencies. The peak at 450–510 cm^{-1} signifies the presence of titanate TiO_3^{2-}, indicating that the transition of TiO_2 involves lattice distortion [30]. It can be deduced that the periodic potential field distortion generated by the modification of CdS and C_{60} caused lattice defects in the TiO_2. The TiO_2 band gap decreases, leading to an enhanced ability to catalyze visible light. Although the sample contains a small amount of titanate TiO_3^{2-}, the photocatalytic activity can be effectively improved by introducing more crystal defects [31]. The band of CdS-C_{60}-TiO_2 changes the lattice to produce some characteristic peaks at 1380 cm^{-1}, which further indicates that CdS, C_{60}, and TiO_2 are bonded, changing the original lattice structure of the TiO_2. It may be inferred from band analyses of CdS, CdS-C_{60}, CdS-TiO_2, and CdS-C_{60}/TiO_2 that the same absorption peak, which may represent a Cd-S bond vibration, is present in all of these materials.

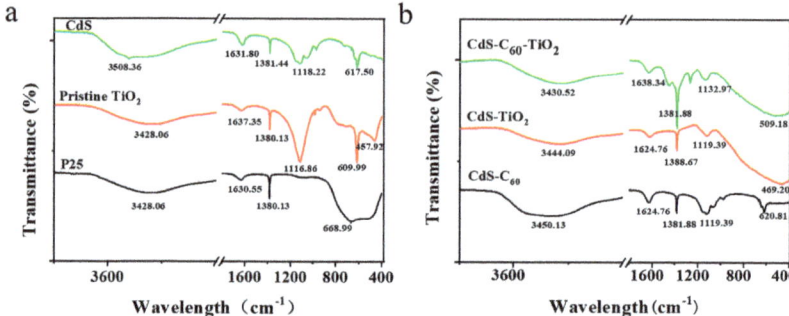

Figure 2. (a) FT–IR spectra of P25, pristine TiO_2, CdS, (b) CdS-C_{60}, CdS-TiO_2, and CdS-C_{60}-TiO_2.

Since the photocatalytic activity of the functionalized materials is largely dependent on their morphological structure, it can be found that P25 (Figure S2a,b, Supporting Information) is a solid microsphere. However, after being modified by CdS, the particles become less uniform, and as seen using XRD, there is an interaction between CdS and TiO_2, making its crystal lattice larger. CdS and C_{60} uniformly cover the surface of TiO_2 in the form of lumps and microspheres, and the aggregation of small particles aggravates the surface roughness (Figure S2c–j, Supporting Information), as is consistent with the results of SEM. By comparing the TEM of the samples CdS-TiO_2 (Figure S3a, Supporting Information) and CdS-C_{60}/TiO_2 (Figure S3b, Supporting Information), we see the crystal plane of the CdS-C_{60}/TiO_2 composite is clearer and regular, the crystal form is more perfect, and the lattice fringes of TiO_2 and CdS can be seen clearly. This indicates that the introduction of C_{60} changed the lattice structure of TiO_2, making the crystal form better. In addition, the connection between the suitable specific surface area and the variables affects the sample's photocatalytic performance. We calculated the pore size and pore distribution of each sample using the Barrett–Joyner–Halenda technique and the Halsey equation, as shown in Table S1 (Supporting Information) [32–34]. From the nitrogen adsorption data of the six catalyst samples, when compared to CdS, the specific surface area of CdS-TiO_2 has increased in comparison to commercial P25 and CdS-C_{60}, but the rate of the change in the pore size is much greater than the change in the specific surface area. This may be a result of the dopant entering the lattice and causing the pore size to decrease, conducive to producing lattice defects in the crystal structure. This improves the photocatalytic activity [35]. As a product of CdS and C_{60} co-modification, in CdS-C_{60}/TiO_2, it is hypothesized that the factors influencing the photocatalytic performance of the sample are related to an appropriate specific surface area and a certain degree of lattice defects. As the specific surface area increases, so does the number of active sites, leading to the higher adsorption performance of the catalyst. Conversely, reducing the specific surface area leads to a decrease in both the pore volume and pore size. The presence of certain lattice defects proves beneficial in enhancing the light-harvesting effect of TiO_2 [36].

By assessing the rate at which MG degrades under 120 min of visible light irradiation, the CdS- and C_{60}-co-modified TiO_2 samples can be evaluated for their visible light catalytic activity. They were put through photocatalytic degradation tests with commercial P25 under identical circumstances to assess their photocatalytic activity. The UV absorption spectra of CdS, CdS-C_{60}, CdS-TiO_2, and CdS-C_{60}/TiO_2 are displayed in Figure 3a–d. These spectra demonstrate that the absorbance values decrease with an increasing visible light irradiation period, indicating a significant and quick degradation of MG. To examine the kinetics of MG's photocatalytic breakdown in more detail, we measured the photocatalytic degradation curves of five catalyst samples, as shown in Figure 3e–f [37]. According to the relevant formula, we calculated the apparent rate constant k_{app} for the catalytic reactions with different catalysts in Table S1 (Supporting information), and it was indicated that the photocatalytic efficiency of all the samples was significantly higher than that of P25 when exposed to visible light. The order of photocatalytic efficiency was CdS-C_{60}/TiO_2 > CdS-C_{60} > CdS > CdS-TiO_2 > P25. It can be seen that the composite materials have significant catalytic activity. Compared with the pure materials, the catalytic activity of the C_{60}-modified materials is improved due to C_{60}'s ability to enhance the quantum efficiency and facilitate charge transfer: C_{60} can enhance the efficiency of separating photogenerated e^--h^+ pairs. Meanwhile, it can also enhance the adsorption efficiency during the degradation process [38,39]. This is also due to the small band gap width (2.4 eV) of CdS, which can induce photocatalysis in the visible light region. CdS can provide excited e^- to TiO_2, while water and oxygen are used to generate hydroxyl radicals (·OH) and superoxide anion radicals (·O_2^-). These free radicals contribute to the photooxidation of the adsorbed organic matrix. The CdS-C_{60}/TiO_2 sample exhibits both the characteristics of C_{60} and CdS, thus exhibiting the highest photocatalytic activity.

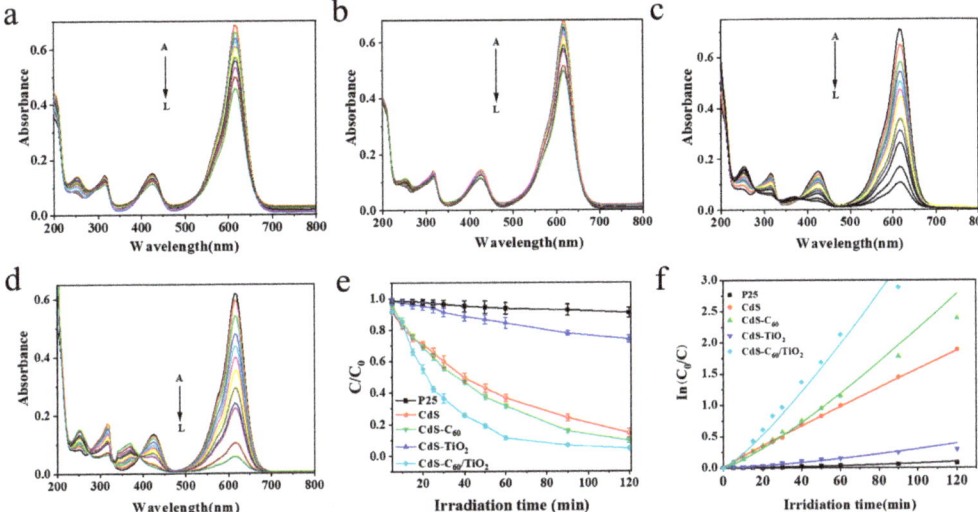

Figure 3. Spectrum of adsorption of MG solution in the presence of (**a**) CdS; (**b**) CdS-TiO_2; (**c**) CdS-C_{60}; and (**d**) CdS-C_{60}/TiO_2 under different irradiation times when exposed to halogen tungsten lamp. For A–G, each absorbance spectrum was recorded over a 5 min interval; for G–J, each absorbance spectrum was recorded over a 10 min interval; for J–L, each absorbance spectrum was recorded over a 30 min interval with visible light illumination. (**e**) Absorbance variations as a function of irradiation time (**f**), ln (c_0/c), and the linear of control for P25, CdS, CdS-C_{60}, CdS-TiO_2, CdS-C_{60}/TiO_2 in MG deterioration after 120 min exposure to radiation at ambient temperature. [MG] = 4 mg/L; [P25] (CdS, CdS-C_{60}, CdS-TiO_2, CdS-C_{60}/TiO_2) = 0.6 g/L.

Moreover, we conducted adsorption mechanical analysis on the different catalysts. The pore size and pore distribution can be calculated in Table S2 (Supporting Information). The factors influencing the photocatalytic performance of the sample are connected to the combined impact of an appropriate specific surface area and a certain level of lattice defects, which is beneficial for improving the optical effect of TiO_2. The results of our adsorption-based linear fitting analysis are consistent with the desorption equilibrium data for MG using different catalysts, as shown in Figure S4 (Supporting Information). One can enhance the TiO_2 by increasing its specific surface area and modifying its crystal structure; TiO_2 co-modified with C_{60} can be expected to improve in its photocatalytic performance; and given sufficient time, CdS-C_{60} is comparable to CdS-C_{60}-TiO_2 in terms of its catalytic performance.

To further analyze the catalytic performance of each catalyst, we conducted a comparison of the photocatalytic hydrogen production yield over time using 10% lactic acid as a sacrificial reagent. This comparison included various catalysts under visible light conditions, as illustrated in Figure 4a. We tested the average hydrogen production of different catalysts per hour under continuous illumination for 5 h in Figure 4b. As illustrated in Figure 4a,b, the hydrogen production performance of these photocatalysts in order from best to worst is CdS-C_{60}/TiO_2 > CdS-C_{60} > CdS > CdS-TiO_2 = P25. The visible light hydrogen production effect of pure P25 and CdS-modified TiO_2 is basically zero, and the average hydrogen production of C_{60}-modified CdS is 5.648 µmol/h. The photocatalytic activity of CdS when co-modified with TiO_2 and C_{60} is approximately twice as high as that of pure CdS, whereas the average yield of TiO_2 co-modified with CdS and C_{60} is 34.025 µmol/h, which is roughly 17 times higher than that of pure CdS. The narrow band gap of C_{60} enables the photocatalyst to have a significantly broader absorption band range, thereby enhancing its light absorption capability in the visible light spectrum. C_{60} has good conductivity and can detach photogenerated e^- and promote good separation of the photogenerated e^--h^+, so the visible light hydrogen production performance in CdS can be enhanced through modification with C_{60}. Furthermore, both CdS and C_{60} demonstrate significant light absorption capabilities within the visible spectrum, and C_{60} as an electronic relay can promote the flow of CdS and TiO_2 photogenerated e^-. C_{60} can be used as an active site for photocatalytic hydrogen production, specifically the catalytic cleavage of water to produce H_2. This indicates that TiO_2 co-modified with CdS and C_{60} has a high photocatalytic activity under visible light conditions.

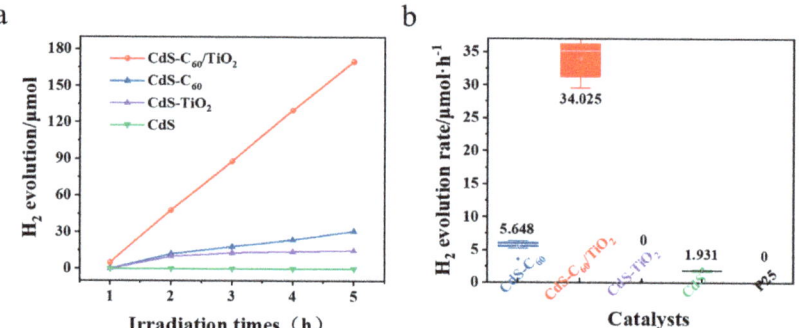

Figure 4. (**a**) Several kinds of hydrogen production with a time change map; (**b**) different catalysts per hour of hydrogen production.

A typical impedance equivalent plot is shown in Figure S4a (Supporting Information) to demonstrate the impact of the charge transfer on the activity of catalysts. The EIS Nyquist plots of different catalysts are shown in Figure S5b (Supporting Information). In the Nyquist plots of EIS, the semicircle's diameter shows that CdS, CdS-C_{60}, CdS-TiO_2, and CdS-C_{60}/TiO_2 have a lower charge transfer resistance compared to pure TiO_2. The order of decreasing resistance values is R(A) < R(B) < R(C) < R(D) < R(E) < R(F). It can speed up the charge separation and promote e^- transfer. This helps to decrease the likelihood of recombining e^--h^+ pairs and enhances the photocatalytic activity [40–42]. From the results of electrochemical AC impedance, we see the effective charge transfer rate of CdS is weaker than that of CdS-TiO_2, but in the photodegradation experiment, the photodegradation rate is higher than that of CdS-TiO_2. The possible reasons for this are that CdS can provide excitation of e^- to TiO_2; within 18 h after the modification of the empty gold sheet, the pure CdS is unstable on its own, and it may have developed photocorrosion, which, in turn, reduced the conductivity.

To illustrate the photodegradation mechanism of the catalyst CdS-C_{60}-TiO_2, we give a schematic diagram of the corresponding photodegradation mechanism, as shown in Figure 5. On the one hand, C_{60} has a good electron transport performance, which can accelerate the photoinduced charge separation and slow down charge recombination. Under natural light irradiation, the valence electrons (e⁻) in TiO_2 are excited to the conduction band (CB), and holes (h⁺) are generated in the valence band (VB). However, these photo-generated electrons–holes can recombine quickly in the photocatalytic reaction, but since C_{60} is a good electron acceptor and can receive electrons from the titanium dioxide, the lifetime of the photogenerated e⁻ and h⁺ will be extended during the transfer process. The photogenerated h⁺ can form hydroxyl radicals, with a strong oxidizing activity and OH⁻ adsorbed on the surface of the TiO_2. The photogenerated e⁻ form oxyanion radicals with the oxygen molecules adsorbed on the surface of the TiO_2, which can effectively photolyze organic radicals or water under the action of free radicals. The resulting active substances (•O^{2-} and •OH) have a high degradation activity for MG.

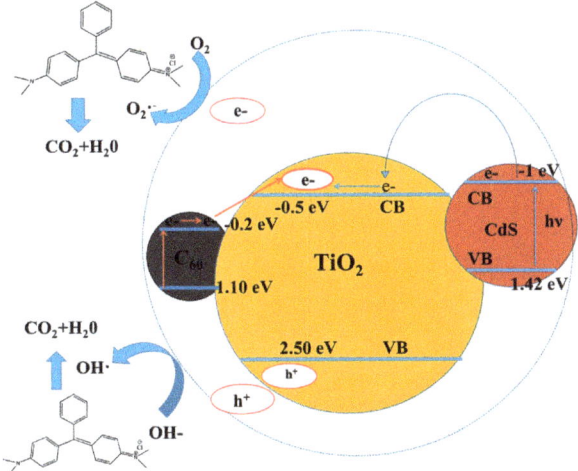

Figure 5. The photodegradation mechanism schematic for CdS-C_{60}-TiO_2.

On the other hand, TiO_2 and CdS can transition under illumination simultaneously, with the latter's conduction band position being higher than the former, so there is a misalignment of the two in the schematic diagram of the mechanism. Generally, when TiO_2 absorbs light with an energy greater than its band gap, the e⁻ in the valence band can be excited and move to the conduction band. This leads to the formation of highly active e⁻ in the conduction band and corresponding h⁺ in the valence band [43]. The photogenerated e⁻ and h⁺ are separated and moved toward the surface of the semiconductor particles due to the presence of an electric field, causing highly active photogenerated e⁻-h⁺ pairs. This indicates that the transition of the electrons from the excited state of CdS to the conduction band of TiO_2 occurs, and then these e⁻ are absorbed by O_2 molecules to form the superoxide anion radical O_2^- [44–46]. Consequently, the adsorbed organic matrix's photooxidation characteristics are improved by this shift in vector transfer, to effectively enhance the photooxidation performance of the adsorbed organic matrix. Due to the photogenerated h⁺'s valence potential, which prevents them from oxidizing hydroxyl groups into hydroxyl radicals, CdS photocorrodes to create Cd^{2+}. In short, both C_{60} and CdS components enhance the catalytic effect of TiO_2; therefore, CdS-C_{60}/TiO_2 exhibits a high catalytic activity.

4. Conclusions

In summary, the preparation of CdS- and C_{60}-co-modified TiO_2 photocatalysts via the sol-gel method was followed by their comprehensive characterization using various techniques. Our rigorous investigation into their photocatalytic performance and hydrogen production capacity showcased the exceptional effectiveness of the ternary composite catalyst CdS-C_{60}/TiO_2 in oxidizing the targeted degradants. The results demonstrated the impressive average hydrogen production of 34.025 µmol/h for TiO_2 co-modified with CdS and C_{60}, marking a substantial 17-fold increase when compared to pure CdS. In parallel, the average hydrogen generation for C_{60}-modified CdS surged to 5.648 µmol/h, representing a notable two-fold enhancement over pure CdS. This remarkable performance can be attributed to the collaborative impact of CdS and C_{60}, which displayed synergistic effects when combined with TiO_2. The co-modifying strategy effectively minimized the likelihood of photogenerated electron–hole recombination while simultaneously reducing the material's band gap. These findings underscore the vast potential of $CdS-C_{60}/TiO_2$ as a high-performance photocatalytic system, with versatile application fields such as renewable energy and environmental remediation. We have compiled a comprehensive overview of highly efficient catalysts for the degradation of MG (refer to Table S3 in the Supporting Information). Although our designed catalysts may not exhibit an optimal performance, our work offers a practical approach to enhancing the effectiveness of photocatalytic degradation by utilizing novel ternary materials.

Supplementary Materials: The following supporting information can be downloaded at https://www.mdpi.com/article/10.3390/ma17051206/s1. Figures S1–S5: XPS, SEM, TEM, N_2 adsorption and desorption isotherms and pore size distributions and electrochemical impedance spectroscopy measurement of different samples; Tables S1–S3: Kinetics parameters of photocatalytic degradation of MG, physical parameters of the adsorption and the photocatalytic degradation of MG at different catalysts. References [47–55] are cited in the supplementary materials.

Author Contributions: Y.L. and M.Z. conceived the idea and performed the experiments. X.L. and Y.G. contributed to the interpretation of the results. X.L. and M.Z. wrote and revised the manuscript. All authors discussed the results and commented on the manuscript. All authors have read and agreed to the published version of the manuscript.

Funding: We gratefully thank Yuzhang Normal University's 2021 University-Level Scientific Research Project (Grant No. YZYB-21-16) and the Science and Technology Research Project of Jiangxi Province Education Department (Grant Nos. GJJ213108, 20202ACB202004, 20212BBE53051, 20213BCJ22024) for their financial support.

Institutional Review Board Statement: Not applicable.

Informed Consent Statement: Not applicable.

Data Availability Statement: Data are contained within the article and supplementary materials.

Conflicts of Interest: The authors declare no conflicts of interest.

References

1. Lei, Q.; Yuan, H.; Du, J.; Ming, M.; Yang, S.; Chen, Y.; Lei, J.; Han, Z. Photocatalytic CO_2 reduction with aminoanthraquinone organic dyes. *Nat. Commun.* **2023**, *14*, 1087. [CrossRef]
2. Huang, W.; Su, C.; Zhu, C.; Bo, T.; Zuo, S.; Zhou, W.; Ren, Y.; Zhang, Y.; Zhang, J.; Rueping, M.; et al. Isolated electron trap-induced charge accumulation for efficient photocatalytic hydrogen production. *Angew. Chem. Int. Ed.* **2023**, *62*, e202304634. [CrossRef]
3. Barrocas, B.T.; Ambroová, N.; Koí, K. Photocatalytic reduction of carbon dioxide on TiO_2 heterojunction photocatalysts—A review. *Materials* **2022**, *15*, 967. [CrossRef]
4. Singh, V.; Rao, A.; Tiwari, A.; Yashwanth, P.; Lal, M.; Dubey, U.; Aich, S.; Roy, B. Study on the effects of Cl and F doping in TiO_2 powder synthesized by a sol-gel route for biomedical applications. *J. Phys. Chem. Solids* **2019**, *134*, 262–271. [CrossRef]
5. Aich, S.; Mishra, M.K.; Sekhar, C.; Satapathy, D.; Roy, B. Synthesis of Al-doped nano Ti-O scaffolds using a hydrothermal route on titanium foil for biomedical applications. *Mater. Lett.* **2016**, *178*, 135–139. [CrossRef]
6. Caglar, A.; Aktas, N.; Kivrak, H. The role and effect of CdS-based TiO_2 photocatalysts enhanced with a wetness impregnation method for efficient photocatalytic glucose electrooxidation. *Surf. Interfaces* **2022**, *33*, 102250. [CrossRef]
7. Chen, W.; Tian, Y.; Wang, X.; Ma, R.; Ding, H.; Zhang, H. Preparation and characterization of Zr-containing silica residue purification loaded nano-TiO_2 composite photocatalysts. *Chem. Phys.* **2023**, *570*, 111889. [CrossRef]
8. Chen, B.C.; Li, P.P.; Wang, B.; Wang, Y.D. Flame-annealed porous TiO_2/CeO_2 nanosheets for enhenced CO gas sensors. *Appl. Surf. Sci.* **2022**, *593*, 153418. [CrossRef]
9. Chen, Y.J.; Luo, X.; Luo, Y.; Xu, P.W.; He, J.; Jiang, L.; Li, J.J.; Yan, Z.Y.; Wang, J.Q. Efficient charge carrier separation in l-alanine acids derived N-TiO_2 nanospheres: The role of oxygen vacancies in tetrahedral Ti^{4+} sites. *Nanomaterials* **2019**, *9*, 698. [CrossRef] [PubMed]

10. Deng, L.; Liu, Y.; Zhao, G.; Chen, J.H.; He, S.F.; Zhu, Y.C.; Chai, B.; Ren, Z.D. Preparation of electrolyzed oxidizing water by TiO_2 doped IrO_2-Ta_2O_5 electrode with high selectivity and stability for chlorine evolution. *J. Electroanal. Chem.* **2019**, *832*, 459–466. [CrossRef]
11. Li, S.Q.; Li, J.Z.; Zhang, H.X.; Luo, B.; Cai, H.; Nie, S.X.; Sha, J.L. Ultraviolet-light-driven electricity generation by a TiO_2/graphene composite in water. *ACS Mater. Lett.* **2023**, *5*, 2862–2869. [CrossRef]
12. Li, G.Q.; Xu, T.; He, R.F.; Li, C.P.; Bai, J. Hollow cadmium sulfide tubes with novel morphologies for enhanced stability of the photocatalytic hydrogen evolution. *Appl. Surf. Sci.* **2019**, *495*, 143642. [CrossRef]
13. Goud, B.S.; Suresh, Y.; Annapurna, S.; Singh, A.K.; Bhikshamaiah, G. Green synthesis and characterization of cadmium sulphide nanoparticles. *Mater. Today Proc.* **2016**, *3*, 4003–4008. [CrossRef]
14. Jang, J.S.; Kim, H.G.; Joshi, U.A.; Jang, J.W.; Lee, J.S. Fabrication of CdS nanovires decorated with TiO_2 nanopartieles for photocatalytic hydrogen production under visible light irradiation. *Int. J. Hydrogen Energ.* **2008**, *33*, 5975–5980. [CrossRef]
15. Ma, J.Y.; An, Z.H.; Zhang, W.W.; Shen, J.; Qi, Y.L.; Chen, D.H. TiO_2/CdS composite photocathode improves the performance and degradation of wastewater in microbial fuel cells. *Anal. Sci. Adv.* **2022**, *3*, 188–197. [CrossRef]
16. Huo, G.N.; Zhang, S.S.; Li, Y.L.; Li, J.X.; Zhao, Y.; Huang, W.P.; Zhang, S.M.; Zhu, B.L. CdS-modified TiO_2 nanotubes with heterojunction structure: A photoelectrochemical sensor for glutathione. *Nanomaterials* **2023**, *13*, 13. [CrossRef] [PubMed]
17. Xing, Z.; Li, S.H.; Hui, Y.; Wu, B.S.; Chen, Z.C.; Yun, D.Q.; Deng, L.L.; Zhang, M.L.; Mao, B.W.; Xie, S.Y.; et al. Star-like hexakis[di(ethoxycarbonyl)methano]-C_{60} with higher electron mobility: An unexpected electron extractor interfaced in photovoltaic perovskites. *Nano Energy* **2020**, *774*, 104859. [CrossRef]
18. Zieleniewska, A.; Lodermeyer, F.; Rotha, A.; Guldi, D.M. Fullerenes-how 25 years of charge transfer chemistry have shaped our understanding of (interfacial) interactions. *Chem. Soc. Rev.* **2018**, *47*, 702–714. [CrossRef] [PubMed]
19. Xia, D.; Zhang, Z.; Zhao, C.; Wang, J.; Li, W. Fullerene as an additive for increasing the efficiency of organic solar cells to more than 17%. *J. Colloid Interf. Sci.* **2021**, *601*, 70–77. [CrossRef] [PubMed]
20. Fang, Y.; Wei, Z.Y.; Guan, Z.H.; Shan, N.Y.; Zhao, Y.; Liu, F.; Fu, L.L.; Huang, Z.P.; Humphrey, M.G.; Zhang, C. Covalent chemical functionalization of Ti_3C_2Tx MXene nanosheets with fullerenes C_{60} and C_{70} for enhanced nonlinear optical limiting. *J. Mater. Chem. C* **2023**, *11*, 7331–7344. [CrossRef]
21. Kang, S.; Bozhilov, K.; Myung, N.; Mulchandani, A.; Chen, W. Microbial Synthesis of CdS Nanocrystals in Genetically Engineered *E. coli. Angew. Chem. Int. Ed.* **2008**, *47*, 5186–5189. [CrossRef]
22. Xu, M.; Kang, Y.; Jiang, L.; Jiang, L.; Tremblay, P.; Zhang, T. The one-step hydrothermal synthesis of CdS nanorods modified with carbonized leaves from Japanese raisin trees for photocatalytic hydrogen evolution. *Int. J. Hydrogen Energy* **2022**, *47*, 15516–15527. [CrossRef]
23. Li, X.H.; Li, Y.J.; Guo, X.; Jin, Z.L. Design and synthesis of $ZnCo_2O_4$/CdS for substantially improved photocatalytic hydrogen production. *Front. Chem. Sci. Eng.* **2023**, *17*, 606–616. [CrossRef]
24. Li, J.L.; Xu, X.T.; Liu, X.J.; Qin, W.; Wang, M.; Pan, L.K. Metal-organic frameworks derived cake-like anatase/rutile mixed phase TiO_2 for highly efficient photocatalysis. *J. Alloys Compd.* **2017**, *690*, 640–646. [CrossRef]
25. Zhang, K.L.; Liu, C.M.; Huang, F.Q.; Zheng, C.; Wang, W.D. Study of the electronic structure and photocatalytic activity of the BiOCl photocatalyst. *Appl. Catal. B Environ.* **2006**, *68*, 125–130. [CrossRef]
26. Li, Y.; Jiang, J.W. Modulation of thermal conductivity of single-walled carbon nanotubes by fullerene encapsulation: The effect of vacancy defects. *Phys. Chem. Chem. Phys.* **2023**, *11*, 7734–7740. [CrossRef] [PubMed]
27. Zhou, W.R.; Jia, L.B.; Chen, M.Q.; Li, X.C.; Su, Z.H.; Shang, Y.B.; Jiang, X.F.; Gao, X.Y.; Chen, T.; Wang, M.T.; et al. An improbable amino-functionalized fullerene spacer enables 2D/3D hybrid perovskite with enhanced electron transport in solar cells. *Adv. Funct. Mater.* **2022**, *32*, 2201374. [CrossRef]
28. Brik, M.G.; Srivastava, A.M.; Popov, A.I. A few common misconceptions in the interpretation of experimental spectroscopic data. *Opt. Mater.* **1992**, *32*, 249–253. [CrossRef]
29. Zhou, X.A.; Ye, S.T.; Zhao, S.F.; Song, H.H.; Gong, H.T.; Fan, S.R.; Liu, M.J.; Wang, M.L.; Zhou, W.H.; Liu, J.J.; et al. Unraveling structure sensitivity in the photocatalytic dehydrogenative C-C coupling of acetone to 2,5-Hexanedione over Pt/TiO_2 catalysts. *ACS Catal.* **2023**, *13*, 11825–11833. [CrossRef]
30. Hu, S.; Qiao, P.Z.; Zhang, L.P.; Jiang, B.J.; Gao, Y.T.; Hou, F.; Wu, B.G.; Li, Q.; Tian, C.G.; et al. Assembly of TiO_2 ultrathin nanosheets with surface lattice distortion for solar-light-driven photocatalytic hydrogen evolution. *Appl. Catal. B* **2018**, *239*, 317–323. [CrossRef]
31. Sambur, J.; Brgoch, J. Unveiling the hidden influence of defects via experiment and data science. *Chem. Mater.* **2023**, *35*, 7351–7354. [CrossRef]
32. Wang, X.; Liu, Y.; Arandiyan, H.; Yang, H.; Bai, L.; Mujtaba, J.; Wang, Q.; Liu, S.; Sun, H. Uniform Fe_3O_4 microflowers hierarchical structures assembled with porous nanoplates as superior anode materials for lithium-ion batteries. *Appl. Surf. Sci.* **2016**, *389*, 240–246. [CrossRef]
33. Rahmati, M.S.; Fazaeli, R.; Saravani, M.G.; Ghiasi, R. Cu–curcumin/MCM-41 as an efficient catalyst for in situ conversion of carbazole to fuel oxygenates: A DOE approach. *J. Nanostruct. Chem.* **2022**, *12*, 307–327. [CrossRef]
34. Yan, J.; Liu, X.L.; Wang, X.T.; Wang, L.J.; Weng, W.S.; Yu, X.T.; Xing, G.B.; Xie, J.; Lu, C.; Luo, Y.; et al. Influence of nano-attapulgite on compressive strength and microstructure of recycled aggregate concrete. *Cem. Concr. Comp.* **2022**, *134*, 104788. [CrossRef]

35. Kanjan, N.; Laokul, P.; Maiaugree, W. Photocatalytic activity of nanocrystalline Fe^{3+}-doped anatase TiO_2 hollow spheres in a methylene blue solution under visible-light irradiation. *J. Mater. Sci. Mater. Electron.* **2022**, *33*, 4659–4680. [CrossRef]
36. Mohajernia, S.; Andryskova, P.; Zoppellaro, G.; Kment, S.; Zboril, R.; Schmidt, J.; Schmuki, P. Influence of Ti^{3+} defect-type on heterogeneous photocatalytic H_2 evolution activity of TiO_2. *J. Mater. Chem. A.* **2020**, *8*, 1432–1442. [CrossRef]
37. Belghiti, M.; Tanji, K.; El Mersly, L.; Lamsayety, I.; Ouzaouit, K.; Faqir, H.; Benzakour, I.; Rafqah, S.; Outzourhit, A. Fast and non-selective photodegradation of basic yellow malachite green, tetracycline, and sulfamethazine using a nanosized ZnO synthesized from zinc ore. *React. Kinet. Mech. Cat.* **2022**, *135*, 2265–2278. [CrossRef]
38. Causa, M.; Risse, J.D.J.; Scarongella, M.; Brauer, J.C.; Domingo, E.B.; Moser, J.E.; Stingelin, N.; Banerji, N. The fate of electron-hole pairs in polymer: Fullerene blends for organic photovoltaics. *Nat. Commun.* **2016**, *7*, 12556. [CrossRef]
39. Katal, R.; Farahan, M.H.D.A.; Hu, J.Y. Degradation of acetaminophen in a photocatalytic (batch and continuous system) and photoelectrocatalytic process by application of faceted-TiO_2. *Sep. Purif. Technol.* **2020**, *230*, 115859. [CrossRef]
40. Low, J.; Yu, J.; Jaroniec, M.; Wageh, S.; Al-Ghamdi, A.A. Heterojunction photocatalysts. *Adv. Mater.* **2017**, *29*, 160694. [CrossRef] [PubMed]
41. Nagappagari, L.R.; Le, T.D.; Ahemad, M.J.; Oh, G.J.; Shin, G.S.; Lee, K.Y.; Yu, Y.T. Enhancement of bifunctional photocatalytic activity of boron-doped g-C_3N_4/SnO_2 heterojunction driven by plasmonic Ag quantum dots. *Mater. Today Nano* **2023**, *22*, 100325. [CrossRef]
42. Islam, M.R.; Chakraborty, A.K.; Gafur, M.A.; Rahman, M.A.; Rahman, M.H. Easy preparation of recyclable thermally stable visible-light-active graphitic-C_3N_4/TiO_2 nanocomposite photocatalyst for efficient decomposition of hazardous organic industrial pollutants in aqueous medium. *Res. Chem. Intermed.* **2019**, *45*, 11753–11773. [CrossRef]
43. Truc, N.T.T.; Pham, T.D.; Nguyen, M.V.; Thuan, D.V.; Trung, D.Q.; Thao, P.; Trang, H.T.; Nguyen, V.N.; Tran, D.T.; Minh, D.N.; et al. Advanced $NiMoO_4$/g-C_3N_4 Z-scheme heterojunction photocatalyst for efficient conversion of CO_2 to valuable products. *J. Alloys Compd.* **2020**, *842*, 155860. [CrossRef]
44. Boettche, S.W.; Oene, S.Z.; Lonerga, M.C.; Surendranat, Y.; Ardo, S.; Broze, C.; Kempler, P.A. Potentially confusing: Potentials in electrochemistry. *ACS Energy Lett.* **2021**, *6*, 261–266. [CrossRef]
45. Pasarán, A.C.; Luke, T.L.; Zarazúa, I.; Rosa, E.D.; Ramírez, R.F.; Sanal, K.C.; Ordaz, A.A. Co-sensitized TiO_2 electrodes with different quantum dots for enhanced hydrogen evolution in photoelectrochemical cells. *J. Appl. Electrochem.* **2019**, *49*, 475–484. [CrossRef]
46. Mao, Z.; Lin, H.; Xu, M.; Miao, J.; He, S.J.; Li, Q. Fabrication of Co-doped CdSe quantum dot-sensitized TiO_2 nanotubes by ultrasound-assisted method and their photoelectrochemical properties. *J. Appl. Electrochem.* **2018**, *48*, 147–155. [CrossRef]
47. Oh, W.C.; Zhang, F.J.; Chen, M.L. Preparation of MWCNT/TiO_2 composites by using MWCNTs and titanium (IV) alkoxide precursors in benzene and their photocatalytic effect and bactericidal activity. *Bull. Korean Chem. Soc.* **2009**, *30*, 2637–2642.
48. Li, Y.S.; Jiang, F.L.; Xiao, Q.; Li, R.; Li, K.; Zhang, M.F.; Zhang, A.Q.; Sun, S.F.; Liu, Y. Enhanced photocatalytic activities of TiO_2 nanocomposites doped with water-soluble mercapto-capped CdTe quantum dots. *Appl. Catal. B* **2010**, *101*, 118–129. [CrossRef]
49. Al-Ekabi, H.; Serpone, N. Kinetic studies in heterogeneous photocatalysis. 1. Photocatalytic degradation of chlorinated phenols in aerated aqueous solutions over TiO_2 supported on a glass matrix. *J. Phys. Chem.* **1988**, *92*, 5726–5731. [CrossRef]
50. Al-Ekabi, H.; Serpone, N.; Wang, X.H.; Li, J.G.; Kamiyama, H.; Moriyoshi, Y.; Ishigaki, T. Wavelength-sensitive photocatalytic degradation of methyl orange in aqueous suspension over Iron(III) doped TiO_2 nanopowders under UV and visible light irradiation. *J. Phys. Chem. B* **2006**, *110*, 6804–6809.
51. Korshin, G.; Chow, C.W.K.; Fabris, R.; Drikas, M. Absorbance spectroscopy-based examination of effects of coagulation on the reactivity of fractions of natural organic matter with varying apparent molecular weights. *Water Res.* **2009**, *43*, 1541–1548. [CrossRef] [PubMed]
52. Halomoan, I.; Yulizar, Y.; Surya, R.M.; Apriandanu, D.O. Facile preparation of CuO-$Gd_2Ti_2O_7$ using acmella uliginosa leaf extract for photocatalytic degradation of malachite green. *Mater. Res. Bull.* **2022**, *150*, 111726. [CrossRef]
53. Jia, J.; Du, X.; Surya, R.M.; Zhang, Q.Q.; Liu, E.Z.; Fan, J. Z-scheme $MgFe_2O_4$/Bi_2MoO_6 heterojunction photocatalyst with enhanced visible light photocatalytic activity for malachite green removal. *Appl. Surf. Sci.* **2019**, *492*, 527–539. [CrossRef]
54. Wu, D.; Li, C.; Zhang, D.; Wang, L.; Zhang, X.; Shi, Z.; Lin, Q. Enhanced photocatalytic activity of Gd^{3+} doped TiO_2 and Gd_2O_3 modified TiO_2 prepared via ball milling method. *J. Rare Earths* **2019**, *37*, 845–852. [CrossRef]
55. Zhang, M.F.; Liang, X.F.; Liu, Y. Co-CNT/TiO_2 composites efectively improved the photocatalytic degradation of malachite green. *Ionics* **2022**, *30*, 521–527. [CrossRef]

Disclaimer/Publisher's Note: The statements, opinions and data contained in all publications are solely those of the individual author(s) and contributor(s) and not of MDPI and/or the editor(s). MDPI and/or the editor(s) disclaim responsibility for any injury to people or property resulting from any ideas, methods, instructions or products referred to in the content.

Article

CO_2-Switchable Hierarchically Porous Zirconium-Based MOF-Stabilized Pickering Emulsions for Recyclable Efficient Interfacial Catalysis

Xiaoyan Pei *, Jiang Liu, Wangyue Song, Dongli Xu, Zhe Wang and Yanping Xie

College of Chemistry and Chemical Engineering, Xinyang Normal University, Xinyang 464000, China
* Correspondence: xiaoyanpei2009@163.com

Abstract: Stimuli-responsive Pickering emulsions are recently being progressively utilized as advanced catalyzed systems for green and sustainable chemical conversion. Hierarchically porous metal–organic frameworks (H-MOFs) are regarded as promising candidates for the fabrication of Pickering emulsions because of the features of tunable porosity, high specific surface area and structure diversity. However, CO_2-switchable Pickering emulsions formed by hierarchically porous zirconium-based MOFs have never been seen. In this work, a novel kind of the amine-functionalized hierarchically porous UiO-66-(OH)$_2$ (H-UiO-66-(OH)$_2$) has been developed using a post-synthetic modification of H-UiO-66-(OH)$_2$ by (3-aminopropyl)trimethoxysilane (APTMS), 3-(2-aminoethylamino) propyltrimethoxysilane (AEAPTMS) and 3-[2-(2-aminoethylamino)ethylamino]propyl-trimethoxysilane (AEAEAPTMS), and employed as emulsifiers for the construction of Pickering emulsions. It was found that the functionalized H-UiO-66-(OH)$_2$ could stabilize a mixture of toluene and water to give an emulsion even at 0.25 wt % content. Interestingly, the formed Pickering emulsions could be reversibly transformed between demulsification and re-emulsification with alternate addition or removal of CO_2. Spectral investigation indicated that the mechanism of the switching is attributed to the reaction of CO_2 with amino silane on the MOF and the generation of hydrophilic salts, leading to a reduction in MOF wettability. Based on this strategy, a highly efficient and controlled Knoevenagel condensation reaction has been gained by using the emulsion as a mini-reactor and the emulsifier as a catalyst, and the coupling of catalysis reaction, product isolation and MOF recyclability has become accessible for a sustainable chemical process.

Keywords: CO_2-switchable; hierarchically porous zirconium-based MOF; Pickering emulsion; mini-reactor; Knoevenagel condensation

Citation: Pei, X.; Liu, J.; Song, W.; Xu, D.; Wang, Z.; Xie, Y. CO_2-Switchable Hierarchically Porous Zirconium-Based MOF-Stabilized Pickering Emulsions for Recyclable Efficient Interfacial Catalysis. *Materials* **2023**, *16*, 1675. https://doi.org/10.3390/ma16041675

Academic Editor: Lucia Carlucci

Received: 24 December 2022
Revised: 8 January 2023
Accepted: 7 February 2023
Published: 17 February 2023

Copyright: © 2023 by the authors. Licensee MDPI, Basel, Switzerland. This article is an open access article distributed under the terms and conditions of the Creative Commons Attribution (CC BY) license (https://creativecommons.org/licenses/by/4.0/).

1. Introduction

Pickering emulsions, stabilized by solid colloid particles, are versatile systems that show great significance for the pharmaceutical, food and biomedical industries [1,2]. More recently, Pickering emulsions have been considered an ideal platform for biphasic catalytic reactions on the basis of their large organic solvent–aqueous interface [3]. Moreover, Pickering interfacial catalysis has been rapidly developed to provide a more environmentally friendly procedure with good selectivity and conversion [4,5]. Although some achievements have been gained in this regard, extensive efforts are still needed for product isolation and catalyst recyclability from the emulsion system due to the tedious demulsification process [3]. Stimuli-responsive Pickering emulsions as advanced catalyzed systems offer considerable advantages for green and sustainable chemical transformation owing to their switchable emulsification and demulsification as required [6]. In this context, various attempts have been devoted to the development of stimuli-responsive emulsifiers. Up to now, numerous surface-active colloid particles, for instance, silica, cellulose nanocrystal, chitosan, lignin, polymer, latex and graphene oxide, have been presented to respond to external stimuli such as light [7–9], pH [10–13], redox [14–16], temperature [17–21], magnetism [22,23], pH–temperature [24–26], pH–magnetism [27,28] and temperature–magnetism [29,30], which have been used to successfully build switchable Pickering emulsions.

CO_2 is ample, cheap, non-poisonous and environmentally friendly, and thus can be employed as an absorbing trigger to control many procedures [31–33]. Metal–organic frameworks (MOFs), constructed through the exquisite assembly of metal ions or clusters and organic linkers, are booming as a kind of fascinating porous material [34,35]. Their permanent porosity, functional tunability and structural diversity invest MOFs with impressive potential as Pickering emulsifiers. Despite most of the Pickering emulsions formed by MOFs exhibiting excellent stability [36–41], they could not be efficiently emulsified and demulsified on command. Until recently, exceptional efforts have been dedicated to developing pH-, CO_2- and thermal-triggered MOF-stabilized Pickering emulsions [42–44]. However, most of the reported MOF emulsifiers are limited to the specific microporous regime, which may restrict substrate diffusion to the active sites of MOFs and hinder large guest molecules from entering the MOF channels, thus greatly limiting their employment, particularly in catalysis. Therefore, it is of critical importance to develop innovative stimuli-responsive MOFs for the formation of Pickering emulsions to achieve their tunable demulsification and diverse applications. Hierarchically porous MOFs (H-MOFs), integrating the advantages of pores of different sizes, are especially desired for the improvement of the diffusion rate, promotion of mass transport, and enhancement of catalytic active sites [45,46]. Therefore, the design and development of functionalized H-MOF emulsifiers combined with micropores, mesopores and macropores are necessary and urgent, but challenging. To the best of our knowledge, CO_2-switchable Pickering emulsions formed by H-MOFs have not been reported.

Zirconium-based MOFs (UiO-66) are specifically selected on account of their extraordinary thermal and chemical stability in organic solvents and aqueous solutions [47]. In this work, a series of amine-functionalized hierarchically porous UiO-66-(OH)$_2$ was designed and synthesized through the post-synthetic modification of H-UiO-66-(OH)$_2$ by (3-aminopropyl)trimethoxysilane (APTMS), 3-(2-aminoethylamino)propyltrimethoxysilane (AEAPTMS) and 3-[2-(2-aminoethylamino)ethylamino]propyl-trimethoxysilane (AEAEAPTMS), denoted as H-UiO-66-(OAPTMS)$_2$, H-UiO-66-(OAEAPTMS)$_2$ and H-UiO-66-(OAEAEAPTMS)$_2$, respectively, and then used as emulsifiers for the fabrication of Pickering emulsions. It turns out that all the amine-functionalized H-UiO-66-(OH)$_2$ could stabilize organic–aqueous mixtures to form emulsions even at the content of 0.25 wt % (Figure 1). Moreover, the formed emulsions could be reversibly converted between demulsification and re-emulsification with alternate CO_2 and N_2 addition. Mechanism studies suggested that the switchable phase transition is ascribed to the efficient reaction of CO_2 with different types of amino silane and the generation of hydrophilic ammonium salts. Based on this, the CO_2-responsive Pickering emulsions have been employed as a mini-reactor and the hierarchically porous MOFs as a catalyst for the Knoevenagel condensation reaction.

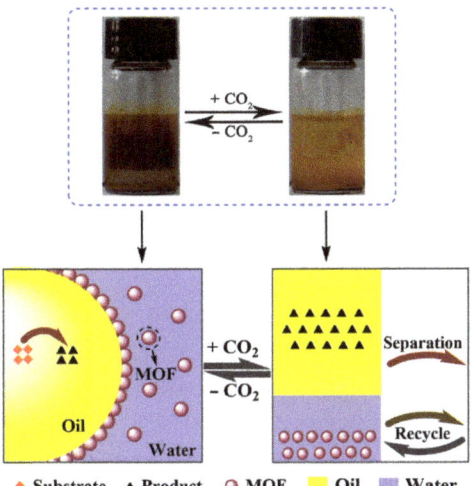

Figure 1. Schematic diagram for CO_2-switchable demulsification and re-emulsification of the functionalized H-UiO-66-(OH)$_2$ stabilized Pickering emulsions.

2. Materials and Methods

2.1. Synthesis of H-UiO-66-(OH)$_2$

H-UiO-66-(OH)$_2$ was synthesized according to the literature [48]. Typically, ZrCl$_4$ (0.120 g, 0.5 mmol), benzoic acid (1.830 g, 15 mmol), Zn(NO$_3$)$_3$·6H$_2$O (0.148 g, 0.5 mmol) and 2,5-dihydroxyterephthalic acid (0.198 g, 1 mmol) were added to 20 mL of N, N-dimethylformamide (DMF) in a Teflon liner vessel. The obtained mixture was sonicated for about 20 min to gain a transparent solution and then heated at 120 °C for 24 h. Upon cooling to room temperature, the obtained powder was separated using centrifugation and washed three times with DMF. The resulting solid was then dispersed in a solution of hydrochloric acid (pH = 1.0) and agitated for about 10 min to break the acid-labile metal–organic assembly template. The final product was isolated using centrifugation and then washed multiple times with DMF and acetone, respectively, to discard the decomposed template residues and then dried overnight at 70 °C under vacuum.

2.2. Synthesis of the Amine-Functionalized H-UiO-66-(OH)$_2$

The amine-functionalized H-UiO-66-(OH)$_2$ was prepared as follows [49]. Briefly, 0.5 g of the as-synthesized H-UiO-66-(OH)$_2$ was firstly dispersed in 60 mL of toluene, and 0.092 g of APTMS, 0.117 g of AEAPTMS or 0.147 g of AEAEAPTMS were subsequently added. The resulting mixtures were stirred at atmospheric temperature for 24 h. The final solids were isolated using centrifugation, washed several times with toluene and ethanol, in turn, and dried at 70 °C overnight, known as H-UiO-66-(OAPTMS)$_2$, H-UiO-66-(OAEAPTMS)$_2$ and H-UiO-66-(OAEAEAPTMS)$_2$, respectively.

2.3. Preparation of the CO_2-Responsive Pickering Emulsions

As an example, the H-UiO-66-(OAEAEAPTMS)$_2$ was added to the toluene–water system (3:2, v/v) followed by high-speed homogenization to give an emulsion. Contents of the H-UiO-66-(OAEAEAPTMS)$_2$ were denoted by mass fraction (wt %) with respect to water. The formed emulsions were found to be able to stabilize for more than one month. The emulsion type was determined using the drop test approach. In addition, CO_2 or N_2 addition was accomplished using a syringe fitted with a needle at the rate of 60 mL min^{-1}. The other emulsions stabilized by H-UiO-66-(OAPTMS)$_2$ and H-UiO-66-(OAEAPTMS)$_2$ were obtained in a similar way.

2.4. General Procedure for Knoevenagel Condensation Reaction

Toluene (3 mL), water (2 mL), the functionalized H-UiO-66-(OH)$_2$ (0.025 g), aldehydes (0.05 mmol) and malononitrile (0.25 mmol) were added to a glass tube and then homogenized at 10,000 rpm for 1 min. The formed emulsion was kept at 25 °C for 2 h in a N$_2$ atmosphere, which was monitored with GC-MS. Once the reaction was completed, demulsification of the emulsion was achieved to separate the product and catalyst with the addition of CO$_2$. At this time, the product was in the upper layer (toluene phase) and the MOF catalyst was in the lower layer (aqueous phase). Then, the product could be acquired with vacuum evaporation, and the toluene phase was then gathered for the next cycles. Once CO$_2$ was driven out, the functionalized H-UiO-66-(OH)$_2$ in the water phase could return to toluene. When reactants were added to the recycled system, followed by re-homogenization, the reaction proceeds as before.

3. Results and Discussion

3.1. The Structure and Morphology of the Amine-Functionalized H-UiO-66-(OH)$_2$

The amine-functionalized H-UiO-66-(OH)$_2$ was produced by anchoring CO$_2$-switchable functional groups onto the pristine H-UiO-66-(OH)$_2$ at ambient temperature. X-ray photoelectron spectrum (XPS), a powerful surface analysis approach, has been applied for the analysis of the compositional and chemical states of H-UiO-66-(OAPTMS)$_2$, H-UiO-66-(OAEAPTMS)$_2$ and H-UiO-66-(OAEAEAPTMS)$_2$, respectively. As shown in Figure S1, all the functionalized H-UiO-66-(OH)$_2$ consisted of zirconium, silicon, carbon, nitrogen and oxygen elements, and the percentages of each element are listed in Table S1. From their high-resolution XPS results, the binding energies of Zr $3d_{5/2}$ and Zr $3d_{3/2}$ were observed at 179.2 and 181.5 eV (Figures S2a–S4a), suggesting the existence of an oxo–zirconium (IV) cluster in the functionalized H-UiO-66-(OH)$_2$. The peaks of 284.2, 285.4 and 287.8 eV were ascribed to C-C/C=C, C-N and C=O (Figures S2b, S3b and 2a) [35], respectively. The N 1s spectrum displayed two peaks at 398.6 eV for C-N and 399.7 eV for N-H (Figures S2c, S3c and S4b). The binding energies of Si $2p_{3/2}$ and Si $2p_{1/2}$ were located at 99.4 and 100.1 eV (Figures S2d, S3d and S4c), respectively. This is good proof for the post-synthetic modification of the pristine H-UiO-66-(OH)$_2$ by various types of amine silane. Nuclear magnetic resonance (NMR) is a useful method that can be used to characterize the structure of H-MOFs at the molecular level in both dry and wet states. Herein, the solid-state ^{13}C NMR spectra of the pristine and functionalized H-UiO-66-(OH)$_2$ were obtained, as presented in Figure S5, showing that except for the pristine H-UiO-66-(OH)$_2$ resonances, other resonances from amine silane were also obvious. It is clear that the peaks of the functionalized H-UiO-66-(OH)$_2$ at about 9.56, 21.67, 35.24 and 42.79 ppm belonged to the CH$_2$ moiety of amine silane, and the signals at nearly 117.32, 133.24 and 150.74 ppm were attributed to the benzene ring, and the carboxyl peaks were in the range of 159.61 to 173.03 ppm. Whereas the weak peaks at 31.43–44.09 ppm from the pristine H-UiO-66-(OH)$_2$ were probably due to residual DMF in its nanopores. Thermogravimetric analysis (TGA) (Figure 2b) showed that the grafting amounts of amine silane on the H-UiO-66-(OH)$_2$ were about 5.10 mmol/g for H-UiO-66-(OAPTMS)$_2$, 3.98 mmol/g for H-UiO-66-(OAEAPTMS)$_2$ and 3.43 mmol/g for H-UiO-66-(OAEAEAPTMS)$_2$. X-ray diffraction (XRD) patterns were employed to evaluate the crystalline structures of H-UiO-66-(OAPTMS)$_2$, H-UiO-66-(OAEAPTMS)$_2$ and H-UiO-66-(OAEAEAPTMS)$_2$. It was found that all the functionalized H-MOFs reserved the basic crystal structure of the pristine H-UiO-66-(OH)$_2$ (Figure 2c), despite the partial amorphization that was found for them owing to the continued exposure of H-UiO-66-(OH)$_2$ to diluted amine silane in the process of functionalization. The morphologies of the pristine and functionalized H-UiO-66-(OH)$_2$ were also obtained using scanning electron microscopy (SEM). As shown in Figures 2d–f and S6, it can be observed that the size of the functionalized H-UiO-66-(OH)$_2$ particles were not obviously changed after modification (about 300 nm).

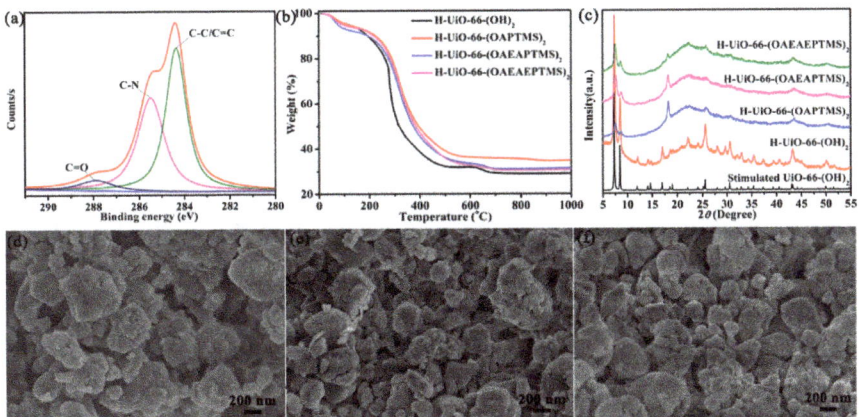

Figure 2. High-resolution C 1s XPS spectra of H-UiO-66-(OAEAEAPTMS)$_2$ (**a**), TGA curves (**b**) and XRD patterns (**c**) of the functionalized H-UiO-66-(OH)$_2$ and SEM images of H-UiO-66-(OAPTMS)$_2$ (**d**), H-UiO-66-(OAEAPTMS)$_2$ (**e**) and H-UiO-66-(OAEAEAPTMS)$_2$ (**f**).

3.2. CO_2-Responsive Demulsification and Re-Emulsification of Pickering Emulsions

In the experimental process, the functionalized H-UiO-66-(OH)$_2$ were employed as emulsifiers for the construction of Pickering emulsions with toluene as the oil phase. It was found that stable Pickering emulsions could be gained by homogenizing the mixture of toluene and water with the added H-UiO-66-(OAPTMS)$_2$, H-UiO-66-(OAEAPTMS)$_2$ and H-UiO-66-(OAEAEAPTMS)$_2$. As an example, Figure 3a presents the Pickering emulsions, formed by homogenizing toluene at 10,000 rpm for 1 min in the water phase containing H-UiO-66-(OAEAEAPTMS)$_2$. It was found that even at the content of 0.25 wt %, H-UiO-66-(OAEAEAPTMS)$_2$ can still effectively emulsify toluene and water to form Pickering emulsions. The morphology and microstructure of the emulsion droplets were recorded with the microscope, as shown in Figure 3b. It is obvious that the droplets of these emulsions were spherical and micrometer-sized. Moreover, their average sizes were found to decrease with the increase in emulsifier content from 0.25 to 1.35 wt %. Similar results were observed for H-UiO-66-(OAPTMS)$_2$ and H-UiO-66-(OAEAPTMS)$_2$ in the same conditions (Figures S7 and S8). In addition, the type of the emulsion was judged to be toluene-in-water (o/w) according to the drop test.

Interestingly, the as-prepared Pickering emulsions were shown to be able to demulsify within 10 min of CO_2 addition. Taking Pickering emulsion stabilized by H-UiO-66-(OAEAEAPTMS)$_2$ as a representative example, it turned out to be very stable even if excess water was removed from the emulsion system. Upon adding CO_2 using a syringe fitted with a needle at an airflow rate of 60 mL min^{-1} for 5 min at normal temperature, the emulsion could be destabilized, resulting in the toluene/water phase separation. Significantly, the Pickering emulsion could be formed again once N_2 was added at a similar rate at 60 °C for about 20 min. Figure 4 provides five cycles of CO_2-responsive demulsification and re-emulsification of the H-UiO-66-(OAEAEAPTMS)$_2$-stabilized Pickering emulsion, indicating excellent reversible switching of Pickering emulsions formed by the functionalized H-UiO-66-(OH)$_2$. In addition, in the optimization process of Pickering emulsions, the common organic solvents such as benzene, ethyl acetate, dichloromethane, trichloromethane, n-hexane and cyclohexane, which were usually used as oil phases to form emulsions, were also selected as representatives to demonstrate the influence of organic solvents on the formation of Pickering emulsions. It was found that benzene, n-hexane and cyclohexane could also be employed as the oil phase to form Pickering emulsions. Moreover, the formed emulsions were capable of transition between emulsification and demulsification with alternate CO_2 and N_2 addition (Figures S9–S17). This means that the functionalized H-UiO-66-(OH)$_2$ reported here may have extensive application prosperity in various oil–water biphasic systems for oil recovery, interface catalysis and as templates for functional material preparation.

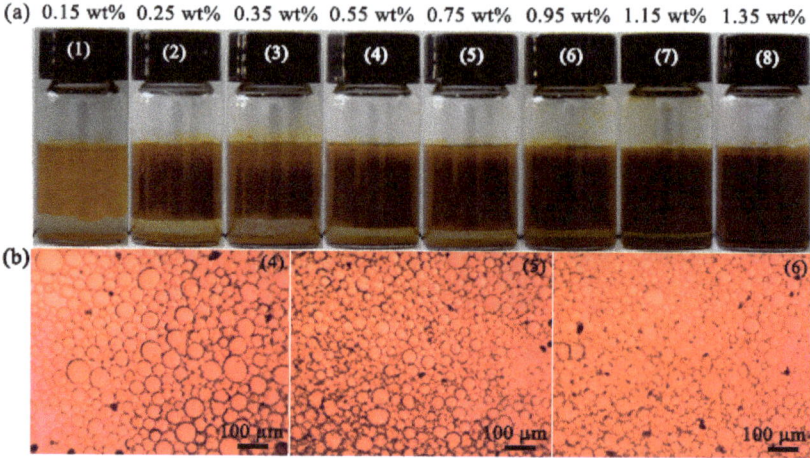

Figure 3. Photographs of toluene-in-water emulsions stabilized by H-UiO-66-(OAEAEAPTMS)$_2$ (**a**) and chosen micrographs for (4), (5) and (6) in (**a**,**b**).

Figure 4. Five cycles of the CO$_2$-triggered demulsification and re-emulsification process of the H-UiO-66-(OAEAEAPTMS)$_2$-stabilized Pickering emulsion.

3.3. Mechanism Analysis for CO$_2$-Responsive Demulsification of Emulsions

To provide insight into the mechanism of CO$_2$-responsive demulsification and re-emulsification of Pickering emulsions, the zeta potential, contact angle, NMR and XPS spectra of the functionalized H-UiO-66-(OH)$_2$ before and after bubbling CO$_2$ were further determined. It was found that the fresh functionalized H-UiO-66-(OH)$_2$ showed the zeta potentials of 18.9 mV for H-UiO-66-(OAPTMS)$_2$ (Figure S18a), 21.8 mV for H-UiO-66-(OAEAPTMS)$_2$ (Figure S19a) and 22.4 mV for H-UiO-66-(OAEAEAPTMS)$_2$ (Figure 5a) pure water. Upon adding CO$_2$ into the system, the zeta potentials were found to increase to 25.9 mV for H-UiO-66-(OAPTMS)$_2$, 32.2 mV for H-UiO-66-(OAEAPTMS)$_2$ and 35.0 mV for H-UiO-66-(OAEAEAPTMS)$_2$. This may be attributed to the fact that the switchable amine groups anchored onto these H-MOFs were ionized by adding CO_2. Ionization of the functionalized H-UiO-66-(OH)$_2$ reduced particle wettability and lowered the emulsion stability. Despite the fact that the electrostatic repulsion appeared between both the dispersed H-MOF particles and emulsion droplets stabilized by the particles, the accessional electronic stabilization via adding CO_2 was inadequate to overcome the destabilization influence of CO_2 on emulsions owing to a lowered wettability of the H-MOF particles. Once CO_2 was removed, the functionalized H-UiO-66-(OH)$_2$ returned back to their original states by a reverse reaction, accompanied by the recovery of the zeta potential values. In addition, the water contact angles of H-UiO-66-(OAPTMS)$_2$, H-UiO-66-(OAEAPTMS)$_2$ and H-UiO-66-(OAEAEAPTMS)$_2$ were determined before and after adding CO_2 for further clarification. It was found that their contact angle values showed an evident decrease from 59° to 25° for H-UiO-66-(OAPTMS)$_2$ (Figure S18a), 55° to 22° for H-UiO-66-(OAEAPTMS)$_2$ (Figure S19a) and 51° to 21° for H-UiO-66-(OAEAEAPTMS)$_2$ upon adding CO_2. Taking into account these results, one can readily conclude that the formation of cationic ammonium on the H-MOFs makes them more hydrophilic, which reduces H-MOF wettability and destabilizes the Pickering emulsions.

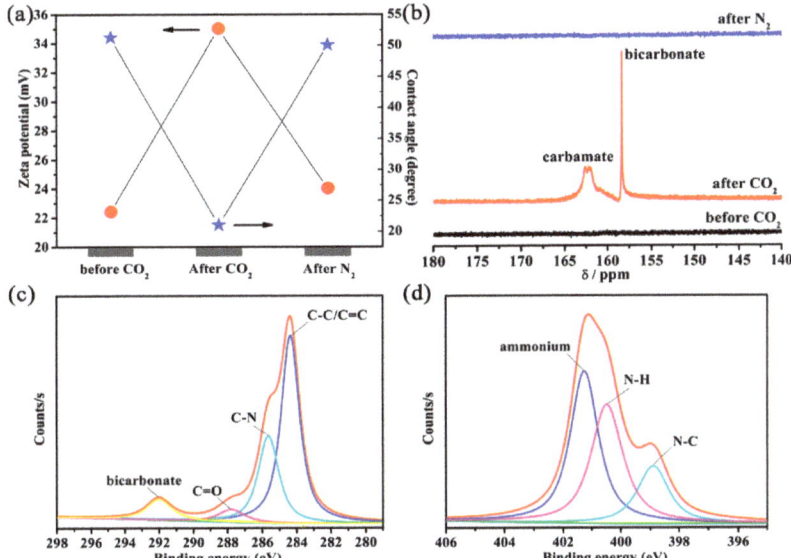

Figure 5. Zeta potentials and water contact angles of H-UiO-66-(OAEAEAPTMS)$_2$ (**a**); ^{13}C NMR spectra of AEAEAPTMS in methanol-d_4 before and after CO_2 bubbling as well as after N_2 bubbling (**b**); and high-resolution C 1s (**c**) and N 1s (**d**) XPS data of H-UiO-66-(OAEAEAPTMS)$_2$ after CO_2 bubbling. Circles in (**a**) stand for zeta potentials, stars stand for water contact angles, and arrows are used to illustrate what the symbol represents.

In an effort to afford further explanations for CO_2-responsive demulsification and re-emulsification of Pickering emulsions, we attempted to obtain solid-state ^{13}C NMR spectroscopy of the functionalized H-UiO-66-(OH)$_2$ before and after adding CO_2 to present their structural changes. However, the unique signals from the reacted H-MOFs were not seen, most likely because of the presence of the carboxylic acid group in the MOFs themselves, overshadowing the newly formed equivalents, and the absence of water. Therefore, the liquid-state NMR spectra of the pure silane coupling agent were measured before and after adding CO_2 (Figures 5b, S18b and S19b). It is clear that two new peaks appeared at about 158.4 and 162.6 ppm for all the amine silanes, ascribing to bicarbonate and carbamate, respectively, implying the production of hydrophilic salts from the reaction of CO_2 with amines of the H-MOFs. In addition, the new signals vanished once CO_2 was expelled by N_2 addition, indicating the reversible reaction process.

XPS, as a favorable approach, was carried out to distinguish the patterns and binding sites of the surface groups of the reacted functionalized H-UiO-66-(OH)$_2$. As a typical example, a new signal at 291.9 eV was observed for the C 1s spectrum of H-UiO-66-(OAEAEAPTMS)$_2$ (Figure 5c), assigning to bicarbonate [50]. The new peak at 401.2 eV was attributed to the N 1s spectra of ammonium (Figure 5d), revealing the generation of ammonium salts [51]. Therefore, it was concluded that the CO_2-responsive demulsification and re-emulsification of Pickering emulsions were involved in the reversible reaction of amino silane with CO_2 and the production of hydrophilic salts, which reduced the H-MOF wettability and lowered the stabilization of emulsions. Once CO_2 was expelled, the functionalized H-MOFs returned back to their original state, and the emulsions could be built again.

3.4. Application of CO_2-Responsive Pickering Emulsions for Knoevenagel Condensation

The Knoevenagel condensation reaction serves as an imperative and valuable condensation approach in organic chemistry because of the synthesis of important intermediates for fine chemicals such as pharmaceuticals, biologically active compounds, natural products and functional polymers through creating new carbon–carbon bonds between carbonyl compounds and active methylene [52–54]. Until now, most of Knoevenagel condensation reactions reported have been carried out under homogeneous conditions. However, it is still challenging for homogeneous systems to efficiently achieve catalyst recovery and product separation from the reaction medium. Hence, it is highly desired but remains an arduous task for the development of new catalysis systems to solve the aforementioned drawbacks. The CO_2-responsive functionalized H-UiO-66-(OH)$_2$-stabilized Pickering emulsions reported here are considered to be promising candidates for the effective coupling of reaction separation and catalyst recovery via CO_2-induced demulsification. As an application example, the switchable Pickering emulsion has been utilized as a mini-reactor and the functionalized H-UiO-66-(OH)$_2$ as catalysts for the efficient integration of Knoevenagel condensation, separation of products and recovery of catalysts.

During the process, Pickering emulsions were initially formed by homogenizing the mixture consisting of toluene, water, reactants (benzaldehyde and malononitrile) and H-UiO-66-(OAEAEAPTMS)$_2$. The reaction for the synthesis of 2-benzylidenemalononitrile in Pickering emulsion progressed well at 25 °C under the catalysis of H-UiO-66-(OAEAEAPTMS)$_2$ (Figure 6a), which was monitored with GC-MS. The yield variation of 2-benzylidenemalononitrile with reaction time in the H-UiO-66-(OAEAEAPTMS)$_2$-stabilized Pickering emulsion is presented in Figure S20. It was found that the yield remained basically the same after 2 h even if the reaction time was extended. Once the reaction was over, the Pickering emulsion was destabilized to separate the product and H-UiO-66-(OAEAEAPTMS)$_2$ by adding CO_2. At this time, the product was in the toluene phase and the H-UiO-66-(OAEAEAPTMS)$_2$ catalyst was dispersed in aqueous solution. GC analysis for the reaction mixture showed that 2-benzylidenemalononitrile was generated at 99% GC yield. The product was then obtained using rotary evaporation under reduced pressure for further purification (Figure S21), toluene was collected for the following cycles and the H-UiO-66-(OAEAEAPTMS)$_2$ in the aqueous phase was utilized directly in the next cycle. As given in Figure 6b, when new reactant was added to the mixture of toluene and water followed by homogenization, along with the removal of CO_2, the reaction proceeded well again. After three cycles, the yield could still be 97% (Figure S22), and the crystalline structure of H-UiO-66-(OAEAEAPTMS)$_2$ did not change markedly (Figure S23).

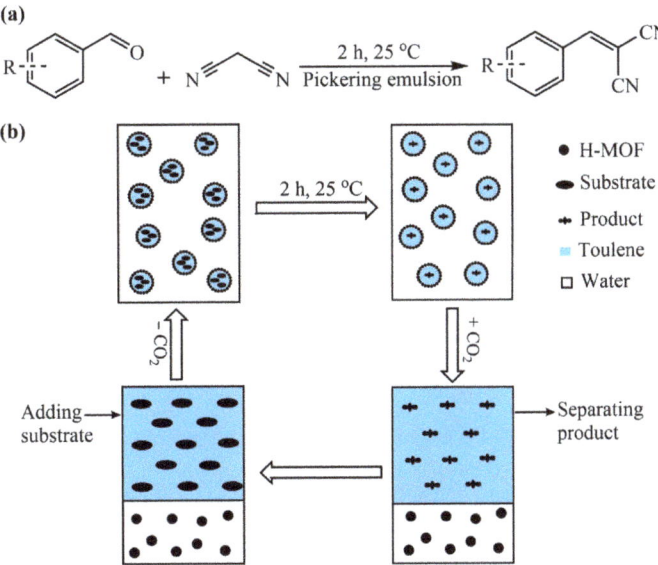

Figure 6. Reaction between aldehydes and malononitrile catalyzed by H-UiO-66-(OAEAEAPTMS)$_2$ (**a**), and the reaction separation procedure for (**a**,**b**).

To extend the range of reaction substrates in the Pickering emulsion, various aldehyde derivatives were selected as reactants, and medium to superior yields were achieved under the same circumstances (Table 1, Figures S24–S28). Moreover, it was found that these yields almost remained the same after three cycles (Table S3). By comparing with other catalyzed systems [55], the CO$_2$-responsive Pickering emulsion proves special catalytic performance for Knoevenagel condensation, and isolation of products from the reaction system could be readily realized by the destabilization of emulsion, and the functionalized H-MOFs could be directly employed for the following cycles. Therefore, an effective combination of catalysis reaction, product isolation and catalyst recovery was obtained to give rise to a sustainable reaction process.

Table 1. Knoevenagel reaction of various aromatic aldehydes.

Entry	Substrate	Product	GC Yield (%)
1	R = *p*-H	R = *p*-H	99
2	R = *o*-NO$_2$	R = *o*-NO$_2$	99
3	R = *p*-NO$_2$	R = *p*-NO$_2$	99
4	R = *p*-CH$_3$	R = *p*-CH$_3$	98
5	R = *p*-F	R = *p*-F	64
6	R = *p*-Br	R = *p*-Br	74

4. Conclusions

In summary, a series of amine-functionalized H-MOFs were prepared and used to form toluene-in-water Pickering emulsions. It was found that the formed emulsions could be switched between demulsification and re-emulsification by alternately adding CO$_2$ and N$_2$ at atmospheric pressure. Mechanism studies showed that the reaction of CO$_2$ with amine anchored on the H-MOFs results in the production of hydrophilic salts, lowering the MOF wettability and reducing the emulsion stability. Once CO$_2$ was expelled, a stable Pickering emulsion could be rebuilt after homogenization through a reverse reaction. Based on the strategy, a highly effective Knoevenagel condensation reaction was gained, and a sustainable synthesis procedure was achieved to couple the catalysis reaction, product separation and MOF recovery. This work provides us with insight into the reversible switching of H-MOF-stabilized Pickering emulsions and paves the way for the design and preparation of other types of emulsion reactors, thus significantly prompting further investigation of green and sustainable procedures for various chemical reactions.

Supplementary Materials: The following supporting information can be downloaded at: https://www.mdpi.com/article/10.3390/ma16041675/s1.

Author Contributions: Conceptualization, X.P.; methodology, D.X., J.L. and Y.X.; validation, X.P., D.X. and Y.X.; investigation, X.P., J.L., W.S. and Z.W.; data curation, X.P.; writing—original draft preparation, X.P.; writing—review and editing, X.P. and D.X.; supervision, D.X. and Y.X.; project administration, X.P.; funding acquisition, X.P. All authors have read and agreed to the published version of the manuscript.

Funding: This research was funded by the National Natural Science Foundation of China (No. 22003055), and the Nanhu Scholars Program for Young Scholars of Xinyang Normal University.

Institutional Review Board Statement: Not applicable.

Informed Consent Statement: Not applicable.

Data Availability Statement: Data available on request from the authors.

Acknowledgments: This work is supported by the National Natural Science Foundation of China (No. 22003055), and the Nanhu Scholars Program for Young Scholars of Xinyang Normal University.

Conflicts of Interest: The authors declare no conflict of interest.

References

1. Xu, M.; Jiang, J.; Pei, X.; Song, B.; Cui, Z.; Binks, B.P. Novel oil-in-water emulsions stabilised by ionic surfactant and similarly charged nanoparticles at very low concentrations. *Angew. Chem. Int. Ed.* **2018**, *57*, 7738–7742. [CrossRef] [PubMed]
2. Binks, B.P.; Rodrigues, J.A. Double inversion of emulsions by using nanoparticles and a di-chain surfactant. *Angew. Chem. Int. Ed.* **2007**, *46*, 5389–5392. [CrossRef] [PubMed]
3. Rodriguez, A.M.B.; Binks, B.P. Catalysis in Pickering emulsions. *Soft Matter* **2020**, *16*, 10221–10243. [CrossRef] [PubMed]
4. Zhang, M.; Wei, L.; Chen, H.; Du, Z.; Binks, B.P.; Yang, H. Compartmentalized Droplets for Continuous Flow Liquid-Liquid Interface Catalysis. *J. Am. Chem. Soc.* **2016**, *138*, 10173–10183. [CrossRef]
5. Chang, F.; Vis, C.M.; Ciptonugroho, W.; Bruijnincx, P.C.A. Recent developments in catalysis with Pickering Emulsions. *Green Chem.* **2021**, *23*, 2575–2594. [CrossRef]
6. Tang, J.; Quinlan, P.J.; Tam, K.C. Stimuli-responsive Pickering emulsions: Recent advances and potential applications. *Soft Matter* **2015**, *11*, 3512–3529. [CrossRef]
7. Li, Z.; Shi, Y.; Zhu, A.; Zhao, Y.; Wang, H.; Binks, B.P.; Wang, J. Light-responsive, reversible emulsification and demulsification of oil-in-water Pickering emulsions for catalysis. *Angew. Chem. Int. Ed.* **2021**, *60*, 3928–3933. [CrossRef]
8. Chen, Z.; Zhou, L.; Bing, W.; Zhang, Z.; Li, Z.; Ren, J.; Qu, X. Light controlled reversible inversion of nanophosphor-stabilized Pickering emulsions for biphasic enantioselective biocatalysis. *J. Am. Chem. Soc.* **2014**, *136*, 7498–7504. [CrossRef]
9. Jiang, J.; Ma, Y.; Cui, Z.; Binks, B.P. Pickering emulsions responsive to CO_2/N_2 and light dual stimuli at ambient temperature. *Langmuir* **2016**, *32*, 8668–8675. [CrossRef]
10. Morse, A.J.; Armes, S.P.; Thompson, K.L.; Dupin, D.; Fielding, L.A.; Mills, P.; Swart, R. Novel Pickering emulsifiers based on pH-responsive poly(2-(diethylamino)ethyl methacrylate) latexes. *Langmuir* **2013**, *29*, 5466–5475. [CrossRef]
11. Liu, H.; Wang, C.; Zou, S.; Wei, Z.; Tong, Z. Simple, reversible emulsion system switched by pH on the basis of chitosan without any hydrophobic modification. *Langmuir* **2012**, *28*, 11017–11024. [CrossRef] [PubMed]
12. Yang, H.; Zhou, T.; Zhang, W. A strategy for separating and recycling solid catalysts based on the pH-triggered Pickering-emulsion inversion. *Angew. Chem. Int. Ed.* **2013**, *52*, 7455–7459. [CrossRef] [PubMed]
13. Kim, J.; Cote, L.J.; Kim, F.; Yuan, W.; Shull, K.R.; Huang, J. Graphene oxide sheets at interfaces. *J. Am. Chem. Soc.* **2010**, *132*, 8180–8186. [CrossRef] [PubMed]
14. Gu, X.; Yang, Y.; Hu, M.; Hu, Y.; Wang, C. Redox responsive diselenide colloidosomes templated from Pickering emulsions for drug release. *J. Control. Release* **2015**, *213*, e119–e120. [CrossRef]
15. Quesada, M.; Muniesa, C.; Botella, P. Hybrid PLGA-organosilica nanoparticles with redox-sensitive molecular gates. *Chem. Mater.* **2013**, *25*, 2597–2602. [CrossRef]
16. Jiang, Q.; Sun, N.; Li, Q.; Si, W.; Li, J.; Li, A.; Gao, Z.; Wang, W.; Wang, J. Redox-responsive Pickering emulsions based on silica nanoparticles and electrochemical active fluorescent molecules. *Langmuir* **2019**, *35*, 5848–5854. [CrossRef]

17. Saigal, T.; Dong, H.; Matyjaszewski, K.; Tilton, R.D. Pickering emulsions stabilized by nanoparticles with thermally responsive grafted polymer brushes. *Langmuir* **2010**, *26*, 15200–15209. [CrossRef]
18. Binks, B.P.; Murakami, R.; Armes, S.P.; Fujii, S. Temperature-induced inversion of nanoparticle-stabilized emulsions. *Angew. Chem. Int. Ed.* **2005**, *44*, 4795–4798. [CrossRef]
19. Zoppe, J.O.; Venditti, R.A.; Rojas, O.J. Pickering emulsions stabilized by cellulose nanocrystals grafted with thermo-responsive polymer brushes. *J. Colloid Interface Sci.* **2012**, *369*, 202–209. [CrossRef]
20. Wang, X.; Zeng, M.; Yu, Y.-H.; Wang, H.; Mannan, M.S.; Cheng, Z. Thermosensitive ZrP-PNIPAM Pickering emulsifier and the controlled-release behavior. *ACS Appl. Mater. Interfaces* **2017**, *9*, 7852–7858. [CrossRef]
21. Thompson, K.L.; Fielding, L.A.; Mykhaylyk, O.O.; Lane, J.A.; Derry, M.J.; Armes, S.P. Vermicious thermo-responsive Pickering emulsifiers. *Chem. Sci.* **2015**, *6*, 4207–4214. [CrossRef] [PubMed]
22. Zhou, J.; Qiao, X.; Binks, B.P.; Sun, K.; Bai, M.; Li, Y.; Liu, Y. Magnetic Pickering emulsions stabilized by Fe_3O_4 nanoparticles. *Langmuir* **2011**, *27*, 3308–3316. [CrossRef] [PubMed]
23. Teixeira, I.F.; da Silva Oliveira, A.A.; Christofani, T.; Camilo Moura, F.C. Biphasic oxidation promoted by magnetic amphiphilic nanocomposites undergoing a reversible emulsion process. *J. Mater. Chem. A* **2013**, *1*, 10203–10208. [CrossRef]
24. Richtering, W. Responsive emulsions stabilized by stimuli-sensitive microgels: Emulsions with special non-Pickering properties. *Langmuir* **2012**, *28*, 17218–17229. [CrossRef]
25. Tang, J.; Lee, M.F.X.; Zhang, W.; Zhao, B.; Berry, R.M.; Tam, K.C. Dual responsive Pickering emulsion stabilized by poly[2-(dimethylamino)ethyl methacrylate] grafted cellulose nanocrystals. *Biomacromolecules* **2014**, *15*, 3052–3060. [CrossRef]
26. Yi, C.; Liu, N.; Zheng, J.; Jiang, J.; Liu, X. Dual-responsive poly(styrene-alt-maleic acid)-graft-poly(N-isopropyl acrylamide) micelles as switchable emulsifiers. *J. Colloid Interfaces Sci.* **2012**, *380*, 90–98. [CrossRef]
27. Wang, X.; Shi, Y.; Graff, R.W.; Lee, D.; Gao, H. Developing recyclable pH-responsive magnetic nanoparticles for oil-water separation. *Polymer* **2015**, *72*, 361–367. [CrossRef]
28. Yoon, K.Y.; Li, Z.; Neilson, B.M.; Lee, W.; Huh, C.; Bryant, S.L.; Bielawski, C.W.; Johnston, K.P. Effect of adsorbed amphiphilic copolymers on the interfacial activity of superparamagnetic nanoclusters and the emulsification of oil in water. *Macromolecules* **2012**, *45*, 5157–5166. [CrossRef]
29. Chen, Y.; Bai, Y.; Chen, S.; Ju, J.; Li, Y.; Wang, T.; Wang, Q. Stimuli-responsive composite particles as solid-stabilizers for effective oil harvesting. *ACS Appl. Mater. Interfaces* **2014**, *6*, 13334–13338. [CrossRef]
30. Brugger, B.; Richtering, W. Magnetic, thermosensitive microgels as stimuli-responsive emulsifiers allowing for remote control of separability and stability of oil in water-emulsions. *Adv. Mater.* **2007**, *19*, 2973–2978. [CrossRef]
31. Jessop, P.G.; Heldebrant, D.J.; Li, X.; Eckert, C.A.; Liotta, C.L. Reversible nonpolar-to-polar solvent. *Nature* **2005**, *436*, 1102. [CrossRef] [PubMed]
32. Pei, X.; Li, Z.; Wang, H.; Zhou, Q.; Liu, Z.; Wang, J. CO_2-Switchable phase separation of nonaqueous surfactant-free ionic liquid-based microemulsions. *ACS Sustain. Chem. Eng.* **2022**, *10*, 1777–1785. [CrossRef]
33. Pei, X.; Liu, J.; Rao, W.; Ma, X.; Li, Z. CO_2-Switchable reversible phase transfer of carbon dot. *Ind. Eng. Chem. Res.* **2021**, *60*, 9296–9303. [CrossRef]
34. Shi, Y.; Wang, Z.; Li, Z.; Wang, H.; Xiong, D.; Qiu, J.; Tian, X.; Feng, G.; Wang, J. Anchoring LiCl in the nanopores of metal-organic frameworks for ultra-high uptake and selective separation of ammonia. *Angew. Chem. Int. Ed.* **2022**, *61*, e202212032. [CrossRef] [PubMed]
35. Pei, X.; Tian, C.; Wang, Y.; Li, Z.; Xiong, Z.; Wang, H.; Ma, X.; Cao, X.; Li, Z. CO_2-Driven reversible transfer of amine-functionalized ZIF-90 between organic and aqueous phases. *Chem. Commun.* **2022**, *58*, 10372–10375. [CrossRef]
36. Xiao, B.; Yuan, Q.; Williams, R.A. Exceptional function of nanoporous metal organic framework particles in emulsion stabilisation. *Chem. Commun.* **2013**, *49*, 8208–8210. [CrossRef]
37. Huo, J.; Marcello, M.; Garai, A.; Bradshaw, D. MOF-polymer composite microcapsules derived from Pickering emulsions. *Adv. Mater.* **2013**, *25*, 2717–2722. [CrossRef]
38. Zhang, B.; Zhang, J.; Liu, C.; Peng, L.; Sang, X.; Han, B.; Ma, X.; Luo, T.; Tan, X.; Yang, G. High-internal-phase emulsions stabilized by metal-organic frameworks and derivation of ultralight metal-organic aerogels. *Sci. Rep.* **2016**, *6*, 21401–21409. [CrossRef]
39. Jin, P.; Tan, W.; Huo, J.; Liu, T.; Liang, Y.; Wang, S.; Bradshaw, D. Hierarchically porous MOF/polymer composites via interfacial nanoassembly and emulsion polymerization. *J. Mater. Chem. A* **2018**, *6*, 20473–20479. [CrossRef]
40. Bian, Z.; Xu, J.; Zhang, S.; Zhu, X.; Liu, H.; Hu, J. Interfacial growth of metal organic framework/graphite oxide composites through Pickering emulsion and their CO_2 capture performance in the presence of humidity. *Langmuir* **2015**, *31*, 7410–7417. [CrossRef]
41. Li, Z.; Zhang, J.; Luo, T.; Tan, X.; Liu, C.; Sang, X.; Ma, X.; Han, B.; Yang, G. High internal ionic liquid phase emulsion stabilized by metal-organic frameworks. *Soft Matter* **2016**, *12*, 8841–8846. [CrossRef] [PubMed]
42. Jiang, W.-L.; Fu, Q.-J.; Yao, B.-J.; Ding, L.-G.; Liu, C.-X.; Dong, Y.-B. Smart pH-responsive polymer-tethered and Pd NP-loaded NMOF as the Pickering interfacial catalyst for one-pot cascade biphasic reaction. *ACS Appl. Mater. Interfaces* **2017**, *9*, 36438–36446. [CrossRef] [PubMed]
43. Shi, Y.; Xiong, D.; Li, Z.; Wang, H.; Qiu, J.; Zhang, H.; Wang, J. Ambient CO_2/N_2 switchable Pickering emulsion emulsified by TETA-functionalized metal-organic frameworks. *ACS Appl. Mater. Interfaces* **2020**, *12*, 53385–53393. [CrossRef] [PubMed]

44. Yao, B.-J.; Fu, Q.-J.; Li, A.-X.; Zhang, X.-M.; Li, Y.-A.; Dong, Y.-B. A thermo-responsive polymer-tethered and Pd NP loaded UiO-66 NMOF for biphasic CB dichlorination. *Green Chem.* **2019**, *21*, 1625–1634. [CrossRef]
45. Cai, G.; Yan, P.; Zhang, L.; Zhou, H.-C.; Jiang, H.-L. Metal-organic framework-based hierarchically porous materials: Synthesis and applications. *Chem. Rev.* **2021**, *121*, 12278–12326. [CrossRef] [PubMed]
46. Yue, Y.; Qiao, Z.-A.; Fulvio, P.F.; Binder, A.J.; Tian, C.; Chen, J.; Nelson, K.M.; Zhu, X.; Dai, S. Template-free synthesis of hierarchical porous metal−organic frameworks. *J. Am. Chem. Soc.* **2013**, *135*, 9572–9575. [CrossRef]
47. Venturi, D.M.; Campana, F.; Marmottini, F.; Costantino, F.; Vaccaro, L. Extensive screening of green solvents for safe and sustainable UiO-66 synthesis. *ACS Sustain. Chem. Eng.* **2020**, *8*, 17154–17164. [CrossRef]
48. Huang, H.; Li, J.-R.; Wang, K.; Han, T.; Tong, M.; Li, L.; Xie, Y.; Yang, Q.; Liu, D.; Zhong, C. An in situ self-assembly template strategy for the preparation of hierarchical-pore metal-organic frameworks. *Nat. Commun.* **2015**, *6*, 8847–8854. [CrossRef]
49. Liang, C.; Liu, Q.; Xu, Z. Surfactant-free switchable emulsions using CO_2-responsive particles. *ACS Appl. Mater. Interfaces* **2014**, *6*, 6898–6904. [CrossRef]
50. Lin, J.; Huang, Y.; Wang, S.; Chen, G. Microwave-assisted rapid exfoliation of graphite into graphene by using ammonium bicarbonate as the intercalation agent. *Ind. Eng. Chem. Res.* **2017**, *56*, 9341–9346. [CrossRef]
51. Ye, X.; Qin, X.; Yan, X.; Guo, J.; Huang, L.; Chen, D.; Wu, T.; Shi, Q.; Tan, S.; Cai, X. π–π conjugations improve the long-term antibacterial properties of graphene oxide/quaternary ammonium salt nanocomposites. *Chem. Eng. J.* **2016**, *304*, 873–881. [CrossRef]
52. Antonangelo, A.R.; Hawkins, N.; Tocci, E.; Muzzi, C.; Fuoco, A.; Carta, M. Tröger's base network polymers of intrinsic microporosity (TB-PIMs) with tunable pore size for heterogeneous catalysis. *J. Am. Chem. Soc.* **2022**, *144*, 15581–15594. [CrossRef] [PubMed]
53. Mancipe, S.; Castillo, J.-C.; Brijaldo, M.H.; López, V.P.; Rojas, H.; Macías, M.A.; Portilla, J.; Romanelli, G.P.; Martínez, J.J.; Luque, R. B-Containing hydrotalcites effectively catalyzed synthesis of 3-(Furan-2-yl)acrylonitrile derivatives via the Knoevenagel condensation. *ACS Sustain. Chem. Eng.* **2022**, *10*, 12602–12612. [CrossRef]
54. Sutar, P.; Maji, T.K. Bimodal self-assembly of an amphiphilic gelator into a hydrogel-nanocatalyst and an organogel with different morphologies and photophysical properties. *Chem. Commun.* **2016**, *52*, 13136–13139. [CrossRef]
55. Tuci, G.; Luconi, L.; Rossin, A.; Berretti, E.; Ba, H.; Innocenti, M.; Yakhvarov, D.; Caporali, S.; Pham-Huu, C.; Giambastiani, G. Aziridine-Functionalized Multiwalled Carbon Nanotubes: Robust and Versatile Catalysts for the Oxygen Reduction Reaction and Knoevenagel Condensation. *ACS Appl. Mater. Interfaces* **2016**, *8*, 30099–30106. [CrossRef]

Disclaimer/Publisher's Note: The statements, opinions and data contained in all publications are solely those of the individual author(s) and contributor(s) and not of MDPI and/or the editor(s). MDPI and/or the editor(s) disclaim responsibility for any injury to people or property resulting from any ideas, methods, instructions or products referred to in the content.

Article

Modified Bamboo Charcoal as a Bifunctional Material for Methylene Blue Removal

Qian Liu [1], Wen-Yong Deng [1], Lie-Yuan Zhang [2], Chang-Xiang Liu [1,*], Wei-Wei Jie [1], Rui-Xuan Su [1], Bin Zhou [1], Li-Min Lu [1], Shu-Wu Liu [1] and Xi-Gen Huang [1]

1. Key Laboratory of Chemical Utilization of Plant Resources of Nanchang, College of Chemistry and Materials, Jiangxi Agricultural University, Nanchang 330045, China
2. Technical Center of Nanchang Customs, Nanchang 330038, China
* Correspondence: lchxiang@126.com; Tel.: +86-791-83813574

Abstract: Biomass-derived raw bamboo charcoal (BC), NaOH-impregnated bamboo charcoal (BC-I), and magnetic bamboo charcoal (BC-IM) were fabricated and used as bio-adsorbents and Fenton-like catalysts for methylene blue removal. Compared to the raw biochar, a simple NaOH impregnation process significantly optimized the crystal structure, pore size distribution, and surface functional groups and increase the specific surface area from 1.4 to 63.0 m^2/g. Further magnetization of the BC-I sample not only enhanced the surface area to 84.7 m^2/g, but also improved the recycling convenience due to the superparamagnetism. The maximum adsorption capacity of BC, BC-I, and BC-IM for methylene blue at 328 K was 135.13, 220.26 and 497.51 mg/g, respectively. The pseudo-first-order rate constants k at 308 K for BC, BC-I, and BC-IM catalytic degradation in the presence of H$_2$O$_2$ were 0.198, 0.351, and 1.542 h^{-1}, respectively. A synergistic mechanism between adsorption and radical processes was proposed.

Keywords: bamboo charcoal; methylene blue; adsorption; degradation; kinetics

Citation: Liu, Q.; Deng, W.-Y.; Zhang, L.-Y.; Liu, C.-X.; Jie, W.-W.; Su, R.-X.; Zhou, B.; Lu, L.-M.; Liu, S.-W.; Huang, X.-G. Modified Bamboo Charcoal as a Bifunctional Material for Methylene Blue Removal. *Materials* 2023, 16, 1528. https://doi.org/10.3390/ma16041528

Academic Editor: Maria Giorgia Cutrufello

Received: 24 December 2022
Revised: 19 January 2023
Accepted: 7 February 2023
Published: 11 February 2023

Copyright: © 2023 by the authors. Licensee MDPI, Basel, Switzerland. This article is an open access article distributed under the terms and conditions of the Creative Commons Attribution (CC BY) license (https:// creativecommons.org/licenses/by/ 4.0/).

1. Introduction

Nowadays, organic wastewater discharge has caused very serious world environmental problems with the rapid development of chemical production in the textile, paper, plastic, rubber, and cosmetics industries. Methylene blue (MB) dye is a commonly used organic pollutant, which has high stability and low biodegradability and is thus difficult to degrade [1]. Various strategies, such as membrane filtration [2], advanced oxidation process (AOPs) [3], photocatalysis [4–6], and adsorption [7] have been reported for the treatment of organic wastewater. Among them, adsorption and the heterogeneous Fenton-like method, an AOP based on the activation of H$_2$O$_2$ by a solid catalyst, are gaining importance in dye wastewater decontamination in recent excellent reviews due to their advantages of simple operation, high removal efficiency, low operating cost, and good recycling performance [8,9].

Biochar is carbon-enriched solid residues commonly produced by biomass carbonization at high temperatures under hypoxic or anaerobic conditions without complex activation processes [10]. Biochar is well-known in wastewater treatment applications due to its abundant pore structure and surface oxygen-containing functional groups [11]. However, the raw biomass-derived biochar always possesses low specific surface area and poor adsorption capacity. For example, Ji et al. reported that fallen leaf-derived biochar had an adsorption capacity for MB of 78.6 mg/g [12]. Sun et al. prepared biochars by pyrolysis at 673 K from eucalyptus, palm bark, and anaerobic digestion residues, with adsorption capacities for MB of 2.0, 2.95, and 9.77 mg/g, respectively [13]. Therefore, modification of carbon materials such as chemical activation, functional surfaces by surface grafting, or deposition with metal nanoparticles via impregnation is an efficient approach to improve adsorption capacity [14–16]. However, in the chemical activation reported in literature, biochar is often soaked with acid, alkali, and salt solution and then calcined at high temperatures [17–20]. Recently, Xu et al. found that only impregnating bamboo char with HNO_3 without secondary high temperature calcination can improve its adsorption of mercury by introducing functional groups onto the surface of the biochar [21]. Wu et al. explored the effects of acid/alkali (HCl, HNO_3 or NaOH) impregnation on the physicochemical properties and adsorption behavior of sludge-based carbon [22] providing a new energy-saving and convenient idea for improving the properties of biochars. On the other hand, biochar is a good support media to disperse metal oxides [23]. Magnetization of Fe_xO_y particles onto biochars has attracted much attention due to its easy recovery performance and good adsorption capacities for organic contaminants [24,25] or heavy metals [26]. However, Liu et al. has pointed out that a large number of uncertainties still need to be explored, such as the synthesis route and potential applications [27].

Reports of biochar for wastewater treatment by AOPs are increasing year by year. Khataee et al. found that Cu_2O–CuO@BC has better photocatalytic degradation efficiency of RO29 than that of single-component Cu_2O–CuO and biochar [28]. Zhai et al. used TiO_2/MgO/ZnO@BC for methylene blue oxidation and found that it can be a bi-functional adsorption and catalytic material with excellent performance [29]. Mian et al. had also reported that MnO_x–N@BC complexes showed good activity and stability for dye degradation in the presence of PMS [30]. Therefore, loading metal oxides on the surface of biochar can increase new active components and promote catalytic activity. Magnetic biochar–iron oxide composites can not only increase the surface active sites but also promote the convenience of recovery. However, only 8.4% of the published papers about magnetic carbon applications are related to dye removal, and most of them are processed by physical adsorption [31]. Thus, it is necessary to investigate the adsorption and catalytic ability of magnetic biochar in the removal of dye from wastewater.

Bamboo is a cheap and fast growing woody plant biomass resource and is a sustainable and suitable source of C-based raw materials [32]. In this paper, raw bamboo charcoal, NaOH-impregnated bamboo charcoal, and magnetic modified bamboo charcoal were successively fabricated and used as bio-adsorbents and Fenton-like catalysts for MB removal. Then, the adsorption kinetics, isotherm, and thermodynamics were analyzed and compared in detail. The catalytic activity and reusability as Fenton-like catalysts were studied and a mechanism was proposed. The object of this work is to explore the modified preparation strategy on the pore texture, surface chemistry, and adsorption/catalytic degradation capacities of bamboo charcoal.

2. Materials and Methods
2.1. Materials

The bamboo (Phyllostachys edulis) were taken from Nanchang, Jiangxi province. Sodium hydroxide (NaOH, 96%, AR), hydrogen peroxide (H_2O_2, 30 wt.%, AR), methylene blue ($C_{16}H_{18}ClN_3S$, MB, 98.5%, AR), ferrous chloride tetrachloride ($FeCl_2 \cdot 4H_2O$, 99%, AR), and ferric chloride hexahydrate ($FeCl_3 \cdot 6H_2O$, 99%, AR) were purchased from Sinopharm Chemical Reagent Co., Ltd., China. All the chemicals were used without any purification.

2.2. Sample Preparation

The bamboo were first cut into pieces and subsequently dried for two days at 373 K, and then were crushed into powder. The bamboo charcoal (BC) was produced by calcination of the bamboo powder at 923 K for 2 h under an atmosphere of N_2. The obtained BC sample was then impregnated with 6 mol/L NaOH solution for 24 h. The impregnated BC sample was then centrifuged, washed, dried, and named BC-I. Next, a certain amount of BC-I (1.15 g), $FeCl_2 \cdot 4H_2O$ (1.837 g) and $FeCl_3 \cdot 6H_2O$ (5 g) were dissolved in 200 mL ultrapure water. The solution was heated to 343 K for 10 min. Then, an NaOH solution was slowly dripped into the mixture and the pH was adjusted to the range of 10~11. The mixture was stirred for 4 h. After filtering, the samples were washed and dried for 24 h at 373 K. The final product was named BC-IM. General preparation mechanism was proposed in Scheme 1.

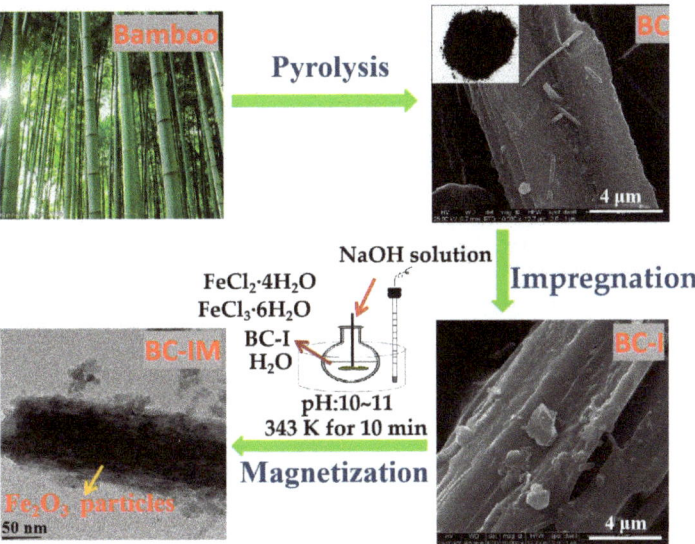

Scheme 1. General preparation mechanism of BC, BC-I, and BC-IM.

2.3. Sample Characterization

The crystal phase composition of the three adsorbents were analyzed by XRD (Bruker AXS D8). The pore structures and BET surface were obtained from the adsorption and desorption of N_2 (Micromeritics TriStar 3000, Micromeritics, USA). The characteristics of the pore size distribution were determined by the Barrett–Joyner–Halenda (BJH) and Horvath–Kawazoe (HK) methods. The morphologies and images of the three adsorbents were investigated by SEM (Philips XL 30, Philips, Czech Republic) and TEM (JEOL 2011, JEOL, Japan), respectively. The element composition analysis and element distribution map of the samples were obtained by EDS coupled to the SEM. FT-IR (Perkinelmer C107996, PerkinElmer, USA) was applied to explore the types of surface functional groups. The magnetic properties were monitored by VSM at room temperature. Finally, XPS (PHI 5000C, PHI, USA) was used to investigate the valence state of surface iron.

2.4. Batch Adsorption Experiments

A total of 100 mg of biochar was poured into 100 mL of a 10 mg/L MB solution at 308 K in a 250 mL glass flask. The adsorption process was investigated by a 722-UV spectrophotometer at 665 nm. Each parameter of the adsorption experiment was repeated three times. The removal efficiency of MB is expressed as $(A_0 - A)/A_0 \times 100\%$, where A_0 and A are the initial absorbance and the absorbance at a given reaction time, respectively. The recycling experiments of the best BC-IM were carried out under similar reaction conditions. The sample was collected by centrifugation, washed, and dried overnight in a vacuum oven at 373 K after each recycle during the recycling experiment.

Furthermore, a series of adsorption tests were carried out with the initial MB concentration varying from 20 to 50 mg/L and the dosage of adsorbent was 100 mg at 308 K. The influence of adsorption temperature on adsorption properties was also studied at 308, 318, and 328 K. The adsorption capacity Q_e (mg/g) at adsorption equilibrium and Q_t (mg/g) at the selected reaction time were obtained from the following equations:

$$Q_e = \frac{(c_o - c_e)V}{m} \quad (1)$$

$$Q_t = \frac{(c_o - c_t)V}{m} \quad (2)$$

where c_o (mg/L), c_e (mg/L), and c_t (mg/L) represent the MB concentration at the initial state, adsorption equilibrium, and a selected time t, respectively. The m (g) and V (L) are the mass of adsorbent and the volume of solution used, respectively.

2.5. Catalytic Degradation Experiments

MB degradation by the catalyst/H_2O_2 system was used as the model experiment to study the catalytic activity of the prepared catalysts. In a 250 mL conical flask, 0.1 g of catalyst, 20 mL H_2O_2, and 100 mL (10 mg/L) MB solution were mixed and placed at 308 K. The absorbance of the solution was measured using a 722-UV spectrophotometer at 665 nm. The catalytic activity of the catalyst was also determined by the removal efficiency of MB, which was calculated as $(A_0 - A)/A_0 \times 100\%$. The recycling experiments were carried out under similar reaction conditions. After reacting for 2 h, the sample was collected by centrifugation, washed, and dried overnight in a vacuum oven at 373 K after each recycle during the recycling experiment.

3. Results

3.1. Characterization Results of the Samples

As demonstrated in Figure 1, the BC, BC-I, and BC-IM samples all exhibited a broad band around 22°, corresponding to the (002) diffraction of the carbon structure, which demonstrates the conversion of lignin–cellulose to a more carbonaceous structure during pyrolysis [33]. For the raw BC, there were many sharp narrow peaks, which may be related to the formation of inorganic ash (various oxides or carbonates) during the pyrolysis process of the bamboo [33,34]. It is worth noting that after NaOH-impregnation, a slight diffraction shift of the (002) plane ranging from 21.7 to 24.5° was observed for the BC-I sample, which has been reported to be related to the formation of more defects [35] and more ordered graphite structures with smaller layer spacing [36]. In addition, a peak around 2θ value of 43° was also found for the BC-I sample, which belongs to the (100) diffraction plane of carbons [37]. The appearance of the (100) plane had been reported to be evidence of the formation of graphitic structures [35], which is consistent with the red shift phenomenon of the (002) plane. Therefore, it indicates that the simple NaOH-modification treatment is an efficient way to optimize the crystal structure of biochar materials. After further magnetization, a diffraction peak centered at 35.6° assignable to the iron oxide Fe_2O_3 was observed, which indicates that iron oxide was successfully introduced to the surface of bamboo charcoal through a facile one-pot in situ precipitation. Rietveld analysis showed that the Fe_2O_3 (PDF-39-1346) is a cubic system with a cell parameter of 0.8832 nm.

Figure 2 presents the SEM images and EDS patterns of the three adsorbents, as well as the elemental mappings of the BC-IM sample. As shown in the SEM figures, BC-I had a rough surface and a large number of irregular channels and folds compared to the raw BC, providing a wide field for the adsorption of organic pollutants [25]. The SEM image of BC-IM showed not only more pore structures but also a large number of particle aggregates distributed around the pores, indicating the formation of magnetic particles on the surface of BC-IM. To identify the differences in the elemental composition of the adsorbents, the EDS patterns were also displayed. The results showed that the raw BC sample contained C, O, K, and Cl elements, with estimated atomic contents of 88.29%, 9.44%, 1.99%, and 0.28%, respectively. The element composition of the BC-I sample was C (89.43%), O (9.16%), and Na (1.41%). The composition of BC-IM was C (33.02%), O (45.61%), Na (3.41%), Cl (58.6%), and Fe (17.58%), which proves that iron was successfully loaded onto the surface of the bamboo charcoal. The elemental mappings of the BC-IM sample further displayed the distribution of iron on the surface of the BC-IM.

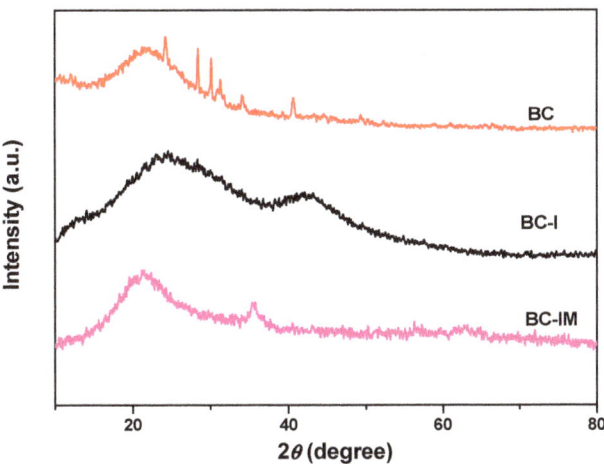

Figure 1. X-ray powder diffraction patterns of the synthesized BC, BC-I, and BC-IM.

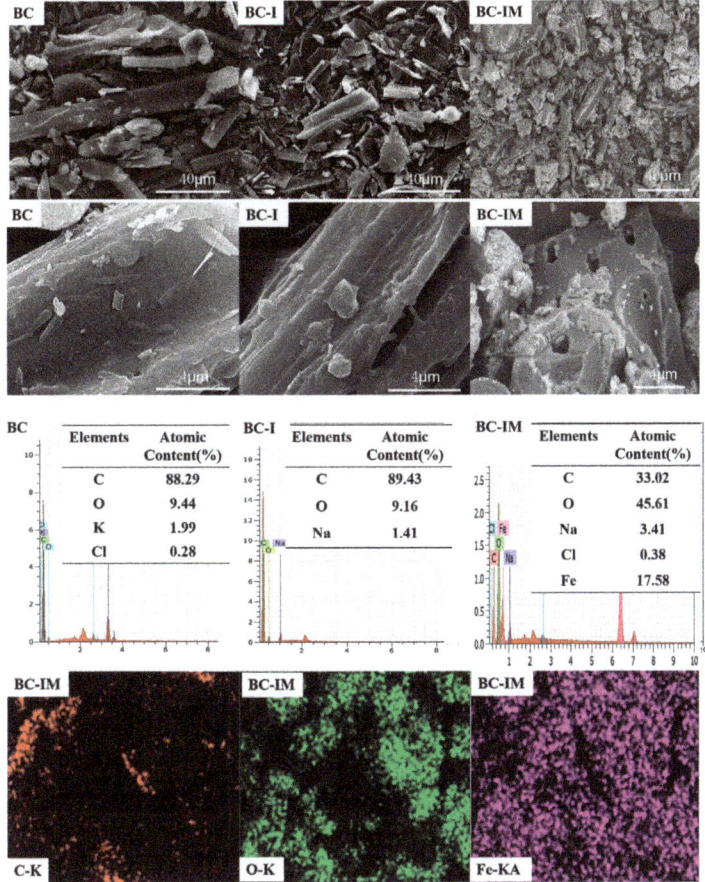

Figure 2. SEM images and EDS patterns of all the samples and elemental mappings of BC-IM.

The nitrogen adsorption–desorption isotherms and corresponding pore size distribution curves of BC-I and BC-IM are summarized in Figure 3. The pore size, pore volume, and BET surface area are displayed in Table 1. For the raw BC sample, a low BET surface area value of 1.4 m^2/g and small pore volume of 0.001 cm^3/g were obtained. The above XRD results demonstrated that the raw BC sample contained some inorganic ash, which may cover the pore structure and surface active sites of BC. The isotherms of both BC-I and BC-IM samples exhibited a rapid nitrogen uptake at the low relative pressure region of less than 0.06, followed by a continuous increase in the rest relative pressure with a large hysteresis loop. In general, the relative pressure in the adsorption–desorption isotherm at 0~0.1, 0.1~0.8, and 0.8~1 is regarded as associated with micropores, mesopores, and macropores, respectively [25]. It shows that high proportion of micropores accompanied with partial mesopores or even macropores appeared in the BC-I and BC-IM samples, which is believed to be beneficial for the transport and diffusion of dyes during adsorption [35].

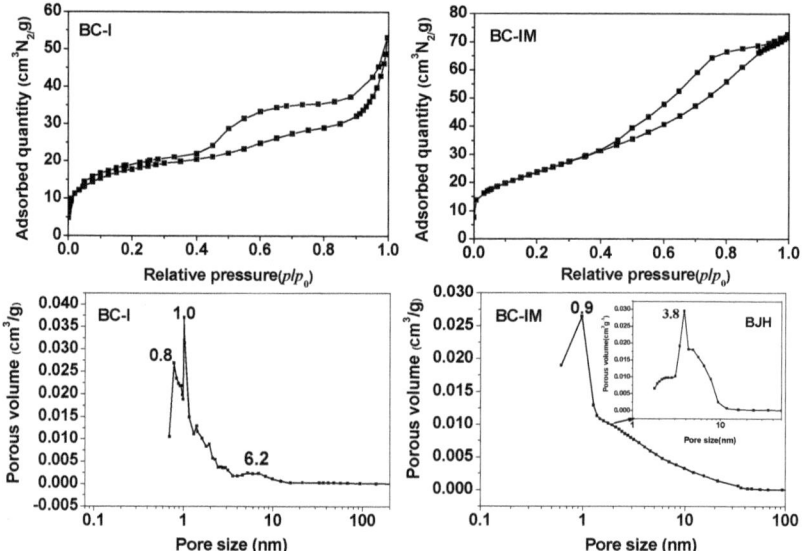

Figure 3. Nitrogen adsorption–desorption isotherms and Barret–Joyner–Halenda (BJH) or Horvath–Kawazoe (HK) pore size distribution curves of BC-I and BC-IM samples.

Table 1. Physicochemical properties of all the adsorbents.

Catalyst	S_{BET} (m^2/g)	Pore Size (nm)	Pore Volume (cm^3/g)
BC	1.4	\	0.001
BC-I	63.0	0.8/1.0/6.2	0.082
BC-IM	84.7	0.9/3.8	0.112

The detailed pore characteristics can be further observed by the pore size distribution curves. The BC-I sample contained pores 0.7–17 nm in size with concentrated pore sizes of 0.8, 1, and 6.2 nm. The BET surface area and pore volume of BC-I were 63.0 m^2/g and 0.082 cm^3/g, respectively. Li et al. had also synthesized a series of bamboo hydrochars with BET surface areas ranging from 2.6 to 43.1 m^2/g [38]. Therefore, it is obvious that the impregnation method using NaOH can effectively remove inorganic ash in the raw BC sample and improve the surface area and pore structure. After the magnetizing treatment, the BC-IM showed a large number of micropores centered at 0.9 nm, partial mesopores centered at 3.8 nm, and small number of macropores. The dimensions of the MB molecule in water is reported as 0.400 × 0.793 × 1.634 nm; Santoso et al. reported that micropores and mesopores less than 6 nm are crucial to increase the amount of adsorbed MB molecules [39]. Furthermore, the BC-IM sample had a higher specific area (84.7 m^2/g) and larger pore volume (0.112 cm^3/g). Therefore, a more suitable pore size, higher surface area, and larger pore volume will make the BC or BC-IM samples have more excellent adsorption capacity for MB.

Figure 4 shows the FT-IR diagrams of BC, BC-I, and BC-IM. Compared to the raw BC, the stretching vibration peak of O-H at 3497 cm^{-1} and C=C bond of the vibration of the sp^2 carbon skeletal network at 1640 cm^{-1} were significantly enhanced in the NaOH-impregnated BC-I sample [40]. After magnetization, more surface functional groups appeared. The obvious bands between 1408 and 1662 cm^{-1} can be ascribed to C=O stretching modes in aromatic ring structures or ring vibrations in a large aromatic skeleton [41,42]. In addition, the band around 760 cm^{-1} is related to the C-C and C-H stretching vibration in aromatic rings [43]. The band at 1020 cm^{-1} is attributed to the lactone structure of C-O-C [25]. The unique band observed at about 517 cm^{-1} in the BC-IM sample is assigned to the stretching vibration of Fe-O, which proves the appearance of iron oxide [44]. This indicates that the oxygen-containing functional groups on the surfaces of BC-I and BC-IM can be greatly increased by modification.

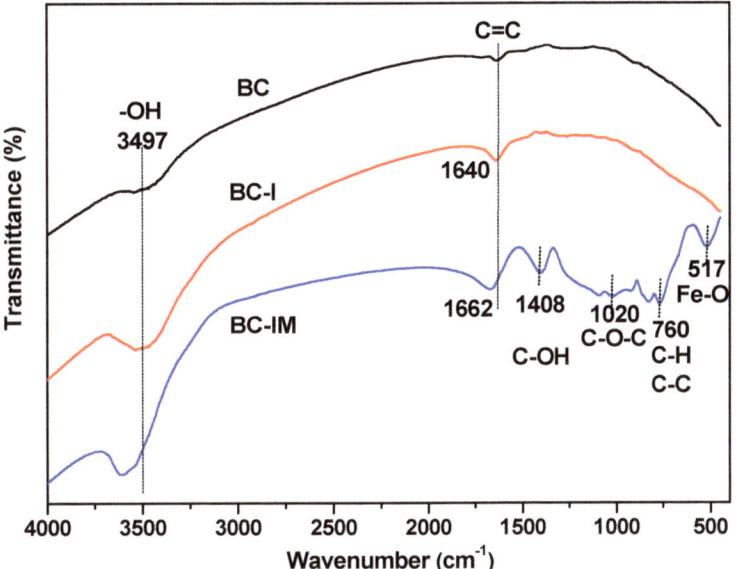

Figure 4. FT-IR spectra of BC, BC-I, and BC-IM.

TEM, HRTEM, and SAED images and VSM magnetization curves of the BC-IM sample are shown in Figure 5. Apparently, there are many agglomerated particles on the surface of the rod-like bamboo charcoal structure with diameter of about 65 nm, owing to the load of iron oxide (Figure 5a). The d-spacing of 0.649 nm was observed in the HR-TEM patterns, which corresponds to the (200) lattice planes of the Fe$_2$O$_3$ phase. It proves that Fe$_2$O$_3$ was successfully loaded on the surface of BC-IM, which is in accordance with the XRD results. The corresponding SAED pattern exhibited a polycrystalline structure pattern. The hysteresis loop at room temperature of BC-IM is shown in Figure 6d. The absence of residual matter and coercivity in the magnetic ring proves that the synthesized BC-IM is superparamagnetic. In addition, the saturation magnetization of BC-IM was 16.4 emu/g. It shows that BC-IM can be easily separated from the solution [45].

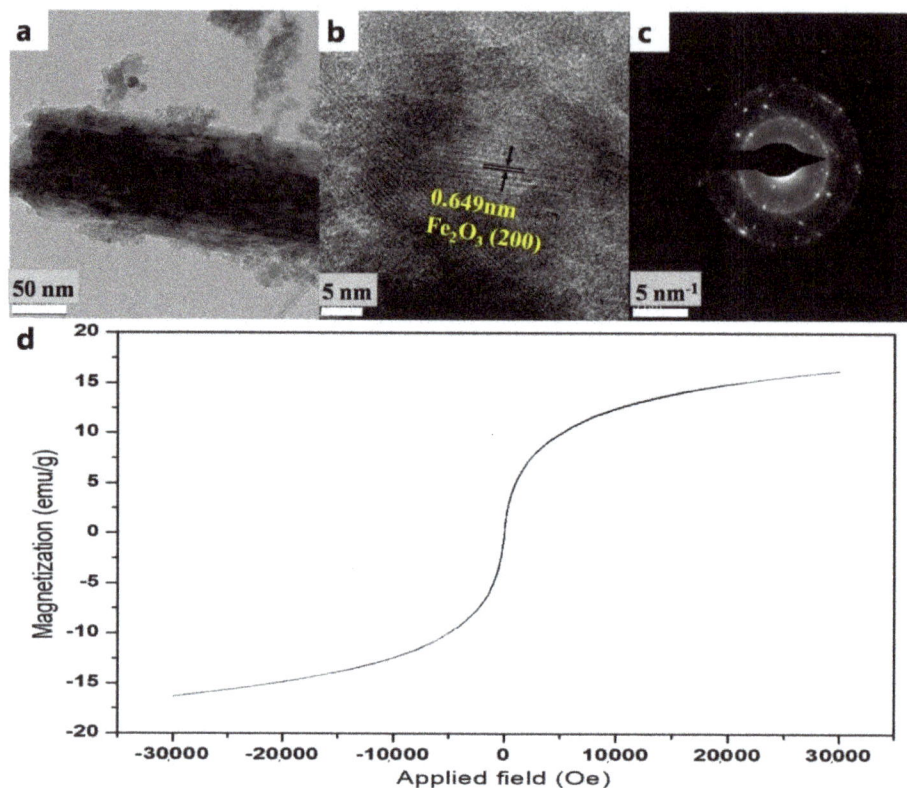

Figure 5. TEM (**a**), HRTEM (**b**), SAED (**c**) images and VSM magnetization curves (**d**) of BC-IM sample.

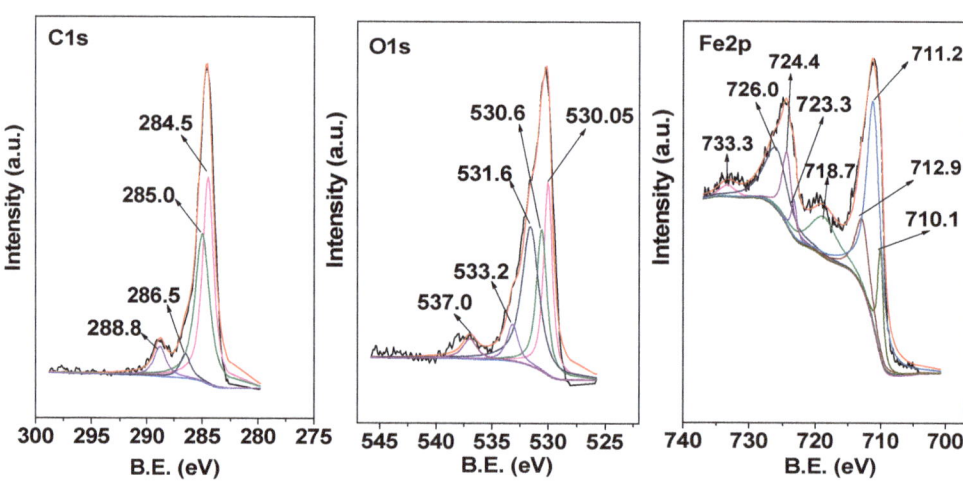

Figure 6. C1s, O1s, and Fe2p spectra of BC-IM.

The XPS technique was also employed to explore the surface species of BC-IM (Figure 6). The C1s spectrum can be fitted into four peaks. The binding energy peak at 288.8 eV of C1s contains both CO_3^{2-} and –COOH(R) oxidation states. The peaks at 286.5,

285.0, and 284.5 eV belong to the functional groups C=O, C-O and C-C, respectively [21,22]. The O1s spectrum can be fitted into five peaks located at the binding energy of 530.05, 530.6, 531.6, 533.2, and 537.0 eV, which are ascribed to Fe-O-Fe, C-O, C=O/Fe-OH, COOH(R), and O/H_2O functional groups [46,47]. All the above analysis indicates that there are abundant functional groups in BC-IM, which is consistent with the FT-IR results. Next, the Fe2p peaks could be deconvoluted into some characteristic peaks, among which the four peaks at 726.0, 724.4, 712.9, and 711.2 eV belong to Fe^{3+} and the two peaks at 723.3 and 710.1 eV belong to Fe^{2+} [47,48]. The results showed that BC-IM has multivalent iron elements.

3.2. Adsorption Properties of Samples

The Performances of the synthesized adsorbents were measured for the adsorption of methylene blue. As can be seen from Figure 7a, the adsorption capacity of the BC-IM composites was superior to that of BC-I and BC. For BC-IM, the MB removal efficiency reached 83% after 11 h, which was higher than that of BC-I (69%) and BC (60%). MB adsorption by the three adsorbents increased rapidly in the first 10 h, then decreased gradually until the equilibrium state was reached at around 58 h. The removal efficiency at equilibrium of BC, BC-I, and BC-IM were 98%, 93%, and 88%, respectively. Combined with the above characterization results, impregnation and magnetization can enlarge the specific surface area of bamboo charcoal, improve the pore structure, and form more surface oxygen-containing functional groups, which may be favorable factors for the improvement of adsorption capacity. Figure 7a shows the images of the solutions before and after the reaction, showing that BC-IM has a good ability to treat methylene blue in water and can be easily recovered under external magnetic action. As can be seen from the figure, the solution was relatively clear after reaching the adsorption equilibrium, and the BC-IM sample could be recovered by an external magnetic field.

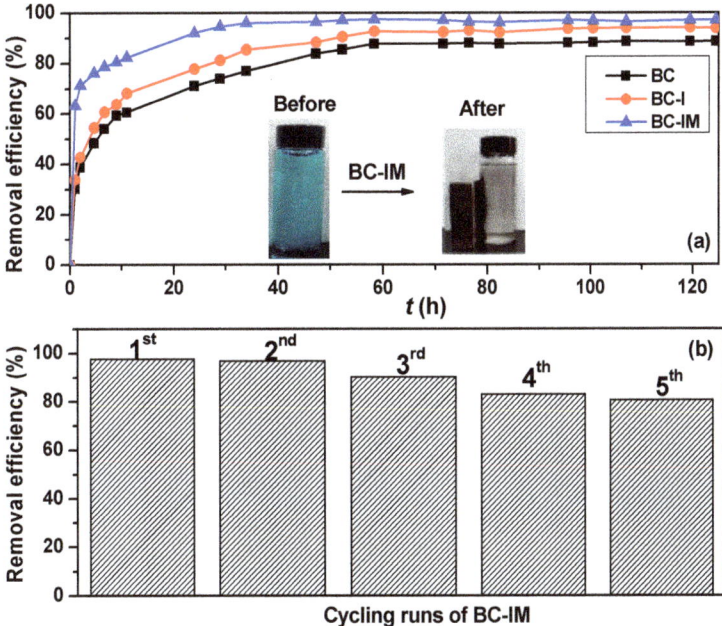

Figure 7. Plots of removal efficiency versus contact time for BC, BC-I, and BC-IM (**a**). Cycling runs for MB adsorption by BC-IM (**b**). Reaction conditions: MB (10 mg/L, 100 mL), contact temperature (308 K), adsorbent dose (0.1 g).

The recycling of biochar used for MB removal was investigated using BC-IM and the results shown in Figure 7b. The adsorption amounts were gradually decreased during the cycling process, which may be attributed to the blockage of pores by adsorbed MB molecules. The removal efficiency of 80% after five cycles showed that the BC-IM sample can act as a potential adsorbent for MB removal.

The point of zero charge (pHpzc) of samples can be determined using a previously reported method [7]. As can be seen in Figure 8, the pHpzc of the three adsorbents were all around 9.4. This indicates that at pH above 9.4, the surface of the adsorbent is considered to have a positive charge, which can adsorb cationic species (like MB). Therefore, the effect of pH on MB removal by BC-IM was examined by varying the pH values from 5 to 12. The initial MB concentration was 10 mg/L at room temperature and reaction lasted for 24 h. At pH > pHpzc, there was an electrostatic attraction between the negative loads from the adsorbent surface and the positive loads from the MB molecules. Therefore, higher removal of MB was achieved in the pH range of 10–12 (Figure 8d). Meanwhile, the electrostatic repulsion, chemical reaction, and Fe leaching in acidic conditions led to the very poor adsorption efficiency of BC-IM.

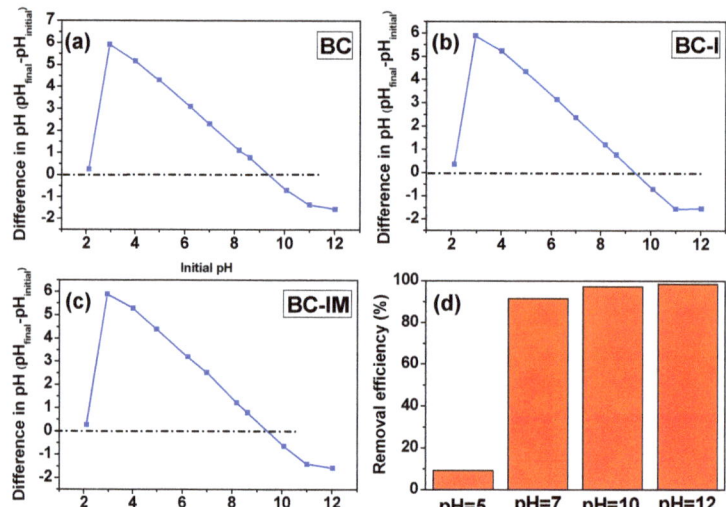

Figure 8. pHpzc of BC, BC-I, and BC-IM (**a**–**c**) and effect of pH on MB removal by BC-IM (**d**).

3.3. Adsorption Kinetics, Adsorption Isotherm, and Adsorption Thermodynamics

To explain the MB adsorption mechanism, pseudo-first-order, pseudo-second-order, and intra-particle diffusion models were first applied to explain the MB adsorption kinetics using the raw BC, BC-I, and BC-IM samples. The equations of these kinetic models are as follows [22,43,49]:

$$\lg(Q_e - Q_t) = \lg Q_e - k_1 t \tag{3}$$

$$\frac{t}{Q_t} = \frac{1}{k_2 Q_e^2} + \frac{t}{Q_e} \tag{4}$$

$$Q_t = k_{ip} t^{1/2} + C_i \tag{5}$$

where k_1 (h^{-1}) and k_2 (g/(mg·h)) obtained from the linear fitting slope of the equations are the adsorption rate constant of the pseudo-first-order or pseudo-second-order models, respectively. k_{ip} (mg/(g·h$^{0.5}$)) is the rate constant of the intra-particle diffusion model and C_i (mg/g) represents the intercept reflecting the boundary layer effect.

The effect of contact time on MB adsorption capacity is shown in Figure 9. It can be seen that adsorption capacity increased with the contact time. The corresponding fitting results and kinetic parameters are shown in Figure 10 and Table 2. A higher correlation coefficient value (R^2) was observed in the pseudo-second-order model for the three adsorbents, indicating that the adsorption system conforms to the pseudo-second-order kinetic model. The basic assumption of the pseudo-second-order model is that chemisorption is the rate-controlled step and chemisorption originates from valence forces between adsorbate and adsorbent by electron exchange or sharing [50]. Therefore, chemisorption is the rate-determining step for MB adsorption, limiting the mass transfer induced participation in the solution [51].

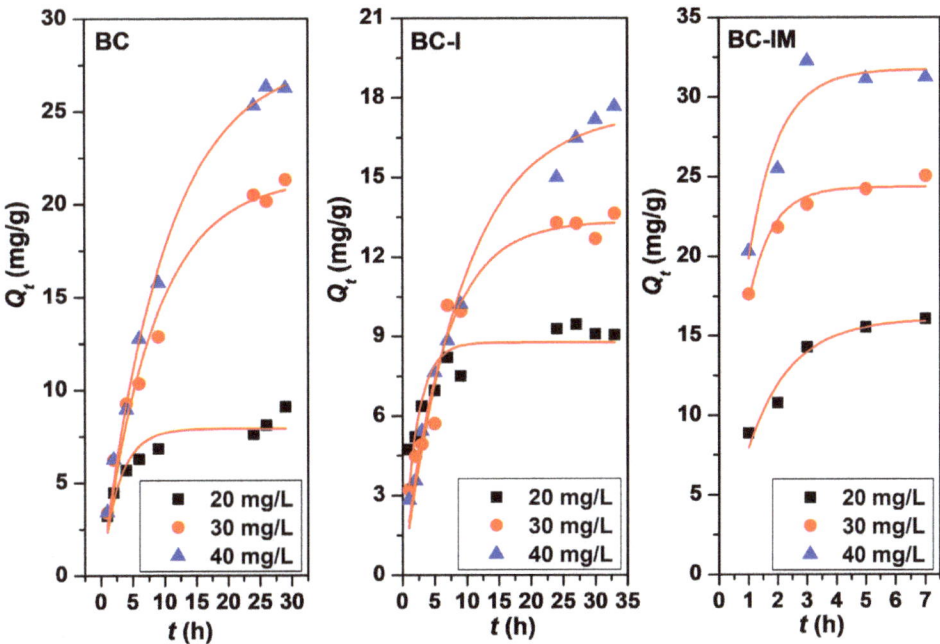

Figure 9. The influence of contact time at 308 K for MB adsorption by BC, BC-I, and BC-IM at different initial MB concentrations.

Then, the Langmuir, Freundlich, and Temkin adsorption isotherm models were applied to evaluate the mechanism during the adsorption process using BC, BC-I, and BC-IM at different temperature. In addition, these model equations are as follows [22,52]:

$$\frac{c_e}{Q_e} = \frac{c_e}{Q_m} + \frac{1}{Q_m k_L} \tag{6}$$

$$\ln Q_e = \ln k_F + \frac{1}{n} \ln c_e \tag{7}$$

$$Q_e = B_T \ln A_T + B_T \ln c_e \tag{8}$$

where Q_m (mg/g) represents the amount of MB at the complete cover state, which displays the maximum adsorption capacity; and k_L (L/mg) is the Langmuir constant related to the adsorption energy; k_F is the Freundlich constant related to adsorption capability, and $1/n$ represents the adsorption intensity; and A_T (L/mg) represents the maximum binding energy and B_T (mg/g) is the heat change during the adsorption, which are the Temkin model constants.

Adsorption isotherms of MB on the three samples at different temperature are shown in Figure 11. Apparently, MB adsorption capacity of all samples increased as the MB concentration increased, resulting from the increased mass transfer driving force during adsorption. The adsorption isotherms were nearly linear, which suggested a complex adsorption mechanism involving partitioning for the adsorption of MB by the bamboo charcoals [38].

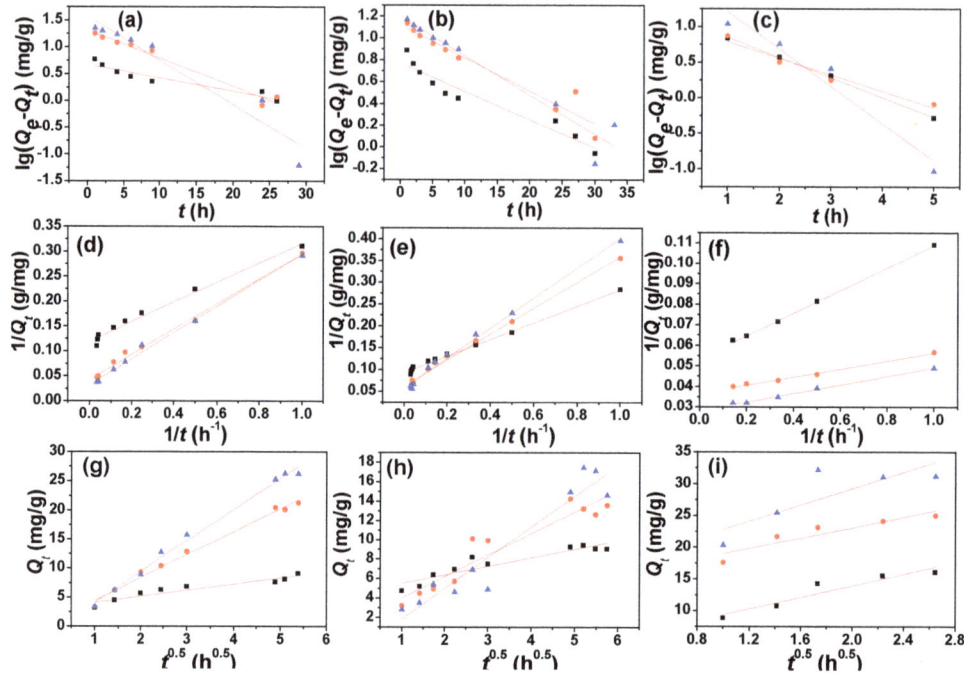

Figure 10. Kinetic model for MB adsorption: the pseudo-first-order model for BC (**a**), BC-I (**b**), and BC-IM (**c**); the pseudo-second-order model for BC (**d**), BC-I (**e**), and BC-IM (**f**); the intra-particle diffusion model for BC (**g**), BC-I (**h**), and BC-IM (**i**) (■ 20 mg/L; ● 30 mg/L; ▲ 40 mg/L).

Table 2. Kinetic model constants and correlation coefficients for MB adsorption by BC, BC-I and BC-IM.

Adsorbents	c_o (mg/L)	Pseudo-First-Order		Pseudo-Second-Order		Intra-Particle Diffusion	
		k_1 (h^{-1})	R^2	k_2 (g/(mg·h))	R^2	k_{ip} (mg/(g·h$^{0.5}$))	R^2
BC	20	0.0246	0.8843	0.0735	0.9812	1.0560	0.8909
	30	0.0522	0.9723	0.0074	0.9928	3.9489	0.9893
	40	0.0823	0.9145	0.0041	0.9944	5.3319	0.9916
BC-I	20	0.0260	0.9209	0.0458	0.9931	0.8832	0.8515
	30	0.0302	0.9281	0.0134	0.9931	2.2269	0.8910
	40	0.0359	0.9261	0.0087	0.9921	3.1691	0.9070
BC-IM	20	0.0280	0.9979	0.0523	0.9990	4.5554	0.8668
	30	0.0232	0.9477	0.0699	0.9910	4.1223	0.7974
	40	0.0525	0.9301	0.0389	0.9936	6.4335	0.5708

The corresponding linear fitting results from the Langmuir, Freundlich, and Temkin models are listed in Figure 12 and Table 3. Compared with the Temkin isotherms, the Langmuir and Freundlich isotherms fit the equilibrium data well. Therefore, the Langmuir and Freundlich models can explain the adsorption mechanism of the three adsorbents. For the raw BC and BC-IM, the Langmuir isotherm model was more suitable for describing the MB adsorption process than the Freundlich model on the whole. It shows that the surface of the BC and BC-IM adsorbents is homogeneous and the adsorption takes place in the monolayer. Meanwhile, for the BC-I sample, the value of R^2 of the Langmuir isotherm model became lower than that of the Freundlich model with the increase in adsorption temperature, indicating that the Freundlich model has higher accuracy. Moradi et al. had also reported the transition of the adsorption isotherm to the Freundlich model at higher temperatures [53]. It shows that the heterogeneous surface may play an important role in multi-layer adsorption for MB removal [22]. Moreover, the 1/n values in the Freundlich isotherm varied from 0.1 to 1, which shows that the sorption process is favorable [54]. The maximum adsorption capacity (Qm) of BC, BC-I, and BC-IM obtained by the Langmuir isotherm model at 328 K was 135.13, 220.26, and 497.51 m^2/g, respectively. Compared to the adsorption capacity for MB from numerous adsorbents reported in the literature (36.25~384.61 mg/g) [43], bamboo charcoal can be used as efficient adsorbents of MB from wastewater.

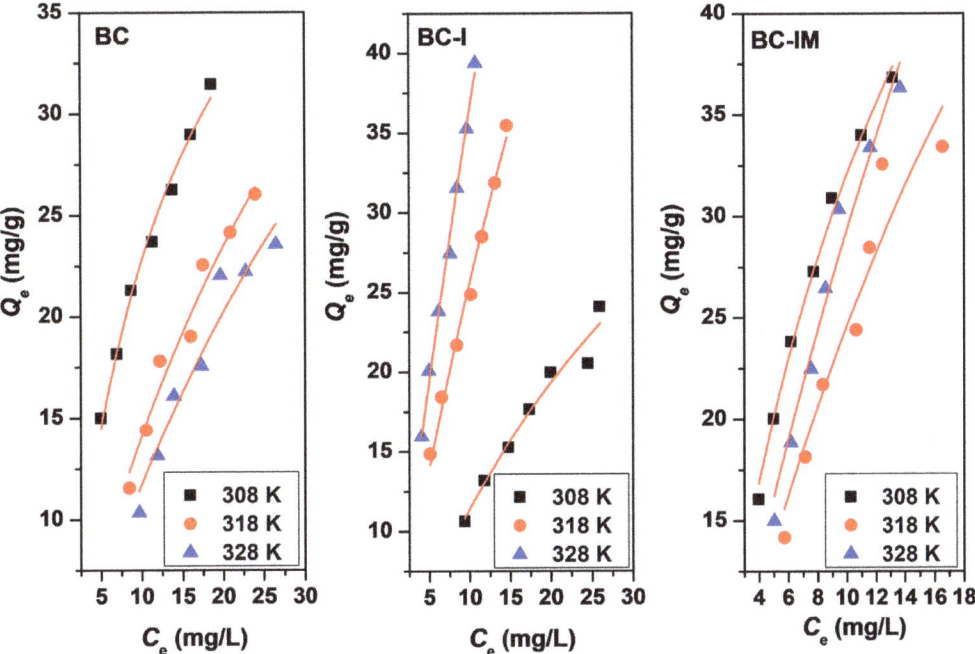

Figure 11. Experimental isotherm data of MB adsorption by BC, BC-I, and BC-IM at different temperatures.

Finally, the thermodynamic parameters of MB adsorption by the three bamboo charcoals including the Gibbs free energy change (ΔG^0), enthalpy change (ΔH^0), and entropy change (ΔS^0) were calculated from the following equations:

$$\Delta G^0 = -RT \ln K_c \tag{9}$$

$$K_c = \frac{Q_e}{c_e} \tag{10}$$

$$\ln K_c = \frac{\Delta S^0}{R} - \frac{\Delta H^0}{RT} \tag{11}$$

where K_c (L/g) represents the equilibrium constant for the MB adsorption process. The ΔH^0 and ΔS^0 were evaluated by plotting $\ln K_c$ against $1/T$ (Figure 13).

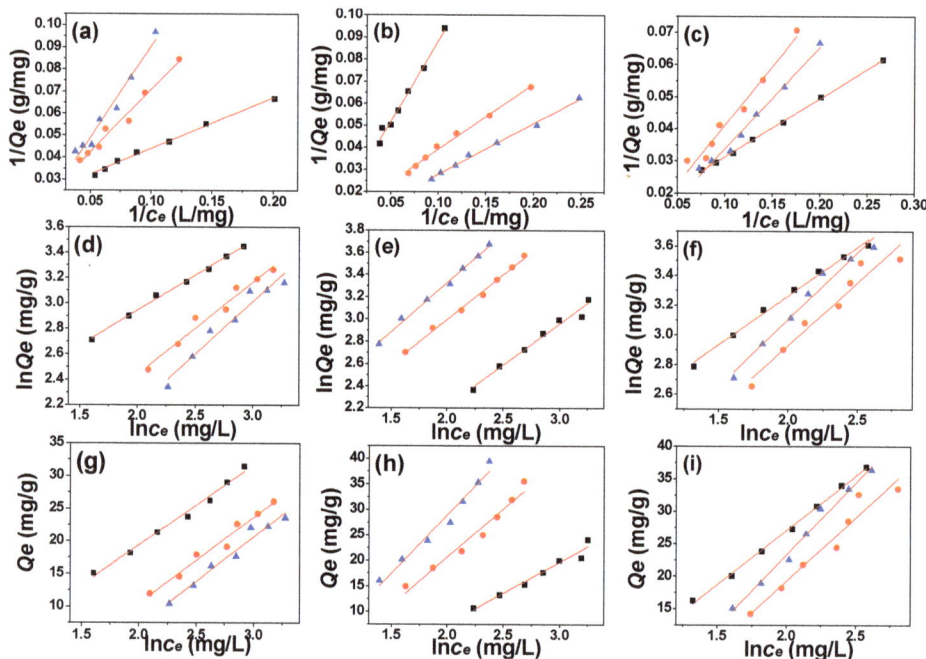

Figure 12. Isotherm model for MB adsorption: Langmuir isotherm model for BC (**a**), BC-I (**b**), and BC-IM (**c**); Freundlich isotherm model for BC (**d**), BC-I (**e**), and BC-IM (**f**); Temkin isotherm model for BC (**g**), BC-I (**h**), and BC-IM (**i**) (■ 308 K; ● 318 K; ▲ 328 K).

Table 3. Langmuir, Freundlich, and Temkin isotherm model constants and correlation coefficients for adsorption of MB by BC, BC-I, and BC-IM.

Adsorbent	T (K)	Langmuir			Freundlich			Temkin		
		Q_m (mg/g)	k_L (L/mg)	R^2	k_F	$1/n$	R^2	A_T (L/mg)	B_T (mg/g)	R^2
BC	308	48.92	0.08777	0.9937	6.2660	0.5518	0.9964	0.65082	12.2765	0.9866
	318	71.48	0.02473	0.9741	2.6739	0.7261	0.9641	0.29622	13.2230	0.9695
	328	135.13	0.00894	0.9633	1.7027	0.8259	0.9424	0.22519	13.6403	0.9593
BC-I	308	65.75	0.02086	0.9866	2.0719	0.7433	0.9726	0.25067	12.1212	0.9571
	318	107.76	0.03123	0.9917	3.9911	0.8051	0.9936	0.40348	18.9115	0.9594
	328	220.26	0.01968	0.9929	4.7295	0.8861	0.9936	0.47785	22.9148	0.9703
BC-IM	308	77.46	0.07074	0.9992	6.9693	0.6600	0.9920	0.67924	16.7576	0.9967
	318	227.79	0.01196	0.9671	3.4210	0.8490	0.9354	0.36229	19.5187	0.9381
	328	497.51	0.00633	0.9831	3.6430	0.9058	0.9689	0.38632	22.1840	0.9841

The values of ΔG^0, ΔH^0, and ΔS^0 are summarized in Table 4. All the ΔG^0 values were negative for the three adsorbents at different temperatures, which indicates that MB adsorption is spontaneous and does not require external energy input. Generally, the physical adsorption energy is from 0 to −20 kJ/mol, while that of chemical adsorption is from −80 to −400 kJ/mol [55]. The ΔG^0 values of the three adsorbents shown in Table 4 at different temperatures are in the range of −3.75~−0.97 kJ/mol, indicating that the adsorption can be regarded as physical adsorption. At the same time, the values of ΔH^0 were less than 25 kJ/mol, further proving that the MB adsorption process belongs to physical adsorption. For BC and BC-IM samples, the negative values of ΔH^0 and ΔS^0 showed that MB adsorption is a randomness decrease and exothermic process. Meanwhile, for the BC-I sample, the positive ΔH^0 and ΔS^0 values denotes that the adsorption of MB by BC-I is endothermic and the randomness is raised at the solid–liquid interface.

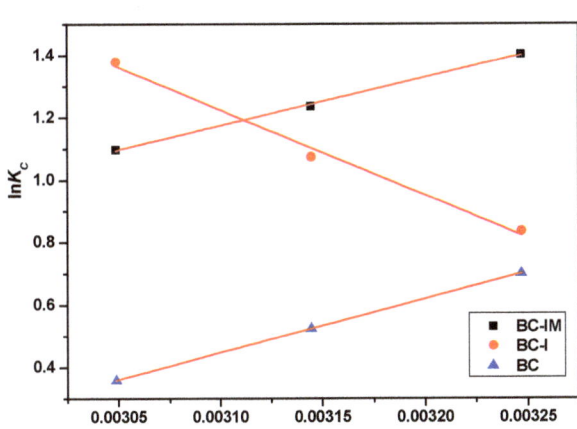

Figure 13. Thermodynamic model plots for MB adsorption by BC, BC-I, and BC-IM.

Table 4. Thermodynamic parameters for MB adsorption by BC, BC-I, and BC-IM.

Adsorbents	T (K)	K_c	ΔG^0 (kJ/mol)	ΔH^0 (kJ/mol)	ΔS^0 (J/(mol K))
BC	308	2.01	−1.79	−0.21	−0.59
	318	1.69	−1.38		
	328	1.43	−0.97		
BC-I	308	2.30	−2.13	0.33	1.17
	318	2.92	−2.83		
	328	3,96	−3.75		
BC-IM	308	4.06	−3.59	−0.18	−0.43
	318	3.44	−3.27		
	328	3.00	−2.99		

The adsorption kinetics suggests that the obtained adsorption data fit best with the pseudo-second order model, which indicates a chemical adsorption. However, it is apparent from the thermodynamics and adsorption isotherm results that physical reaction also occurred in the MB adsorption process. In conclusion, MB adsorption by bamboo charcoals is a complex physical and chemical reaction process.

3.4. Catalytic Properties and Reusability of Samples

The catalysis experiments of the three samples were conducted at 308 K and are summarized in Figure 14a. The removal efficiency of MB in the presence of H_2O_2 by BC, BC-I, and BC-IM was 58.85%, 70.95%, and 98.43%, respectively. For comparison, the pure adsorption experiment was also performed. The removal efficiency of MB in the absence of H_2O_2 by BC, BC-I, and BC-IM was 38.65%, 42.57%, and 71.46%, respectively. Meanwhile, the degradation efficiency was less than 14% only in the presence of hydrogen peroxide and no catalyst. The significant improvement of removal efficiency after adding hydrogen peroxide indicates that Fenton-like catalysis also plays an important role in methylene blue removal, besides adsorption. Pseudo first-order kinetics was used to obtain the kinetic rates of MB degradation (Figure 14b). The fitted rate constants at 308 K of BC, BC-I, and BC-IM were 0.198 h^{-1}, 0.351 h^{-1}, and 1.542 h^{-1}, respectively. Apparently, the degradation capabilities of the modified bamboo charcoals were higher than that of the raw BC. In particular, the BC-IM sample showed the best rate constants, which was also better than other Fenton-like catalytic systems used in methylene blue degradation reactions (Table 5).

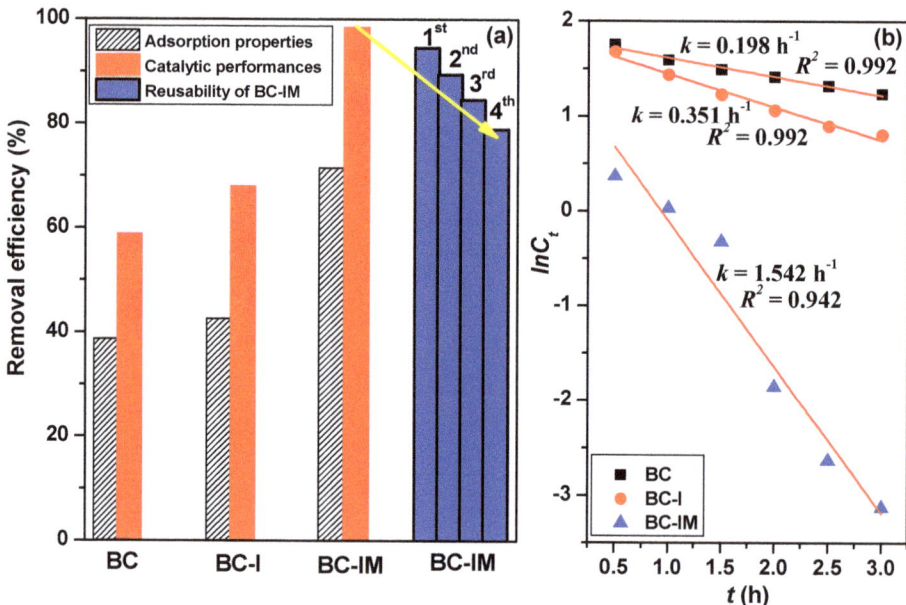

Figure 14. Removal efficiency of MB by the three samples in the absence or presence of H_2O_2 at 2 h. The reusability of BC-IM (**a**) and the pseudo-first-order rate constant k (h^{-1}) was calculated from the slope of lnC_t versus reaction time (**b**). Reaction conditions: MB (10 mg/L, 100 mL), sample dose (0.1 g), H_2O_2 (30 wt%, 20 mL), room temperature.

Table 5. Comparison of catalytic performance of BC-IM for MB degradation with cited literature.

Catalyst	Degradation Time and Temperature	Degradation Efficiency	k (h^{-1})	Reference
$Ca_{0.5}Pb_{0.4}Yb_{0.1}Zn_{1.0}Fe_{11}O_{19}$	120 min	96.1%	1.0788	[56]
$Ca_{0.5}Pb_{0.4}Dy_{0.1}Ni_{1.0}Fe_{11.9}O_{19}$	105 min	85%	0.801	[57]
$CoMn_{30}Ce_{10}$	360 min, 308 K	90%	0.251	[1]
CuO	210 min, 308 K	96%	0.8112	[58]
BC-IM	120 min, 308 K	98.43%	1.542	This work

The reusability of the best sample (BC-IM) is also shown in Figure 14a. The removal efficiency after four cycles was 94.6%, 89.5%, 84.7%, and 79.0%, respectively. The loss of activity may be caused by the slight leaching of iron ions from the catalyst and the surface active site being blocked by byproducts. The obtained results showed that the Fenton process can stably remove most of the MB in dye-polluted waters, suggesting that the BC-IM sample can act as a potential Fenton catalyst for the treatment of actual wastewater.

3.5. Mechanism Study

The MB degradation process is proposed to be a synergistic mechanism involving adsorption and radical processes. For the raw BC sample, the MB and H_2O_2 molecules are first adsorbed on the surface of BC, the persistent free radicals (PFRs) existed on the biomass charcoal will facilitate the decomposition of H_2O_2 into hydroxyl radical [59], which can further attack and degrade MB molecules. Finally, the byproducts produced by degradation are desorbed. As described above, a simple NaOH impregnation process can significantly optimize the crystal structure, pore size distribution, and surface functional groups and increase the specific surface area from 1.4 to 63.0 m^2/g. Therefore, the enhancement of degradation efficiency is related to these factors, which may produce more active sites.

Further magnetization of the BC-I sample by one-pot in situ precipitation can not only continuously improve physicochemical properties and facilitate the recycling convenience, but also introduce Fe oxide as a new activator (Equations (1) and (2)).

$$Fe^{2+} + H_2O_2 \rightarrow Fe^{3+} + \cdot OH + {}^-OH \quad (12)$$

$$OH + MB \rightarrow intermediates \rightarrow CO_2 + H_2O \quad (13)$$

Therefore, the monolayer adsorbed MB on BC and BC-IM or multilayer adsorbed MB on BC-I would be degraded by the hydroxyl radical originating from PFRs and Fe oxide-activated H_2O_2. Then, the active sites on the surface of BC-IM can be reused to adsorb MB molecules again, which could improve the MB removal capacity. Magnetic bamboo charcoal can offer a potential opportunity in the field of organic pollutants treatment.

4. Conclusions

In this study, a simple and feasible modified strategy of bamboo charcoal without secondary high temperature pyrolysis was studied. In particular, the magnetic modification not increases the maximum sorption capacity and degradation rate constants of BC-IM 3.7 times or 7.8 times that of the raw BC, but also makes it more convenient to quickly recover under an external magnetic field. The pseudo-first-order rate constants k at room temperature for BC, BC-I, and BC-IM for catalytic degradation in the presence of H_2O_2 were 0.198, 0.351, and 1.542 h^{-1}, respectively. A synergistic mechanism involving adsorption and radical processes was proposed, which is expected to act as an adsorption/catalytic bifunctional catalyst in organic pollutants treatment. The findings provide insight into the MB adsorption/catalytic degradation mechanism by bamboo charcoal, and gives guidance to the design of modification strategies for biochars.

Author Contributions: Conceptualization, Q.L. and C.-X.L.; methodology, Q.L. and L.-Y.Z.; validation, Q.L., W.-Y.D., W.-W.J., R.-X.S. and B.Z.; formal analysis, L.-Y.Z. and S.-W.L.; investigation, W.-W.J., R.-X.S. and B.Z.; writing—original draft preparation, Q.L., W.-Y.D. and C.-X.L.; writing—review and editing, Q.L., L.-M.L. and C.-X.L.; supervision, X.-G.H. and C.-X.L.; project administration, X.-G.H. and C.-X.L.; funding acquisition, Q.L., L.-M.L., S.-W.L. and X.-G.H. All authors have read and agreed to the published version of the manuscript.

Funding: This work was supported by the National Natural Science Foundation of China (22064010, 51862014), Foundation of Jiangxi Educational Committee (GJJ200430), and Jiangxi Province Key Research and Development Project (20192BBEL50029).

Institutional Review Board Statement: Not applicable.

Informed Consent Statement: Not applicable.

Data Availability Statement: Not applicable.

Conflicts of Interest: The authors declare no conflict of interest.

References

1. Wang, Q.W.; Deng, W.Y.; Lin, X.C.; Huang, X.G.; Wei, L.; Gong, L.; Liu, C.X.; Liu, G.B.; Liu, Q. Solid-state preparation of mesoporous Ce-Mn-Co ternary mixed oxide nanoparticles for the catalytic degradation of methylene blue. *J. Rare Earth.* **2021**, *31*, 826–834. [CrossRef]
2. Fu, W.; Zhang, W. Microwave-enhanced membrane filtration for water treatment. *J. Membr. Sci.* **2018**, *568*, 97–104. [CrossRef]
3. Zupko, R.; Kamath, D.; Coscarelli, E.; Rouleau, M.; Minakata, D. Agent-Based model to predict the fate of the degradation of organic compounds in the aqueous-phase UV/H$_2$O$_2$ advanced oxidation process. *Process. Saf. Environ.* **2020**, *136*, 49–55. [CrossRef]
4. Malik, M.; Len, T.; Luque, R.; Osman, S.M.; Paone, E.; Khan, M.I.; Wattoo, M.A.; Jamshaid, M.; Anum, A.; Rehman, A.U. Investigation on synthesis of ternary g-C$_3$N$_4$/ZnO-W/M nanocomposites integrated heterojunction II as efficient photocatalyst for environmental applications. *Environ. Res.* **2023**, *217*, 114621. [CrossRef] [PubMed]
5. Jamshaid, M.; Khan, M.I.; Fernandez, J.; Shanableh, A.; Hussaine, T.; Rehman, A.U. Synthesis of Ti^{4+} doped Ca-BiFO$_3$ for the enhanced photodegradation of moxifloxacin. *New J. Chem.* **2022**, *46*, 19848. [CrossRef]

6. Haounati, R.; Alakhras, F.; Ouachtak, H.; Saleh, T.A.; Al-Mazaideh, G.; Alhajri, E.; Jada, A.; Hafifid, N.; Addi, A.A. Synthesized of zeolite@Ag$_2$O nanocomposite as superb stability photocatalysis toward hazardous rhodamine B dye from water. *Arab. J. Sci. Eng.* **2023**, *48*, 169–179. [CrossRef]
7. Ouachtak, H.; Guerdaoui, A.E.; Haounati, R.; Akhouairi, S.; Haouti, R.E.; Hafifid, N.; Addi, A.A.; Šljukic, B.; Santos, D.M.F.; Taha, M.L. Highly effifficient and fast batch adsorption of orange G dye from polluted water using superb organo-montmorillonite: Experimental study and molecular dynamics investigation. *J. Mol. Liq.* **2021**, *335*, 116560. [CrossRef]
8. Osagie, C.; Othmani, A.; Ghosh, S.; Malloum, A.; Esfahani, Z.K.; Ahmadi, S. Dyes adsorption from aqueous media through the nanotechnology: A review. *J. Mater. Res. Technol.* **2021**, *14*, 2195–2218. [CrossRef]
9. Li, T.; Ge, L.; Peng, X.; Wang, W.; Zhang, W. Enhanced degradation of sulfamethoxazole by a novel Fenton-like system with significantly reduced consumption of H_2O_2 activated by g-C_3N_4/MgO composite. *Water Res.* **2021**, *190*, 116777. [CrossRef]
10. Li, X.P.; Wang, C.B.; Zhang, J.G.; Liu, J.P.; Liu, B.; Chen, G.Y. Preparation and application of magnetic biochar in water treatment: A critical review. *Sci. Total Environ.* **2020**, *711*, 134847. [CrossRef]
11. Oleszczuk, P.; Godlewska, P.; Reible, D.; Kraska, P. Bioaccessibility of polycyclic aromatic hydrocarbons in activated carbon or biochar amended vegetated (Salix viminalis) soil. *Environ. Pollution.* **2017**, *227*, 406–413. [CrossRef] [PubMed]
12. Ji, B.; Wang, J.; Song, H.; Chen, W. Removal of methylene blue from aqueous solutions using biochar derived from a fallen leaf by slow pyrolysis: Behavior and mechanism. *J. Environ. Chem. Eng.* **2019**, *7*, 103036. [CrossRef]
13. Sun, L.; Wan, S.; Luo, W. Biochars prepared from anaerobic digestion residue, palm bark, and eucalyptus for adsorption of cationic methylene blue dye: Characterization, equilibrium, and kinetic studies. *Bioresour. Technol.* **2013**, *140*, 406–413. [CrossRef] [PubMed]
14. Wu, F.C.; Tseng, R.L. High adsorption capacity NaOH-activated carbon for dye removal from aqueous solution. *J. Hazard. Mater.* **2008**, *152*, 1256–1267. [CrossRef]
15. Jia, Z.; Zeng, W.; Xu, H.; Li, S.; Peng, Y. Adsorption removal and reuse of phosphate from wastewater using a novel adsorbent of lanthanum-modified platanus biochar. *Process. Saf. Environ.* **2020**, *140*, 221–232. [CrossRef]
16. Anastopoulos, I.; Pashalidis, I.; Hosseini-Bandegharaei, A.; Giannakoudakis, D.A.; Robalds, A.; Usman, M.; Escudero, L.B.; Zhou, Y.; Colmenares, J.C.; Núñez-Delgado, A.; et al. Agricultural biomass/waste as adsorbents for toxic metal decontamination of aqueous solutions. *J. Mol. Liq.* **2019**, *295*, 111684. [CrossRef]
17. Popa, N.; Visa, M. The synthesis, activation and characterization of charcoal powder for the removal of methylene blue and cadmium from wastewater. *Adv. Powder Technol.* **2017**, *28*, 1866–1876. [CrossRef]
18. Lalhruaitluanga, H.; Prasad, M.N.V.; Radha, K. Potential of chemically activated and raw charcoals of Melocannabaccifera for removal of Ni(II) and Zn(II) from aqueous solutions. *Desalination* **2011**, *271*, 301–308. [CrossRef]
19. Islama, M.A.; Ahmed, M.J.; Khanday, W.A.; Asif, M.; Hameed, B.H. Mesoporous activated carbon prepared from NaOH activation of rattan (Lacospermasecundiflorum) hydrochar for methylene blue removal. *Ecotox. Environ. Safe.* **2017**, *138*, 279–285. [CrossRef]
20. Oginni, O.; Singh, K.; Oporto, G.; Dawson-Andoh, B.; McDonald, L.; Sabolsky, E. Influence of one-step and two-step KOH activation on activated carbon characteristics. *Bioresour. Technol.* **2019**, *7*, 100266. [CrossRef]
21. Xu, H.; Shen, B.; Yuan, P.; Lu, F.; Tian, L.; Zhang, X. The adsorption mechanism of elemental mercury by HNO_3-modified bamboo char. *Fuel Process. Technol.* **2016**, *154*, 139–146. [CrossRef]
22. Wu, C.; Li, L.; Zhou, H.; Ai, J.; Zhang, H.; Tao, J.; Wang, D.; Zhang, W. Effects of chemical modification on physicochemical properties and adsorption behavior of sludge-based activated carbon. *J. Environ. Sci.* **2021**, *100*, 340–352. [CrossRef] [PubMed]
23. Guo, F.; Peng, K.; Liang, S.; Jia, X.; Jiang, X.; Qian, L. One-step synthesis of biomass activated char supported copper nanoparticles for catalytic cracking of biomass primary tar. *Energy* **2019**, *180*, 584–593. [CrossRef]
24. Thines, K.R.; Abdullah, E.C.; Mubarak, N.M.; Ruthiraan, M. Synthesis of magnetic biochar from agricultural waste biomass to enhancing route for waste water and polymer application: A review. *Renew. Sust. Energ. Rev.* **2017**, *67*, 257–276. [CrossRef]
25. Hao, Z.; Wang, C.; Yan, Z.; Jiang, H.; Xu, H. Magnetic particles modification of coconut shell-derived activated carbon and biochar for effective removal of phenol from water. *Chemosphere* **2018**, *211*, 962–969. [CrossRef]
26. Zhang, Q.; Wang, Y.; Wang, Z.; Zhang, Z.; Wang, X.; Yang, Z. Active biochar support nano zero-valent iron for efficient removal of U(VI) from sewage water. *J. Alloy Compd.* **2021**, *852*, 156993. [CrossRef]
27. Liu, J.W.; Jiang, J.G.; Meng, Y.; Aihemaiti, A.; Xu, Y.W.; Xiang, H.L.; Gao, Y.; Chen, X.C. Preparation, environmental application and prospect of biochar-supported metal nanoparticles: A review. *J. Hazard. Mater.* **2020**, *388*, 122026. [CrossRef]
28. Khataee, A.; Kalderis, D.; Gholami, P.; Fazli, A.; Moschogiannaki, M.; Binas, V.; Lykaki, M.; Konsolakis, M. Cu_2O-CuO@biochar composite: Synthesis, characterization and its efficient photocatalytic performance. *Appl. Surf. Sci.* **2019**, *498*, 143846. [CrossRef]
29. Zhai, S.; Li, M.; Wang, D.; Zhang, L.; Yang, Y.; Fu, S. In situ loading metal oxide particles on bio-chars: Reusable materials for efficient removal of methylene blue from wastewater. *J. Clean. Prod.* **2019**, *220*, 460–474. [CrossRef]
30. Mian, M.M.; Liu, G.; Fu, B.; Song, Y. Facile synthesis of sludge-derived MnO_x-N-biochar as an efficient catalyst for peroxymonosulfate activation. *Appl. Catal. B Environ.* **2019**, *255*, 117765. [CrossRef]
31. Yi, Y.; Huang, Z.; Lu, B.; Xian, J.; Tsang, E.P.; Cheng, W.; Fang, J.; Fang, Z. Magnetic biochar for environmental remediation: A review. *Bioresour. Technol.* **2020**, *298*, 122468. [CrossRef]
32. Oyedum, A.O.; Gebreegziabher, T.; Hui, C.W. Mechanism and modeling of bamboo pyrolysis. *Fuel Process. Technol.* **2013**, *106*, 595–604. [CrossRef]

33. Regmi, P.; Moscoso, J.L.G.; Kumar, S.; Cao, X.; Mao, J.; Schafran, G. Removal of copper and cadmium from aqueous solution using switchgrass biochar produced via hydrothermal carbonization process. *J. Environ. Manag.* **2012**, *109*, 61–69. [CrossRef]
34. Karunanayake, A.G.; Navarathna, C.M.; Gunatilake, S.R.; Crowley, M.; Anderson, R.; Mohan, D.; Perez, F.; Pittman Jr, C.U.; Mlsna, T. Fe_3O_4 Nanoparticles Dispersed on Douglas Fir Biochar for Phosphate Sorption. *ACS Appl. Nano. Mater.* **2019**, *2*, 3467–3479. [CrossRef]
35. Hou, Y.; Yan, S.; Huang, G.; Yang, Q.; Huang, S.; Cai, J. Fabrication of N-doped carbons from waste bamboo shoot shell with high removal efficiency of organic dyes from water. *Bioresour. Technol.* **2020**, *303*, 122939. [CrossRef] [PubMed]
36. Hou, Y.R.; Huang, G.G.; Li, J.H.; Yang, Q.P.; Huang, S.R.; Cai, J.J. Hydrothermal conversion of bamboo shoot shell to biochar: Preliminary studies of adsorption equilibrium and kinetics for rhodamine B removal. *J. Anal. Appl. Pyrol.* **2019**, *143*, 104694. [CrossRef]
37. Jena, L.; Soren, D.; Deheri, P.K.; Pattojoshi, P. Preparation, characterization and optical properties evaluations of bamboo charcoal. *Curr. Res. Green Sustain. Chem.* **2021**, *4*, 100077. [CrossRef]
38. Li, Y.; Meas, A.; Shan, S.; Yang, R.; Gai, X. Production and optimization of bamboo hydrochars for adsorption of Congo red and 2-naphthol. *Bioresour. Technol.* **2016**, *207*, 379–386. [CrossRef]
39. Santoso, E.; Ediati, R.; Kusumawati, Y.; Bahruji, H.; Sulistiono, D.O.; Prasetyoko, D. Review on recent advances of carbon based adsorbent for methylene blue removal from waste water. *Mater. Today Chem.* **2020**, *16*, 100233. [CrossRef]
40. Liu, S.S.; Ge, H.Y.; Wang, C.C.; Zou, Y.; Liu, J.Y. Agricultural waste/grapheme oxide 3D bio-adsorbent for highly efficient removal of methylene blue from water pollution. *Sci. Total Environ.* **2018**, *628–629*, 959–968. [CrossRef]
41. Saleh, S.; Kamarudin, K.B.; Ghani, W.A.; Kheang, L.S. Removal of organic contaminant from aqueous solution using magnetic biochar. *Procedia Eng.* **2016**, *148*, 228–235. [CrossRef]
42. Oh, W.D.; Lua, S.K.; Dong, Z.I.; Lim, T.T. Performance of magnetic activated carbon composites as peroxymonosulfate activator and regenerable adsorbent via sulfate radical-mediated oxidation processes. *J. Hazard. Mater.* **2015**, *284*, 1–9. [CrossRef] [PubMed]
43. Idohou, E.A.; Fatombi, J.K.; Osseni, S.A.; Agani, I.; Neumeyer, D.; Verelst, M.; Mauricot, R.; Aminou, T. Preparation of activated carbon/chitosan/carica papaya seeds composite for efficient adsorption of cationic dye from aqueous solution. *Surf. Interfaces* **2020**, *21*, 100741. [CrossRef]
44. Panneerselvam, P.; Morad, N.; Tan, K.A. Magnetic nanoparticle (Fe_3O_4) impregnated onto tea waste for the removal of nickel(II) from aqueous solution. *J. Hazard. Mater.* **2011**, *186*, 160–168. [CrossRef]
45. Xie, Y.; Wu, Y.; Qin, Y.; Yi, Y.; Liu, Z.; Lv, L.; Xu, M. Evalution of perchlorate removal from aqueous solution by cross-linked magnetic chitosan/poly (vinyl alcohol) particles. *J. Taiwan Inst. Chem. Eng.* **2016**, *65*, 295–303. [CrossRef]
46. Daou, T.; Begin-Colin, S.; Greneche, J.; Thomas, F.; Derory, A.; Ernhardt, P.; Legaré, P.; Pourroy, P.G. Phosphate adsorption properties of magnetite-based nanoparticles. *Chem. Mater.* **2007**, *19*, 4494–4505. [CrossRef]
47. Navarathna, C.M.; Karunanayake, A.G.; Gunatilake, S.R.; Pittman Jr, C.U.; Perez, F.; Mohan, D.; Mlsna, T. Removal of Arsenic (III) from water using magnetite precipitated onto Douglas fir biochar. *J. Environ. Manag.* **2019**, *250*, 109429. [CrossRef]
48. Luo, H.Y.; Lin, Q.T.; Zhang, X.F.; Huang, Z.F.; Fu, H.G.; Xiao, R.B.; Liu, S.S. Determining the key factors of nonradical pathway in activation of persulfate by metal-biochar nanocomposites for bisphenol A degradation. *Chem. Eng. J.* **2020**, *391*, 123555.
49. Fan, S.S.; Tang, J.; Wang, Y.; Li, H.; Zhang, H.; Tang, J.; Wang, Z.; Li, X. Biochar prepared from co-pyrolysis of municipal sewage sludge and tea waste for the adsorption of methylene blue from aqueous solutions: Kinetics, isotherm, thermodynamic and mechanism. *J. Mol. Liq.* **2016**, *220*, 432–441. [CrossRef]
50. Bayramoglu, G.; Altintas, B.; Arica, M.Y. Adsorption kinetics and thermodynamic parameters of cationic dyes from aqueous solutions by using a new strong cation-exchange resin. *Chem. Eng. J.* **2009**, *152*, 339–346. [CrossRef]
51. Wu, F.C.; Tseng, R.L.; Juang, R.S. Kinetic modeling of liquid-phase adsorption of reactive dyes and metal ions on chitosan. *Water Res.* **2001**, *35*, 613–618. [CrossRef] [PubMed]
52. Wang, Y.Q.; Pan, J.; Li, Y.H.; Zhang, P.F.; Li, M.X.; Zheng, H.; Zhang, X.P.; Li, H.; Du, Q.J. Methylene blue adsorption by activated carbon, nickel alginate/activated carbon aerogel, and nickel alginate/grapheme oxide aerogel: A comparison study. *J. Mater. Res. Technol.* **2020**, *9*, 12443–12460. [CrossRef]
53. Moradi, H.; Azizpour, H.; Bahmanyar, H.; Mohammadi, M. Molecular dynamics simulation of H_2S adsorption behavior on the surface of activated carbon. *Inorg. Chem. Commun.* **2020**, *118*, 108048. [CrossRef]
54. Rehman, M.U.; Manan, A.; Uzair, M.; Khan, A.S.; Ullah, A.; Ahmad, A.S.; Wazir, A.H.; Qazi, I.; Khan, M.A. Physicochemical characterization of Pakistani clay for adsorption of methylene blue: Kinetic, isotherm and thermodynamic study. *Mater. Chem. Phys.* **2021**, *269*, 124722. [CrossRef]
55. Boparai, H.K.; Joseph, M.; O'Carroll, D.M. Kinetics and thermodynamics of cadmiumion removal by adsorption onto nano zerovalent iron particles. *J. Hazard. Mater.* **2011**, *186*, 458–465. [CrossRef] [PubMed]
56. Jamshaid, M.; Nazir, M.A.; Najam, T.; Shah, S.S.A.; Khan, H.M.; Rehman, A.U. Facile synthesis of Yb^{3+}-Zn^{2+} substituted M type hexaferrites: Structural, electric and photocatalytic properties under visible light for methylene blue removal. *Chem. Phy. Lett.* **2022**, *805*, 139939. [CrossRef]
57. Jamshaid, M.; Rehman, A.U.; Kumar, O.P.; Iqbal, S.; Nazir, M.A.; Anum, A.; Khan, H.M. Design of dielectric and photocatalytic properties of Dy–Ni substituted $Ca_{0.5}Pb_{0.5-x}Fe_{12-y}O_{19}$ M-type hexaferrites. *J Mater Sci: Mater Electron* **2021**, *32*, 16255–16268. [CrossRef]

58. Liu, Q.; Deng, W.; Wang, Q.; Lin, X.; Gong, L.; Liu, C.; Xiong, W.; Nie, X. An efficient chemical precipitation route to fabricate 3D flower-like CuO and 2D leaf-like CuO for degradation of methylene blue. *Adv. Powder Technol.* **2020**, *31*, 1391–1401. [CrossRef]
59. Li, L.; Lai, C.; Huang, F.; Cheng, M.; Zeng, G.; Huang, D.; Li, B.; Liu, S.; Zhang, M.; Qin, L.; et al. Degradation of naphthalene with magnetic bio-char activate hydrogen peroxide: Synergism of bio-char and FeMn binary oxides. *Water Res.* **2019**, *160*, 238–248. [CrossRef]

Disclaimer/Publisher's Note: The statements, opinions and data contained in all publications are solely those of the individual author(s) and contributor(s) and not of MDPI and/or the editor(s). MDPI and/or the editor(s) disclaim responsibility for any injury to people or property resulting from any ideas, methods, instructions or products referred to in the content.

Article

A Novel POP-Ni Catalyst Derived from PBTP for Ambient Fixation of CO_2 into Cyclic Carbonates

Fen Wei [1,2], Jiaxiang Qiu [2], Yanbin Zeng [2], Zhimeng Liu [1], Xiaoxia Wang [1,*] and Guanqun Xie [2,*]

[1] Guangdong Provincial Engineering Technology Research Center of Key Material for High Performance Copper Clad Laminate, School of Materials Science and Engineering, Dongguan University of Technology, Dongguan 523808, China

[2] School of Environment and Civil Engineering, Dongguan University of Technology, Dongguan 523808, China

* Correspondence: wangxx@dgut.edu.cn (X.W.); gqxie@dgut.edu.cn (G.X.)

Abstract: The immobilization of homogeneous catalysts has always been a hot issue in the field of catalysis. In this paper, in an attempt to immobilize the homogeneous [Ni(Me$_6$Tren)X]X (X = I, Br, Cl)-type catalyst with porous organic polymer (POP), the heterogeneous catalyst PBTP-Me$_6$Tren(Ni) (POP-Ni) was designed and constructed by quaternization of the porous bromomethyl benzene polymer (PBTP) with tri[2-(dimethylamino)ethyl]amine (Me$_6$Tren) followed by coordination of the Ni(II) Lewis acidic center. Evaluation of the performance of the POP-Ni catalyst found it was able to catalyze the CO_2 cycloaddition with epichlorohydrin in N,N-dimethylformamide (DMF), affording 97.5% yield with 99% selectivity of chloropropylene carbonate under ambient conditions (80 °C, CO_2 balloon). The excellent catalytic performance of POP-Ni could be attributed to its porous properties, the intramolecular synergy between Lewis acid Ni(II) and nucleophilic Br anion, and the efficient adsorption of CO_2 by the multiamines Me$_6$Tren. In addition, POP-Ni can be conveniently recovered through simple centrifugation, and up to 91.8% yield can be obtained on the sixth run. This research provided a facile approach to multifunctional POP-supported Ni(II) catalysts and may find promising application for sustainable and green synthesis of cyclic carbonates.

Keywords: heterogeneous catalysis; POP-Ni catalyst; carbon dioxide fixation; green chemistry; catalytic material

Citation: Wei, F.; Qiu, J.; Zeng, Y.; Liu, Z.; Wang, X.; Xie, G. A Novel POP-Ni Catalyst Derived from PBTP for Ambient Fixation of CO_2 into Cyclic Carbonates. *Materials* **2023**, *16*, 2132. https://doi.org/10.3390/ma16062132

Academic Editor: Haralampos N. Miras

Received: 11 February 2023
Revised: 2 March 2023
Accepted: 4 March 2023
Published: 7 March 2023

Copyright: © 2023 by the authors. Licensee MDPI, Basel, Switzerland. This article is an open access article distributed under the terms and conditions of the Creative Commons Attribution (CC BY) license (https:// creativecommons.org/licenses/by/ 4.0/).

1. Introduction

In recent years, the issue of persistent increases in CO_2 in the atmosphere caused by the massive combustion of fossil fuels has become one of the most concerns around the world, and solutions to the issue have been paid significant attention to by both governments and researchers all over the world [1–3]. A number of measures have been taken to reduce the release of CO_2. Alternatively, utilization of CO_2 as an abundant, non-toxic, and inexpensive C1 source could provide a practical solution. As has been reported, CO_2 can be converted into alcohols [4,5], acids [6], esters [7,8], polycarbonates [9], and other high-value chemicals [10,11]. It is worth noting that among various chemical fixation methods of CO_2, the cycloaddition of CO_2 and epoxide to produce cyclic carbonate has attracted widespread attention from many researchers [12–14]. On the one hand, the cycloaddition reaction of CO_2 and epoxide is more beneficial to the development of green chemistry due to its 100% atomic economy [15]; on the other hand, as an important type of chemical product, cyclic carbonates have been commonly applied as fuel additives [16], polar non-protonic solvents [17], battery electrolytes [18], useful intermediates of drugs/fine chemicals [19], and the monomers for polycarbonate and polyurethane [20,21]. However, owing to the high thermodynamic and kinetic stability of CO_2, its conversion to cyclic carbonate is highly energy-consuming and requires the use of an active catalyst. Hitherto, there have been many catalytic systems available for the transformation of CO_2 [22,23]. Among them, metal complexes have stood out in many catalytic systems due to their high activities, simple preparative procedures, as well as low-cost starting materials [24,25]. The metal complex catalytic system may be divided into binary catalytic systems and one-component bifunctional catalysts. As for the former, the metal complex catalysts have to be assisted by additional co-catalysts such as tetrabutylammonium bromide to afford satisfactory performance [14,26]. Recently, more and more efforts have been focused on the development of one-component bifunctional metal catalyst systems that integrate metal Lewis sites and nucleophilic sites in one component [27,28]. The development of one-component bifunctional catalysts could not only eliminate the extra addition of co-catalysts, but also increase the intramolecular cooperation between the metal Lewis acidic sites and nucleophilic sites in the catalysts, thereby enhancing the catalytic activity. In this context, Naveen [29] and our group [15] reported a series of one-component bifunctional catalysts [M(Me$_6$Tren)X]X. Among them, [Ni(Me$_6$Tren)X]X (X = I, Br, Cl) exhibited the best catalytic activity for the efficient conversion of CO_2 and epoxide to cyclic carbonate under mild conditions. However, recovery of these homogeneous [Ni(Me$_6$Tren)X]X catalysts was challenging due to its cumbersome recovery procedures having a large amount of organic solvent and incomplete recovery. In view of the recoverability of heterogeneous catalysts [30–33], which could maintain the good catalytic activity of the counterpart homogeneous catalysts and have better potential for practical application, we were inspired to attempt the heterogenization of the [Ni(Me$_6$Tren)X]X catalysts as a novel heterogeneous catalyst.

PBTP-(x)-R, a series of novel amine-containing porous organic copolymers synthesized by Yang and co-workers [34] via copolymerization of 4,4'-bis (chloromethyl) biphenyl (BCMBP) and 1,3,5-tris (chloromethyl) benzene (TCB) followed by reactions with multiamines with benzyl chloride functionality, showed selective-adsorption and high-adsorption capacity for CO_2. In combination with our previous work where [Ni(Me$_6$Tren)X]X (X = I, Br, Cl) could be an efficient homogeneous catalyst for the fixation of CO_2 into cyclic carbonates, herein a novel heterogeneous PBTP-Me$_6$Tren(Ni) (POP-Ni) catalyst was designed and constructed by grafting tri[2-(dimethylamino)ethyl]amine (Me$_6$Tren) on a porous organic polymer PBTP followed by coordination with Ni(II) salt, envisioning that it would be able to enrich CO_2 and catalyze CO_2 cycloaddition reaction efficiently.

2. Materials and Methods

2.1. Chemicals and Instruments

1,3,5-Tris(bromomethyl) benzene (TBB) (98%), 1,3,5-trimethoxybenzene (98%), tri[2-(dimethylamino)ethyl] amine (Me$_6$Tren) (98%), epichlorohydrin (98%), epibromohydrin (97%), styrene oxide (98%), butyl glycidyl ether (98%), tert-butyl glycidyl ether (97%), allyl glycidyl ether (98%), phenyl glycidyl ether (97%), and 1,2-epoxyhexane (97%) were obtained from Energy Chemical (Anhui, China). 4,4'-(Bromomethyl) biphenyl (BBMBP) (97%) was purchased from TCI (Tokyo, Japan). Ferric chloride (95%) was purchased from 3A Chemicals (Shanghai, China). Both 1,2-dichloroethane and nickel acetate tetrahydrate (98%) were purchased from Macklin Chemicals (Shanghai, China). All reagents used in this experiment were directly used without any pretreatment.

The surface morphology of the sample was characterized by means of scanning electron microscopy (SEM) on a JSM-6701F scanning electron microscope produced by Japan Electronics Co., Ltd. (Tokyo, Japan). The N_2 sorption isotherms were performed on Quantachrome-EVO (Quantachrome, Hillsboro, OR, USA), and the sample powder was degassed at 150 °C in a vacuum for 12 h before measurement. By using Thermo Scientific K-Alpha (Thermo Fisher Scientific, Waltham, MA, USA), X-ray photoelectron spectroscopy (XPS) characterization was performed for the sample. Thermogravimetric analysis (TGA) was conducted by using Netzsch TG209-F3 (Netzsch, Selbu, Germany) in an N_2 atmosphere. The reaction yield was detected using the Shimadzu GC2010-QP2010Plus gas chromatography mass spectrometer (GC-MS) (Shimadzu, Japan) with the gas chromatography column of Agilent J&W HP-5, and the GC column temperature was programmed at 32–150 °C at a rate of 8 °C/min.

2.2. Synthesis of PBTP

PBTP was synthesized by referring to Yang's work [34]. A mixture of 4,4'-(bromomethyl) biphenyl (BBMBP, 0.340 g), and 1,3,5-tris (bromomethyl) benzene (TBB, 1.427 g) was completely dissolved in 1,2-dichloroethane (30 mL). Under an N_2 atmosphere, anhydrous $FeCl_3$ (0.811 g) was added swiftly into the above solution. After being stirred at 45 °C for 1 h, the reaction system was heated to 80 °C for another 1 h. After the completion of the reaction, the solid reaction product was thoroughly washed with methanol and dried in a vacuum at 70 °C for 24 h to obtain dark-brown solid powder.

2.3. The Synthesis of PBTP-Me$_6$Tren

PBTP (0.1 g) was added into a Pyrex tube (15 mL), and toluene (2.5 mL) was added for its uniform distribution. Then, tri[2-(dimethylamino)ethyl] amine (Me$_6$Tren, 0.2 mmol, 0.11 mL) was added dropwise under stirring. After the complete dripping of Me$_6$Tren, the screw plug of the Pyrex tube was immediately fitted. Vigorous stirring was performed at 90 °C for 72 h. After the reaction, the reaction product was fully washed with methanol and then dried in a vacuum at 60 °C for 12 h to obtain a dark-brown powder.

2.4. The Synthesis of PBTP-Me$_6$Tren (Ni) (POP-Ni)

Nickel acetate tetrahydrate (2 mmol, 2.488 g) was dissolved in N, N-dimethylformamide (DMF, 25 mL) and filtered to obtain a clear solution. Then, PBTP-Me$_6$Tren (0.1 g) was uniformly distributed in the above solution, and the reaction system was allowed to react under reflux at 90 °C for 6 h. After the reaction, the reaction product was washed thoroughly with a large amount of methanol and dried in a vacuum at 60 °C for 12 h to obtain a dark-brown powder.

2.5. Evaluation of the Catalyst by CO_2 Cycloaddition Reaction

Typically, a Schlenk tube (10 mL) containing the catalyst POP-Ni (80 mg) was emptied and backfilled with CO_2 three times. Under CO_2 balloon pressure, DMF (0.5 mL) and epichlorohydrin (5 mmol, 0.394 mL) were added via a syringe. Then, the mixture was stirred at 80 °C for 24 h. The crude reaction mixture was centrifuged, and the up-layer solution was then diluted with ethyl acetate. The yield was determined by GC-MS using 1,3,5-trimethoxybenzene as an internal standard. The low-layer catalyst was repeatedly washed with ethanol several times. It was used for the next run after drying in a vacuum at 70 °C for 6 h. The same conditions were used for the reusability studies.

3. Results and Discussion

3.1. Catalyst Characterization

Porous organic polymer PBTP was synthesized through the copolymerization of monomer 4,4'-(bromomethyl) biphenyl (BBMBP) and 1,3,5-tris (bromomethyl) (TBB) via a Friedel–Crafts alkylation reaction, and then the bifunctional catalyst PBTP-Me$_6$Tren(Ni) (POP-Ni) was synthesized by following a two-step sequential post-synthetic modification procedure. (Figure 1). Firstly, the surface morphology of the synthesized material was characterized by scanning electron microscopy (SEM). As shown in Figure 2a, the synthetic PBTP is formed by the accumulation of a spherical-like structure with a rich pore structure. After post-synthetic modification, the surface morphology of POP-Ni shows no obvious change, and it remains an accumulation of a spheroid-like structure. However, compared with PBTP, the size of the spheroid-like structural unit of POP-Ni shows a significant increase (increase from 106.7 ± 19.9 nm to 153.7 ± 21.5 nm) (Figures 2b and S1), indicating that the post-synthetic modification strategy follows a core–shell growth process based on PBTP. To further explore the impact of post-synthetic modification strategies on the pore size distribution of materials, the specific surface area and pore size distributions of PBTP and POP-Ni were examined by means of Brunauer–Emmett–Teller (BET) measurements (Figure 2c,d). According to the measurement results, PBTP has a large specific surface area of up to 902.3 m^2g^{-1}. In terms of pore size distribution, the pore sizes of PBTP concentrates were in the range of 0.8–2.0 nm and 3.2–27 nm. Following the process of two-step sequential post-synthetic modification, the specific surface area of material POP-Ni was reduced from 902.3 m^{2}g^{-1} to 576.3 m^{2}g^{-1}. Meanwhile, compared with PBTP, the pore size distribution of POP-Ni decreased significantly in the range of 0.8–2.0 nm, which may be the leading cause for the decrease in the specific surface area. On the other hand, there is no significant difference in the pore size distribution of POP-Ni as compared with that of PBTP in the range of 3.2–27 nm. Overall, the presence of micropores in the POP-Ni structure could facilitate CO$_2$ enrichment in the material, while the extensive distribution of mesoporous could facilitate product transfer.

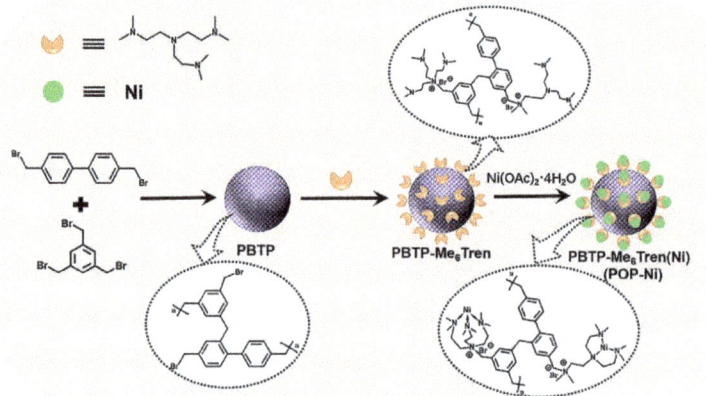

Figure 1. The preparation of PBTP-Me6Tren(Ni) (POP-Ni).

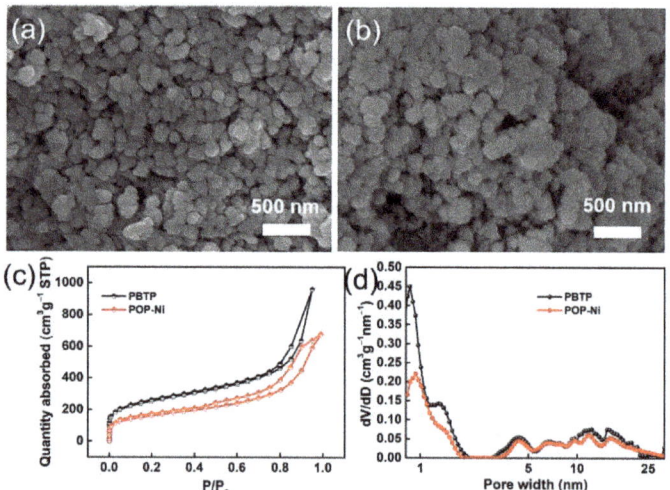

Figure 2. SEM image of POP-Ni (**a**) and POP-Ni (**b**); N$_2$ adsorption/desorption isotherm (**c**) and pore size distribution of PBTP and POP-Ni (**d**).

Furthermore, XPS was employed to examine the element composition and valence distribution of the bifunctional catalysts. As indicated by the XPS spectrum of POP-Ni, there are four different elements in the material, namely Ni, C, N and Br (Figure S2). As shown in the Ni 2p XPS spectrum of POP-Ni (Figure 3a), the two peaks at around 855.68 eV and 873.33 eV are assigned to Ni^{2+} 2p3/2 and Ni^{2+} 2p1/2, respectively, while the satellite peaks of Ni 2p3/2 at 861.21 eV and the peaks at 878.81 eV are assigned to the satellite peak of Ni 2p1/2. In the Br 3d spectrum (Figure 3b), the peak at 68.19 eV is assigned to the quaternary ammonium bromide anion, and the peak with a binding energy of 70.84 eV results from methyl bromide, indicating an incomplete conversion of the benzyl bromide groups in the porous polymer precursor into the nucleophilic Br anion through a quaternization reaction. The incomplete quaternization reaction may be caused by the steric hindrance and pore size limitation of PBTP material. In the C1s spectrum (Figure 3c), there are three different forms of carbon observed, with the binding energy of 283.85 eV and 286.53 eV related to phenyl carbon and aliphatic chain skeleton carbon, respectively. In addition, the peak at 285.49 eV is assigned to the carbon of $-C-N-$. In the N 1s spectrum of POP-Ni (Figure 3d), the peaks at 398.64 eV, 400.55 eV, and 406.21 eV are ascribed to $-N-C-$, N$-$Ni, and quaternary ammonium N cations, respectively. Then, TGA was applied to evaluate the thermal stability of the synthetic bifunctional POP-Ni catalyst (Figure S3b). The weight loss of POP-Ni before 150 °C should result from the volatilization of the DMF solvent, and the catalyst maintains sufficient thermal stability before 250 °C. In summary, the above results demonstrate the successful preparation of the bifunctional catalyst POP-Ni.

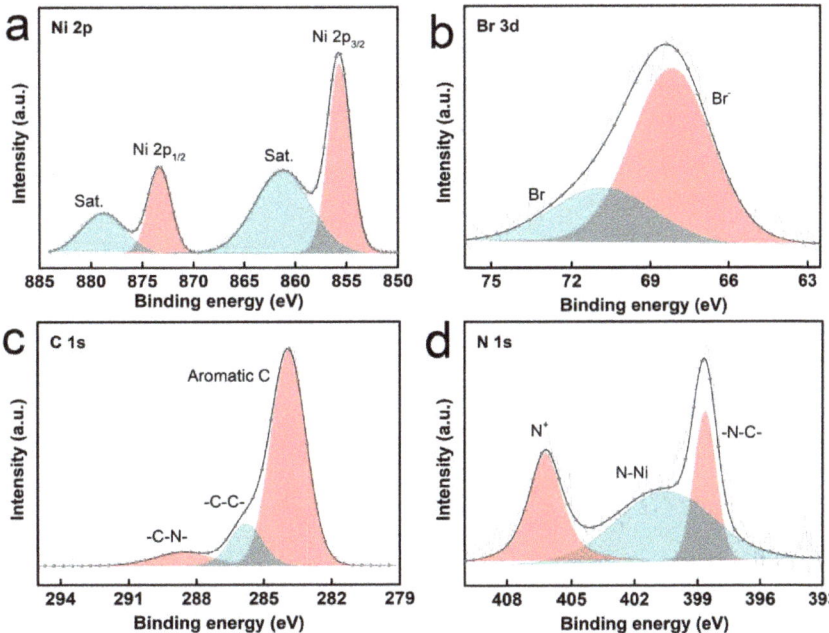

Figure 3. Ni 2p spectrum (**a**); Br 3d spectrum (**b**); C 1s spectrum (**c**); N 1s spectrum of POP-Ni (**d**).

3.2. Investigation of the Catalytic Performance

The bifunctional catalyst has attracted increased attention due to two advantages: acid-base coordination and easy reusability. In addition, there has been some attention paid in recent years to the research of more moderate reaction conditions and a wide range of substrates. Therefore, the conditions of CO_2 cycloaddition reactions were examined in this study by using POP-Ni as a catalyst and epichlorohydrin as a template substrate (Table 1). Firstly, the catalytic activity of catalysts PBTP, PBTP-Me$_6$Tren, and POP-Ni for cycloaddition reaction was explored under solvent-free conditions with a catalyst loading of 40 mg (Table 1, Entries 1–3). The results show that PBTP as a carrier plays little role in the catalytic activity for CO_2 cycloaddition reactions. Different from PBTP, the PBTP-Me$_6$Tren with nucleophilic halogen active sites and multiamine groups converted epichlorohydrin into chloropropene carbonate with a 32.8% yield. In contrast, the bifunctional catalyst POP-Ni afforded epichlorohydrin with a 51.9% yield, suggesting Ni(II) as the active site of Lewis acid to synergize cycloaddition with nucleophilic Br. These results clearly indicate the important role played by both the intramolecular synergy of metal Ni(II) with Lewis acidity and the nucleophilic halogen Br and the effective adsorption of multiamine groups on CO_2 [34] in the occurrence of CO_2 cycloaddition reactions.

Table 1. Evaluation of the catalytic performance of POP-Ni for CO_2 cycloaddition reactions [a].

Entry	Catalyst	Amount of Catalyst	Solvent	Yield (%) [c]	Selectivity (%) [c]
1	PBTP	40 mg	none	2	>99
2	PBTP-Me$_6$Tren	40 mg	none	32.8	>99
3	POP-Ni	40 mg	none	51.9	>99
4	POP-Ni	40 mg	DMSO	58.4	>99
5	POP-Ni	40 mg	DMA	69.6	>99
6	POP-Ni	40 mg	DMF	78.5	>99
7	POP-Ni	80 mg	DMF	97.5	>99
8 [b] [15]	[Ni(Me$_6$Tren)Br]Br	22 mg	none	98.9	>99

[a] Reaction conditions: 5 mmol of epichlorohydrin, 0.5 mL of solvent, CO_2 (balloon), 80 °C, and 24 h. [b] The performance of the homogeneous catalyst. Reaction conditions: 5 mmol of epichlorohydrin, 1 mol% of catalyst, CO_2 (balloon), 80 °C, and 24 h. [c] Determined by GC-MS.

To further improve the catalyst activity, three highly polar solvents, DMSO, DMA, and DMF, were introduced into the CO_2 cycloaddition reaction (Table 1, Entries 4–6) since solvent-free conditions seemed to not disperse the catalyst well. The yield of chloropropene carbonate reached 58.4% when DMSO was used as a solvent, while DMA and DMF as the solvent could afford 69.6% and 78.5% yield, respectively. Although DMSO shows a higher polarity to facilitate the dispersion and swelling of the catalysts, its higher viscosity fails to enhance its catalytic activity significantly compared with DMA and DMF [35]. DMF as a solvent achieves a higher catalytic yield than DMA. The primary reason for this is that the DMF with higher polarity produces a more significant swelling effect on the catalyst, which allows POP-Ni to be better distributed in the reaction system, thus enhancing its catalytic activity.

The influence of the catalyst loading on the product yields was further investigated (Figure S4). According to the results, the catalytic activity was boosted significantly with the increase in catalyst loading: 40 mg (78.5%) < 50 mg (82.3%) < 60 mg (88.2%) < 70 mg (91.2%) < 80 mg (97.5%) < 90 mg (98.7%). The results showed 90 mg of the catalyst did not bring about significant improvement in yields and 80 mg of the catalyst could already afford satisfactory yields (97.5%); thus, 80 mg was set as the optimal catalyst dosage (Table 1, Entry 7). It should be noted that POP-Ni catalyst at 80 °C and CO_2 balloon conditions showed slightly lower catalytic activity (80 mg catalyst loading, 97.5% yield) compared with that of the homogeneous catalyst [Ni(Me$_6$Tren)Br]Br (1 mol% catalyst loading, 98.9% yield), and the selectivity in both cases were excellent (>99%).

In addition, a comparison of POP-Ni with other available POP-related catalysts for the CO_2 cycloaddition reaction of epichlorohydrin was also made (Table 2). Although mild and efficient catalysts involving Zn or Co Lewis centers on the POP have been developed [36–38], the addition of TBAB as a cocatalyst was required, which may make the recovery of catalytic systems tedious and difficult. For other Mg^{2+}-, Zn^{2+}-, Co^{2+}-, or Al^{3+}-involved POP catalysts [39–45], higher temperatures or pressures were usually applied so that comparable yields could be obtained.

Table 2. The comparison of POP-Ni with available POP related catalysts for the CO_2 cycloaddition reaction of epichlorohydrin.

Entry	Catalyst	Cocatalyst	CO_2 (Mpa)	Temp. (°C)	Time (h)	Yield (%)	Number of Recycling (yield, %) [a]	Ref.
1	Co/POP-TPP	TBAB	0.1	29	24	95.6	18 (93.6)	[36]
2	Zn/TPA-TCIF(BD)	TBAB	0.5	40	10	98.8	10 (84)	[37]
3	Co@H-POP-4,4'-bipyridine	TBAB	0.3	30	48	97.1	none	[38]
4	Py-Zn@IPOP$_I$	none	2	120	6	96	5 (94)	[39]
5	NHC-CAP-1(Zn^{2+})	none	2	100	3	97	10 (95)	[40]
6	ZnTPP/QA-azo-PiP$_1$	none	1	80	12	99	7 (92)	[41]
7	POF-Zn^{2+}-I^-	none	1	60	8	92.2	none	[42]
8	AlPor–PIP–Br	none	0.5	40	24	98	6 (97)	[43]
9	Al-CPOP	none	0.1	120	24	95.0	5 (95)	[44]
10	Co-HIP	none	0.1	80	20	96	5 (95)	[45]
11	POP-Ni	none	0.1	80	24	97.5	6 (91.5)	This work

[a] The yields of the cyclic carbonate from the last run are shown in the parentheses.

3.3. Applicability of Catalyst POP-Ni

To illustrate the application scope of POP-Ni, the cycloaddition of CO_2 with other epoxides was examined under the same conditions (CO_2 balloon, 80 °C, Table 3, and Table S1). POP-Ni can also catalyze epibromohydrin with high catalytic activity to generate bromopropylene carbonate with good yields (95.7%) (Table 3, entry 2). For aliphatic epoxides, the yield of cyclic carbonate declines sharply with the increase in the alkyl chain. For 1,2-epoxyhexane, the corresponding cyclic carbonate can be obtained with a medium yield of 62.7% (Table 3, entry 3). With further increases in the length of the alkyl chain, the catalytic activity of POP-Ni was further reduced, as can be seen for the allyl glycidyl ether and butyl glycidyl ether, which afford the yield of 48.0% and 49.2%, respectively (Table 3, entries 4, 5). In addition, the yield of *tert*-butyl glycidyl ethers containing branched alkanes was found to be even lower (35.6%) compared to that of the butyl glycidyl ethers containing straight-chain alkanes (Table 3, entry 6). For phenyl glycidyl ether, it can be transformed into the corresponding cyclic carbonate with a moderate yield of 51.5%. In addition, POP-Ni afforded a yield of 48.2% of the cyclic carbonate when styrene oxide was used as the substrate. It can be seen from the above results that POP-Ni shows good catalytic activity for epichlorohydrin and epichlorohydrin, and its catalytic activity is on a medium level for substrates of relatively larger sizes. This is mainly because the pores in the catalyst are mainly micropores, which restricts the contact between the larger substrate and the catalytic active site of the catalyst and shows certain substrate selectivity.

3.4. Catalyst Reusability

Recycling experiments were conducted to investigate the recoverability and stability of the catalyst since they are considered the most important indicators of excellent heterogeneous catalysts. With epichlorohydrin as the substrate, recycling of the catalyst for six runs led to a slight reduction in the catalytic activity of the catalyst (from 97.5% to 91.8%) while maintaining excellent selectivity (>99%) (Figure 4 and Table S2). At each run of the reaction, the catalyst was easily separated from the reaction system by centrifugation and used for the next run. BET measurement of the recycled catalyst (Figure S5) showed the specific surface area of POP-Ni obviously decreased (403.7 m^2g^{-1}) as compared with the fresh catalyst (576.3 m^2g^{-1}), due mainly to the decrease in micropore numbers, which may probably be caused by blockage of the product molecules in the channel. Overall, the results showed that the mild catalytic condition (80 °C) exerts no obvious effect on the structure and performance of the catalyst, and the catalyst exhibited good potential for practical application.

Table 3. Screening on the substrate scope for CO_2 cycloaddition with the POP-Ni catalyst [a].

Entry	Epoxide	Product	Yield (%) [b]	Selectivity (%) [b]
1	Cl-epoxide	Cl-cyclic carbonate	97.5	>99
2	Br-epoxide	Br-cyclic carbonate	95.7	>99
3	propyl-epoxide	propyl-cyclic carbonate	62.7	>99
4	allyl glycidyl ether	allyl glycidyl carbonate	48.0	89.4
5	butyl glycidyl ether	butyl glycidyl carbonate	49.2	90.3
6	tert-butyl glycidyl ether	tert-butyl glycidyl carbonate	35.6	91.3
7	phenyl glycidyl ether	phenyl glycidyl carbonate	51.5	93.4
8	styrene oxide	styrene carbonate	48.2	87.6

[a] Reaction condition: 5 mmol of epoxide, 80 mg of POP-Ni, 0.5 mL of DMF, CO_2 (balloon), 80 °C, and 24 h.
[b] Determined by GC-MS.

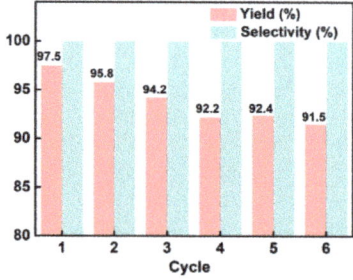

Figure 4. Recycling of the catalyst POP-Ni. Reaction conditions: epichlorohydrin (5 mmol), POP-Ni (80 mg), 0.5 mL of DMF, CO_2 balloon. The mixture was stirred at 80 °C for 24 h. Yield and selectivity were determined by GC-MS.

3.5. Plausible Reaction Pathway

Based on the previous reports [44,46] and our experimental results, the reaction pathway of CO_2 cycloaddition catalyzed by POP-Ni was proposed (Figure 5). Firstly, the epoxide was activated through the Ni^{2+} active site in the catalyst to coordinate with the O atom of the epoxide. Then, the nucleophilic Br anion attacked the carbon atom on the side of the epoxy with less hindrance through a nucleophilic attack to open up the ring (I), thus obtaining the ring-opening O anion intermediate (II). Intermediate II further launches nucleophilic attacks on CO_2 (III) to obtain carbonate intermediate (IV). Finally, the product cyclic carbonate is obtained through intramolecular substitution. At the same time, the catalyst is released for further catalysis.

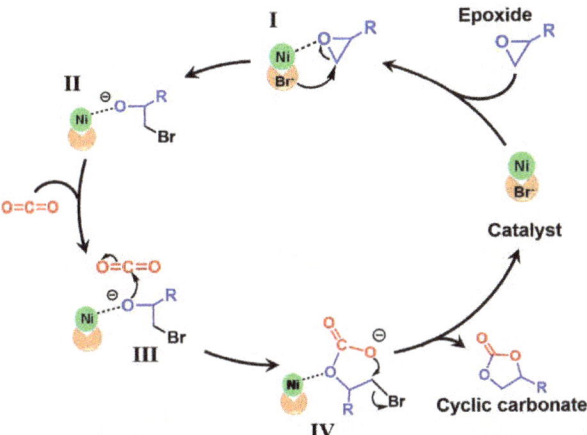

Figure 5. Plausible mechanistic pathway for CO_2 cycloaddition with epoxides.

4. Conclusions

A novel heterogeneous POP-Ni catalyst has been successfully constructed by grafting Me_6Tren on the PBTP, synthesized by copolymerization of BBMBP and TBB in a molar ratio of 1/4 and subsequent coordination of the Ni(II) Lewis acidic center. The POP-Ni not only possessed a large specific surface area (576.3 $m^2 g^{-1}$) and good thermal stability, but also maintained great catalytic activity after heterogenization. It exhibited excellent catalytic performance with a yield of 97.5% in converting epichlorohydrin and CO_2 into propylene chlorocarbonate under mild conditions (80 °C, CO_2 balloon). The excellent catalytic performance of POP-Ni could be attributed to its porous properties, the synergism between the Lewis acid Ni(II) and nucleophilic Br anion, and the efficient adsorption of CO_2 by the multiamines Me_6Tren. In particular, POP-Ni could be effectively recovered through simple centrifugal and still maintain excellent catalytic performance with a yield of 91.5% after six consecutive recycles. The POP-Ni catalyst may find promising application in view of its availability, high activity, and good reusability.

Supplementary Materials: The following supporting information can be downloaded at: https://www.mdpi.com/article/10.3390/ma16062132/s1, Figure S1. SEM image of PBTP (a) and POP-Ni (b); particle size distribution diagram of PBTP (c) and POP-Ni (d). Figure S2. XPS spectrum of POP-Ni. Figure S3. (a) TGA profile of PBTP; (b) TGA profile of POP-Ni. Figure S4. The catalytic activity corresponding to different catalyst amounts. Figure S5. N2 adsorption/desorption isotherm (a) and pore size distribution (b) of POP-Ni (fresh) and POP-Ni (recycled catalyst after the fourth run). Table S1. Screening on the substrate scope for CO_2 cycloaddition with POP-Ni catalyst. Table S2. Recyclability of catalyst POP-Ni.

Author Contributions: F.W.: investigation, experiment, writing—original draft. J.Q.: investigation, experiment. Y.Z.: investigation, experiment. Z.L.: supervision, writing—review and editing. X.W.: supervision, writing—review and editing, project administration. G.X.: supervision, writing—review and editing. All authors have read and agreed to the published version of the manuscript.

Funding: This work was supported by Applied Basic Research Foundation of Guangdong Province (No. 2019A1515110551), Science Foundation for Distinguished Scholars of Dongguan University of Technology (No. 211135283).

Institutional Review Board Statement: Not applicable.

Informed Consent Statement: Not applicable.

Data Availability Statement: All data reported in this paper are contained within the manuscript.

Conflicts of Interest: The authors declare no conflict of interest.

References

1. Gilfillan, D.; Marland, G. CDIAC-FF: Global and national CO_2 emissions from fossil fuel combustion and cement manufacture: 1751–2017. *Earth Syst. Sci. Data* **2021**, *13*, 1667–1680. [CrossRef]
2. Ostovari, H.; Muller, L.; Skocek, J.; Bardow, A. From Unavoidable CO_2 Source to CO_2 Sink? A Cement Industry Based on CO_2 Mineralization. *Environ. Sci. Technol.* **2021**, *55*, 5212–5223. [CrossRef]
3. Wang, Y.L.; Ciais, P.; Broquet, G.; Breon, F.M.; Oda, T.; Lespinas, F.; Meijer, Y.; Loescher, A.; Janssens-Maenhout, G.; Zheng, B.; et al. A global map of emission clumps for future monitoring of fossil fuel CO_2 emissions from space. *Earth Syst. Sci. Data* **2019**, *11*, 687–703. [CrossRef]
4. Han, H.; Noh, Y.; Kim, Y.; Park, S.; Yoon, W.; Jang, D.; Choi, S.M.; Kim, W.B. Selective electrochemical CO_2 conversion to multicarbon alcohols on highly efficient N-doped porous carbon-supported Cu catalysts. *Green Chem.* **2020**, *22*, 71–84. [CrossRef]
5. Lian, Y.; Fang, T.F.; Zhang, Y.H.; Liu, B.; Li, J.L. Hydrogenation of CO_2 to alcohol species over Co@Co_3O_4/C-N catalysts. *J. Catal.* **2019**, *379*, 46–51. [CrossRef]
6. Wang, X.N.; Gao, C.; Low, J.X.; Mao, K.K.; Duan, D.L.; Chen, S.M.; Ye, R.; Qiu, Y.R.; Ma, J.; Zheng, X.S.; et al. Efficient photoelectrochemical CO_2 conversion for selective acetic acid production. *Sci. Bull.* **2021**, *66*, 1296–1304. [CrossRef]
7. Frogneux, X.; von Wolff, N.; Thuery, P.; Lefevre, G.; Cantat, T. CO_2 Conversion into Esters by Fluoride-Mediated Carboxylation of Organosilanes and Halide Derivatives. *Chem. Eur. J* **2016**, *22*, 2930–2934.
8. Lee, H.J.; Choi, J.; Lee, S.M.; Um, Y.; Sim, S.J.; Kim, Y.; Woo, H.M. Photosynthetic CO_2 Conversion to Fatty Acid Ethyl Esters (FAEEs) Using Engineered Cyanobacteria. *J. Agric. Food Chem.* **2017**, *65*, 1087–1092. [CrossRef]
9. Wang, Y.Y.; Darensbourg, D.J. Carbon dioxide-based functional polycarbonates: Metal catalyzed copolymerization of CO_2 and epoxides. *Coord. Chem. Rev.* **2018**, *372*, 85–100. [CrossRef]
10. Chang, K.; Zhang, H.C.; Cheng, M.J.; Lu, Q. Application of ceria in CO_2 conversion catalysis. *ACS Catal.* **2020**, *10*, 613–631. [CrossRef]
11. Guo, Y.; Qian, C.; Wu, Y.L.; Liu, J.; Zhang, X.D.; Wang, D.D.; Zhao, Y.L. Porous catalytic membranes for CO_2 conversion. *J. Energy Chem.* **2021**, *63*, 74–86. [CrossRef]
12. Gupta, A.K.; Guha, N.; Krishnan, S.; Mathur, P.; Rai, D.K. A Three-Dimensional Cu(II)-MOF with Lewis acid-base dual functional sites for Chemical Fixation of CO_2 via Cyclic Carbonate Synthesis. *J. CO_2 Util.* **2020**, *39*, 101173. [CrossRef]
13. Wu, X.; North, M. A Bimetallic Aluminium(Salphen) Complex for the Synthesis of Cyclic Carbonates from Epoxides and Carbon Dioxide. *ChemSusChem* **2017**, *10*, 74–78. [CrossRef] [PubMed]
14. Melendez, D.O.; Lara-Sanchez, A.; Martinez, J.; Wu, X.; Otero, A.; Castro-Osma, J.A.; North, M.; Rojas, R.S. Amidinate Aluminium Complexes as Catalysts for Carbon Dioxide Fixation into Cyclic Carbonates. *ChemCatChem* **2018**, *10*, 2271–2277. [CrossRef]
15. Wei, F.; Tang, J.; Zuhra, Z.; Wang, S.S.; Wang, X.X.; Wang, X.F.; Xie, G.Q. [M(Me_6Tren)X] X complex as efficacious bifunctional catalyst for CO_2 cycloaddition: The synergism of the metal and halogen ions. *J. CO_2 Util.* **2022**, *61*, 102048. [CrossRef]
16. Perez-Sena, W.Y.; Eranen, K.; Kumar, N.; Estel, L.; Leveneur, S.; Salmi, T. New insights into the cocatalyst-free carbonation of vegetable oil derivatives using heterogeneous catalysts. *J. CO_2 Util.* **2022**, *57*, 101879. [CrossRef]
17. Hernandez, E.; Santiago, R.; Belinchon, A.; Vaquerizo, G.M.; Moya, C.; Navarro, P.; Palomar, J. Universal and low energy-demanding platform to produce propylene carbonate from CO_2 using hydrophilic ionic liquids. *Sep. Purif. Technol.* **2022**, *295*, 121273. [CrossRef]
18. He, M.N.; Su, C.C.; Peebles, C.; Zhang, Z.C. The Impact of Different Substituents in Fluorinated Cyclic Carbonates in the Performance of High Voltage Lithium-Ion Battery Electrolyte. *J. Electrochem. Soc.* **2021**, *168*, 010505. [CrossRef]
19. Cristofol, A.; Bohmer, C.; Kleij, A.W. Formal Synthesis of Indolizidine and Quinolizidine Alkaloids from Vinyl Cyclic Carbonates. *Chem. Eur. J.* **2019**, *25*, 15055–15058. [CrossRef]
20. Yu, W.; Maynard, E.; Chiaradia, V.; Arno, M.C.; Dove, A.P. Aliphatic Polycarbonates from Cyclic Carbonate Monomers and Their Application as Biomaterials. *Chem. Rev.* **2021**, *121*, 10865–10907. [CrossRef] [PubMed]
21. Tang, J.; Wei, F.; Ding, S.J.; Wang, X.X.; Xie, G.Q.; Fan, H.B. Azo-Functionalized Zirconium-Based Metal-Organic Polyhedron as an Efficient Catalyst for CO_2 Fixation with Epoxides. *Chem. Eur. J.* **2021**, *27*, 12890–12899. [CrossRef] [PubMed]
22. Liu, J.; Yang, G.Q.; Liu, Y.; Wu, D.S.; Hu, X.B.; Zhang, Z.B. Metal-free imidazolium hydrogen carbonate ionic liquids as bifunctional catalysts for the one-pot synthesis of cyclic carbonates from olefins and CO_2. *Green Chem.* **2019**, *21*, 3834–3838. [CrossRef]
23. Song, H.B.; Wang, Y.J.; Xiao, M.; Liu, L.; Liu, Y.L.; Liu, X.F.; Gai, H.J. Design of Novel Poly(ionic liquids) for the Conversion of CO_2 to Cyclic Carbonates under Mild Conditions without Solvent. *ACS Sustain. Chem. Eng.* **2019**, *7*, 9489–9497. [CrossRef]
24. Della Monica, F.; Buonerba, A.; Paradiso, V.; Milione, S.; Grassi, A.; Capacchione, C. OSSO-type Fe(III) metallate as single-component catalyst for the CO_2 cycloaddition to epoxides. *Adv. Synth. Catal.* **2019**, *361*, 283–288. [CrossRef]
25. Fu, X.Y.; Jing, X.Y.; Jin, L.L.; Zhang, L.L.; Zhang, X.F.; Hu, B.; Jing, H.W. Chiral basket-handle porphyrin-Co complexes for the catalyzed asymmetric cycloaddition of CO_2 to epoxides. *Chin. J. Catal.* **2018**, *39*, 997–1003. [CrossRef]
26. Damiano, C.; Sonzini, P.; Cavalleri, M.; Manca, G.; Gallo, E. The CO_2 cycloaddition to epoxides and aziridines promoted by porphyrin-based catalysts. *Inorg. Chim. Acta* **2022**, *540*, 121065. [CrossRef]

27. Guo, Y.C.; Chen, W.J.; Feng, L.; Fan, Y.C.; Liang, J.S.; Wang, X.M.; Zhang, X. Greenery-inspired nanoengineering of bamboo-like hierarchical porous nanotubes with spatially organized bifunctionalities for synergistic photothermal catalytic CO_2 fixation. *J. Mater. Chem. A* **2022**, *10*, 12418–12428. [CrossRef]
28. Yu, K.; Puthiaraj, P.; Ahn, W.S. One-pot catalytic transformation of olefins into cyclic carbonates over an imidazolium bromide-functionalized Mn(III)-porphyrin metal-organic framework. *Appl. Catal. B* **2020**, *273*, 119059. [CrossRef]
29. Naveen, K.; Ji, H.; Kim, T.S.; Kim, D.; Cho, D.H. C_3-symmetric zinc complexes as sustainable catalysts for transforming carbon dioxide into mono- and multi-cyclic carbonates. *Appl. Catal. B* **2021**, *280*, 119395. [CrossRef]
30. Balas, M.; Beaudoin, S.; Proust, A.; Launay, F.; Villanneau, R. Advantages of Covalent Immobilization of Metal-Salophen on Amino-Functionalized Mesoporous Silica in Terms of Recycling and Catalytic Activity for CO_2 Cycloaddition onto Epoxides. *Eur. J. Inorg. Chem.* **2021**, *2021*, 1581–1591. [CrossRef]
31. Jayakumar, S.; Li, H.; Tao, L.; Li, C.Z.; Liu, L.N.; Chen, J.; Yang, Q.H. Cationic Zn-Porphyrin Immobilized in Mesoporous Silicas as Bifunctional Catalyst for CO_2 Cycloaddition Reaction under Cocatalyst Free Conditions. *ACS Sustain. Chem. Eng.* **2018**, *6*, 9237–9245. [CrossRef]
32. Liu, Y.; Liu, Q.Y.; Sun, K.H.; Zhao, S.F.; Kim, Y.D.; Yang, Y.P.; Liu, Z.Y.; Peng, Z.K. Identification of the Encapsulation Effect of Heteropolyacid in the Si-Al Framework toward Benzene Alkylation. *ACS Catal.* **2022**, *12*, 4765–4776. [CrossRef]
33. Zhang, S.F.; Zhang, B.; Liang, H.J.; Liu, Y.Q.; Qiao, Y.; Qin, Y. Encapsulation of Homogeneous Catalysts in Mesoporous Materials Using Diffusion-Limited Atomic Layer Deposition. *Angew. Chem. Int. Ed.* **2018**, *57*, 1091–1095. [CrossRef]
34. Yang, Y.Q.; Chuah, C.Y.; Bae, T.H. Polyamine-appended porous organic copolymers with controlled structural properties for enhanced CO_2 capture. *ACS Sustain. Chem. Eng.* **2021**, *9*, 2017–2026. [CrossRef]
35. Zhang, F.; Bulut, S.; Shen, X.J.; Dong, M.H.; Wang, Y.Y.; Cheng, X.M.; Liu, H.Z.; Han, B.X. Halogen-free fixation of carbon dioxide into cyclic carbonates via bifunctional organocatalysts. *Green Chem.* **2021**, *23*, 1147–1153. [CrossRef]
36. Dai, Z.F.; Sun, Q.; Liu, X.L.; Bian, C.Q.; Wu, Q.M.; Pan, S.X.; Wang, L.; Meng, X.J.; Deng, F.; Xiao, F.S. Metalated porous porphyrin polymers as efficient heterogeneous catalysts for cycloaddition of epoxides with CO_2 under ambient conditions. *J. Catal.* **2016**, *338*, 202–209. [CrossRef]
37. Zheng, Y.T.; Wang, X.Q.; Liu, C.; Yu, B.Q.; Li, W.L.; Wang, H.L.; Sun, T.T.; Jiang, J.Z. Triptycene-supported bimetallic salen porous organic polymers for high efficiency CO_2 fixation to cyclic carbonates. *Inorg. Chem. Front.* **2021**, *8*, 2880–2888. [CrossRef]
38. Kong, L.Y.; Han, S.L.; Zhang, T.; He, L.C.; Zhou, L.S. Developing hierarchical porous organic polymers with tunable nitrogen base sites via theoretical calculation-directed monomers selection for efficient capture and catalytic utilization of CO_2. *Chem. Eng. J.* **2021**, *420*, 127621. [CrossRef]
39. Zhang, W.W.; Ping, R.; Lu, X.Y.; Shi, H.B.; Liu, F.S.; Ma, J.J.; Liu, M.S. Rational design of Lewis acid-base bifunctional nanopolymers with high performance on CO_2/epoxide cycloaddition without a cocatalyst. *Chem. Eng. J.* **2021**, *420*, 127621. [CrossRef]
40. Puthiaraj, P.; Ravi, S.; Yu, K.; Ahn, W.S. CO_2 adsorption and conversion into cyclic carbonates over a porous $ZnBr_2$-grafted N-heterocyclic carbene-based aromatic polymer. *Appl. Catal. B* **2019**, *251*, 195–205. [CrossRef]
41. Chen, Y.J.; Ren, Q.G.; Zeng, X.J.; Tao, L.M.; Zhou, X.T.; Ji, H.B. Sustainable synthesis of multifunctional porous metalloporphyrin polymers for efficient carbon dioxide transformation under mild. *Chem. Eng. Sci.* **2021**, *232*, 116380. [CrossRef]
42. Ma, D.X.; Li, J.X.; Liu, K.; Li, B.Y.; Li, C.G.; Shi, Z. Di-ionic multifunctional porous organic frameworks for efficient CO_2 fixation under mild and co-catalyst free conditions. *Green Chem.* **2018**, *20*, 5285–5291. [CrossRef]
43. Liu, X.Y.; Yang, Y.Y.; Chen, M.; Xu, W.; Chen, K.C.; Luo, R.C. High-Surface-Area Metalloporphyrin-Based Porous Ionic Polymers by the Direct Condensation Strategy for Enhanced CO_2 Capture and Catalytic Conversion into Cyclic Carbonates. *ACS Appl. Mater. Interfaces* **2023**, *15*, 1085–1096. [CrossRef] [PubMed]
44. Liu, T.T.; Liang, J.; Huang, Y.B.; Cao, R. A bifunctional cationic porous organic polymer based on a Salen-(Al) metalloligand for the cycloaddition of carbon dioxide to produce cyclic carbonates. *Chem. Commun.* **2016**, *52*, 13288–13291. [CrossRef]
45. Li, J.; Han, Y.L.; Ji, T.; Wu, N.H.; Lin, H.; Jiang, J.; Zhu, J.H. Porous Metallosalen Hypercrosslinked Ionic Polymers for Cooperative CO_2 Cycloaddition Conversion. *Ind. Eng. Chem. Res.* **2020**, *59*, 676–684. [CrossRef]
46. Chen, Y.J.; Luo, R.C.; Xu, Q.H.; Jiang, J.; Zhou, X.T.; Ji, H.B. Charged Metalloporphyrin Polymers for Cooperative Synthesis of Cyclic Carbonates from CO_2 under Ambient Conditions. *ChemSusChem* **2017**, *10*, 2534–2541. [CrossRef]

Disclaimer/Publisher's Note: The statements, opinions and data contained in all publications are solely those of the individual author(s) and contributor(s) and not of MDPI and/or the editor(s). MDPI and/or the editor(s) disclaim responsibility for any injury to people or property resulting from any ideas, methods, instructions or products referred to in the content.

Review

New Materials Used for the Development of Anion-Selective Electrodes—A Review

Cecylia Wardak [1,*], Klaudia Morawska [1] and Karolina Pietrzak [2]

[1] Department of Analytical Chemistry, Institute of Chemical Sciences, Faculty of Chemistry, Maria Curie-Sklodowska University, Maria Curie-Sklodowska Sq. 3, 20-031 Lublin, Poland; klaudiamorawska0905@gmail.com

[2] Department of Food and Nutrition, Medical University of Lublin, 4a Chodzki Str., 20-093 Lublin, Poland; karolina.pietrzak@umlub.pl

* Correspondence: cecylia.wardak@mail.umcs.pl; Tel.: +48-815375655

Abstract: Ion-selective electrodes are a popular analytical tool useful in the analysis of cations and anions in environmental, industrial and clinical samples. This paper presents an overview of new materials used for the preparation of anion-sensitive ion-selective electrodes during the last five years. Design variants of anion-sensitive electrodes, their advantages and disadvantages as well as research methods used to assess their parameters and analytical usefulness are presented. The work is divided into chapters according to the type of ion to which the electrode is selective. Characteristics of new ionophores used as the electroactive component of ion-sensitive membranes and other materials used to achieve improvement of sensor performance (e.g., nanomaterials, composite and hybrid materials) are presented. Analytical parameters of the electrodes presented in the paper are collected in tables, which allows for easy comparison of different variants of electrodes sensitive to the same ion.

Keywords: ion-selective electrode; potentiometry; anion ionophore; solid contact; carbon paste ion-selective electrode

1. Introduction

Ion-selective electrodes (ISEs) are sensors used in potentiometric measurements. They are also called membrane electrodes due to the presence of an ion-selective membrane, which is one of the most important elements of each ISE. Because of the presence of the ionophore in the membrane, the electrode is sensitive to changes in the activity of particular ions in solutions [1]. The development of ion-selective sensors began more than 100 years ago when at the beginning of the 20th century, Cramer invented a glass electrode, which, for quite a long time (until the 1930s), was successfully used for analytical measurements. Haber and Klemensiewicz were also working on ISEs at the beginning of the last century [2]. Extremely important for potentiometry was the year 1966, when Frant and Ross announced their pioneering discovery of the fluoride ion-selective electrode [3]. Ion-selective sensors were improved in subsequent years by researchers such as Bloch, Moody, Thomas, Bakker and Sokalski. However, the breakthrough in the field of potentiometry came in 1971 when Freiser and Cattrall invented the first solid contact ion-selective electrode (SC-ISE), which they called a coated wire electrode (CWE). It was the first ion-selective electrode in which there was no internal solution. This completely revolutionized the field of potentiometry and opened the way to a number of new construction possibilities [4]. A brief history of ISEs is presented in Figure 1.

Their discovery contributed to the development of SC-ISEs, which, compared to classical ion-selective electrodes, have a number of advantages undoubtedly due to the elimination of the internal solution. As a result of this action, the design of these electrodes has been simplified and downsized, and, consequently, the production costs of these measuring instruments have been significantly reduced [5,6].

Citation: Wardak, C.; Morawska, K.; Pietrzak, K. New Materials Used for the Development of Anion-Selective Electrodes—A Review. *Materials* **2023**, *16*, 5779. https://doi.org/10.3390/ma16175779

Academic Editors: Jinsheng Zhao and Zhenyu Yang

Received: 24 July 2023
Revised: 13 August 2023
Accepted: 21 August 2023
Published: 23 August 2023

Copyright: © 2023 by the authors. Licensee MDPI, Basel, Switzerland. This article is an open access article distributed under the terms and conditions of the Creative Commons Attribution (CC BY) license (https://creativecommons.org/licenses/by/4.0/).

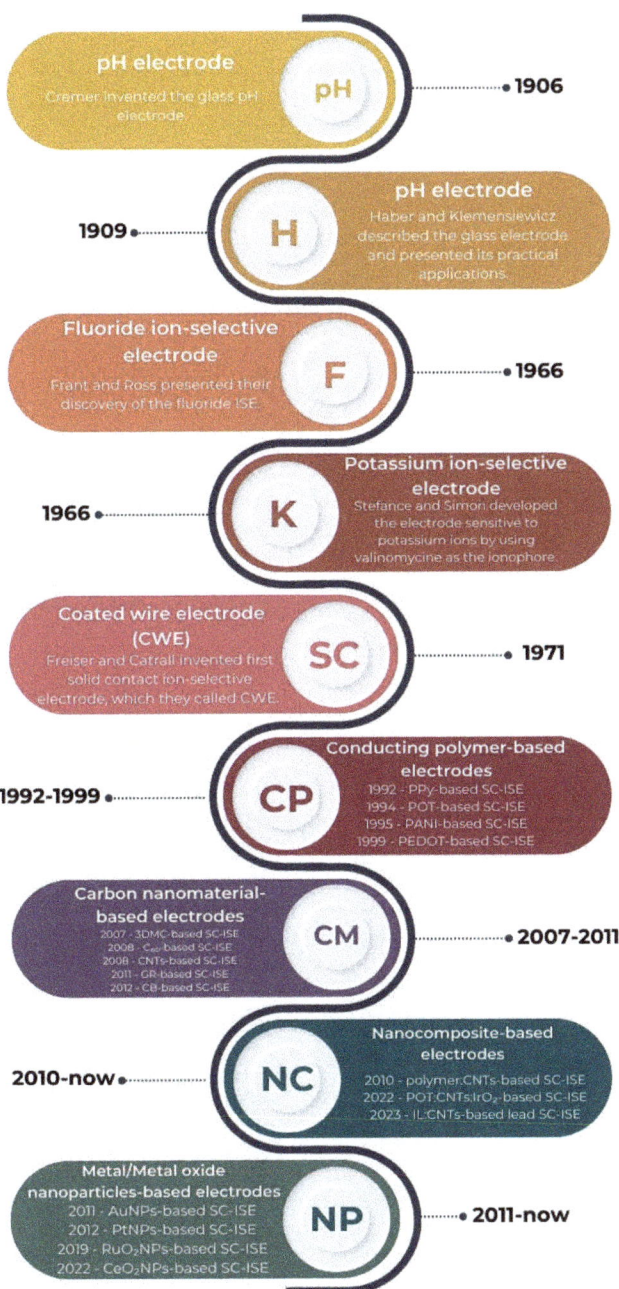

Figure 1. A brief history of ISEs development.

By comparing the differences in the design of classic ion-selective electrodes with a liquid contact and ion-selective electrodes with a solid contact (Figure 2), it can be observed that SC-ISEs have an ion-selective membrane and an inner electrode just like classical ISEs. SC-ISEs differ from classical electrodes because of the absence of an internal electrolyte and the presence of solid contact in the form of an intermediate layer between the internal electrode and the ion-selective membrane. The solid contact that is present in SC-ISEs provides the transport of ions and the conversion of their signal into an electrical signal [7,8]. It fulfills the functions of the internal solution in classical electrodes, the presence of which forces the measurement in a vertical position and is the cause of difficulties in the miniaturization of these electrodes.

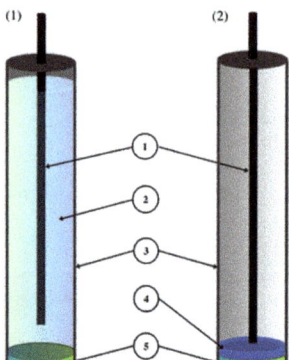

Figure 2. Construction of liquid contact ion-selective electrode (**1**) and solid contact ion-selective electrode (**2**) (1—inner electrode; 2—internal electrolyte; 3—holder; 4—solid contact layer; and 5—ion-selective membrane).

The complete elimination of the internal electrolyte solution in the SC-ISEs enables not only the miniaturization of these sensors but also often obtains a lower limit of detection (LOD) even to nanomole concentrations [9,10]. This is a consequence of the fact that in the absence of an internal electrolyte solution, there is no flow of ions from the internal solution (containing usually a high concentration of these ions) through the membrane into the sample solution. In addition, the lack of an internal solution eliminates the operations related to it: the problem of its leakage, the presence of air in the electrode and the need to work in a vertical position. SC-ISEs show greater mechanical resistance. This is due to the fact that the membrane of the electrodes with solid contact is usually thicker and placed on a hard surface. This makes them easier to store and transport, which is especially important for field measurements. [11,12]. The electrodes are easy to obtain in small sizes and in any shape. This allows the construction of multi-sensor measurement platforms, e.g., electronic tongues [13] or wearable sensors in the form of wristbands or elements of clothing [4]. The most common disadvantage of SC-ISEs is insufficient potential stability related to inadequate mutual adhesion of the used materials [14] and the formation of an unfavorable water layer between the membrane and the inner electrode [15].

The research in the area of ion-selective electrodes is still being intensively conducted and has resulted in the development of sensors with better analytical performance and/or simpler operations. These include two main research directions: the development of new active substances to obtain more selective ion-sensitive membranes and/or new primarily electroconductive materials, which are used in the construction of electrodes without an internal electrolyte solution to improve the process of charge transfer between the membrane and the internal electron conductor. Considering the type of ion to which the electrode is sensitive, there is definitely less work involving anionic electrodes. First and foremost, this is associated with the fact that, for anions, the number of commercially available ionophores is much smaller. This fact is also an inspiration to search for new compounds that can fulfill this function in the membrane.

In the literature on ion-selective electrodes, there are almost no review articles focusing on anion-selective electrodes. Only one short review has been published in the last 20 years in which the authors focused on ionophores used in anion-sensitive membranes [16].

This review presents recent developments in the field of ion-selective electrodes sensitive to inorganic anions. It describes 102 anionic ion-selective electrodes used in potentiometric measurements developed over the last five years. Among them were 67 electrodes with solid contact and 35 with liquid contact. ISEs sensitive to 18 different ions were described, which include anions such as NO_3^-, F^-, Cl^-, Br^-, I^-, ClO_4^-, S^{2-}, SO_3^{2-}, SO_4^{2-}, $H_2PO_4^-$, HPO_4^{2-}, PO_4^{3-}, SCN^-, AsO_4^{3-}, BO_3^{3-}, CH_3COO^-, CO_3^{2-} and SiO_3^{2-}. This paper is a comprehensive overview containing the characteristics of new materials used in the construction of ISEs, including both materials used as active substances of the ion-selective membrane, as well as materials used to improve the efficiency and/or easier and more universal use of ISEs.

2. Constructional Variants of Anionic Ion-Selective Electrodes

The most important component of any ion-selective electrode is the ion-sensitive membrane. The most important component of the membrane from the point of view of electrode operation is the ionophore, which gives it sensitivity and selectivity to a specific ion. The material used as an ionophore should selectively bind a given ion in the membrane environment. In the case of anion-sensitive electrodes, three types of membranes can be distinguished: polymeric membranes based on PVC, crystalline membranes composed of insoluble salts of the determined ion and composite membranes containing the active ingredient combined in various ways with other materials. Polymeric and crystalline membranes are still used in classic liquid-contact electrodes; however, a much larger group includes a variant of electrodes without internal electrolyte solution called all-solid-state ion-selective electrodes (ASS-ISEs). In recent years, they have become, among ISEs, one of the most studied groups of analytical measuring instruments. Depending on their construction, ASS-ISEs can be classified as follows:

- CWE, CDE—a coated wire/coated disc electrode, which was originally a platinum wire coated with a layer of PVC membrane. CWEs/CDEs are created by applying, for example, a polymer membrane or crystal membrane on a wire/disc used as the inner electrode. Such sensors are now rarely used. In publications, they are studied as a comparative system to assess the effectiveness of solid contact or modification of the membrane composition.
- SC-ISE—solid contact ISE is a type of electrode in which an intermediate layer of material, generally called solid contact, is introduced between the ion-selective membrane and the inner electrode. Conductive polymers, carbon nanomaterials, metal and metal oxide nanoparticles and composite materials are most often used as intermediate layers.
- SP-ISE—a single-piece ion-selective electrode with a membrane in which the modifying material is dispersed or dissolved. An ion-selective electrode is created by depositing a membrane cocktail of conductive polymers or other conductive materials, e.g., carbon nanotubes, onto a surface that is usually a glassy carbon electrode [17,18]. Carbon paste ion-selective electrodes can also be included in this group of electrodes in which the active component of the membrane is mixed with graphite powder and binder material, and the mixture is packed in the sensor holder with an internal electrode placed inside (often a copper wire). In order to improve their work, the composition of the paste electrode membrane is also modified by the addition of various materials, which are often carbon nanomaterials. A comparison of the construction of different ASS-ISEs electrodes is shown in Figure 3.

All-solid-state constructions, except CWEs and CDEs, provide a reduction of potential drift and improve stability as well as reproducibility of the potential. The additional solid contact materials and conductive materials used for membrane modification increase the electrical capacitance and improve the ion-electron conductivity. To achieve these goals, a system with solid contact must fulfill three important conditions, which have been defined by Nikolski and Materov: (1) The current passing through the sample during the measurement should be significantly lower than the exchange current, which should be characterized by high values. (2) The chemical indifference towards the other interferents present in the sample—there must be no side reactions taking place alongside the main electrode reaction. (3) The equilibrium of the ion-electron conductivity should be reversible and stable [19].

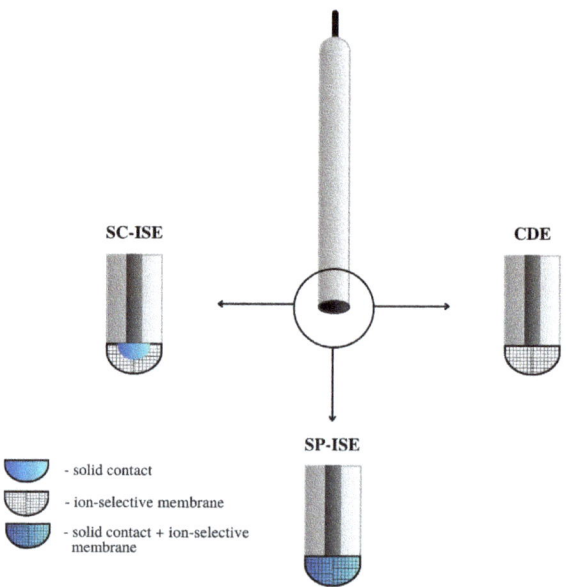

Figure 3. A comparison of the construction of CP-ISE, SP-ISE and SC-ISE electrodes.

Two of the most disadvantageous properties of CWEs and CDEs are their weak reproducibility and stability of the potential. Besides this, the formation of an aqueous layer between the ion-selective membrane and the internal ("lead") electrode is a fairly common problem, which in turn results in the generation of a higher potential drift and a shorter electrode lifetime. This problem can be significantly minimized by the above-mentioned modifications. In this case, the hydrophobic materials prove to be very suitable.

3. Research Methods Used to Assess New ISEs

Newly developed ion-selective electrodes are subjected to numerous tests to determine their parameters and assess their analytical usefulness. The basic measurements carried out in each case include the production of a calibration curve from which the LOD, measuring range and slope of the characteristic are determined. Equally important are the selectivity tests consisting of determining the values of selectivity coefficients towards interfering ions. These measurements are absolutely necessary when a new active substance is introduced into the electrode membrane. The most commonly used methods for determining selectivity coefficients are those recommended by IUPAC, i.e., the separate solution method (SSM) and the fixed interference method (FIM) [20].

In order to check the effectiveness of the introduced electroconductive materials in the form of an intermediate layer or as a membrane component, the potential drift tests are performed: the potential change under zero current conditions and the estimated E^0 potential stability in time [21]. Chronopotentiometry and electrochemical impedance spectroscopy measurements also provide valuable information in this area as they make it possible to determine the electrical capacitance of the electrodes and their resistance, which significantly influence the stability of the electrode potential [15,18,22]. Water layer tests are also performed according to the procedure proposed by the Pretsch group [15].

4. Ion-Selective Electrodes Sensitive to Nitrates Ions

The monitoring of nitrate ions is very important, which is why new tools are being developed to make their identification more accurate and simple. Determining the concentration of nitrate ions is crucial because an excess of nitrates in the human diet is harmful to health and can cause, for example, problems with the cardiovascular system and dysfunction of the intestines and the rest of the gastrointestinal system. It can also lead to pathological conditions such as cancer. We have collected 19 nitrate ion-selective electrodes that were developed over the past five years. Almost all of the nitrate electrodes were solid contact electrodes, with the exception of two that had liquid contact.

For the determination of nitrates, a few groups of scientists proposed electrodes with a solid contact, which had TDMAN (tridodecylmethylammonium nitrate) as the ionophore. They were different from each other, for example, in the inner electrode and material of the intermediate layer. In article [23], the structure of the electrode was developed by using screen-printing technology, in which the working electrode was a graphite electrode deposited from a graphite-based ink, was described. Graphite was used as the transducer layer due to its stability, as well as its good conductive and hydrophobic properties. The slope of the characteristic of −55.4 mV/decade slightly deviated from the Nernst value, and the obtained linearity range for the tested microelectrode was not very wide (2.9×10^{-4}–1.7×10^{-1} M). The limit of detection was quite high, so measurements of low concentrations could be problematic. The advantage of this electrode is its good repeatability.

The second electrode with TDMAN as the ionophore was described in the publication [24], where the ISE based on hydrophobic laser-induced graphene (LIG) as an intermediate layer was presented. The hydrophobic LIG was created using a polyimide substrate and a double lasing process. A study using the XPS (X-ray photoelectron spectroscopy) technique was carried out to check the composition of the LIG, and the static contact water angle of $135.5 \pm 0.7°$ was determined. The slope of the characteristic equaled −58.17 mV/decade and was much closer to the Nernst response in comparison to the electrode described as the first. The ranges of the linearity were comparable, but the detection limit extended to micromole values, which was a significantly better result. This electrode was successfully applied in the monitoring of surface water quality.

The authors of another work [25] proposed a novel electrode in which a new nanocomposite consisting of poly(3-octylthiophene) and molybdenum disulfide (POT-MoS_2) was used as the intermediate layer. A gold electrode (AuE) was used as the basic electrode. The SC-ISE exhibited an over-Nerstian response, while the other parameters were comparable to the ISEs described above. The parameters of the nanocomposite-modified electrodes were not remarkably different from those for ISEs based on MoS_2 or POT. Nevertheless, the combination of POT-MoS_2 gave the best results, i.e., the slope, LOD or potential stability. An additional advantage of the newly proposed composite is its high hydrophobicity and redox properties. By replacing the membrane, this electrode can be used for the determination of other ions, i.e., potassium, phosphate or sulfate ions.

Subsequent electrodes having the same type of ionophore—TDMAN—were presented in the work [26,27]. In the publication [26], the solid contact ion-selective electrodes with different types of transducer layers (TLs) were described. The TL was constituted by polyaniline nanofibers doped with Cl^- (PANINFs-Cl^-) or NO_3^- (PANINFs-NO_3^-) ions (the chemical formulas are shown in Figure 4). Several variations of electrodes were prepared using the polymer conductor—it was used as a suspension dropped directly onto the surface of the inner electrode, which was GCE (glass carbon electrode), and as a component of a membrane cocktail, where it occurred in different amounts (0.5%, 1%, 2%). All polyaniline nanofiber-modified electrodes had better parameters than the unmodified electrode. Electrodes with PANI nanofibers doped with Cl^- ions had a wider range of linearity and a lower detection limit than those where PANINFs were doped with NO_3^-. On the other hand, the second ones exhibited a slope closer to the Nernst slope of −57.8 mV/decade. ISEs with both types of nanofibers did not show potential sensitivity to changes in pH in a very wide range. Furthermore, these types of electrodes showed excellent stability and potential reversibility and also had a lower membrane resistance value and a higher value of double-layer electrical capacitance relative to the unmodified electrode. These electrodes were successfully used to determine nitrate ions in real environmental samples, i.e., drinking, river and groundwater.

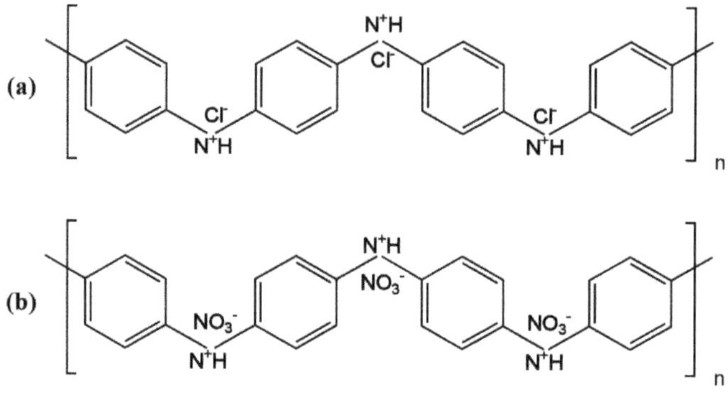

Figure 4. The structure of polyaniline doped with chloride (a) and nitrate (b) ions [26].

The same type of ionophore was also presented in the article [27], where a miniaturized version of the ion-selective electrode was presented in which the solid contact was mesoporous black carbon (MCB) and the conductive electrode was a silver wire placed in a glass capillary. MCB provides good ion-electron conductivity and, in addition, has a large specific surface. An additional advantage is its high availability and low price. The addition of MCB significantly stabilized the electrode's response, ensuring adequate double-layer capacitance and decreasing the resistance more than a hundred times over coated wire electrodes. In comparison to electrodes described in the work of [26], these ISEs had a lower slope of −54.8 mV/decade, an order of magnitude more narrow range of linearity and a similar detection limit. Undoubtedly, the disadvantage of this electrode is its relatively short lifetime oscillating around 20 days. By comparing the electrodes proposed in the publications [26,27], it is clearly preferable to incorporate modifications such as PANINFs rather than MCBs.

Also, in the paper [28], for the determination of nitrate ions, an electrode in which TDMAN performed the function of the ionophore was proposed. The inner electrode was a screen-printed carbon electrode (SPCE), on whose surface cobalt (II, III) oxide nanoparticles (Co_3O_4NPs) were deposited, and acted as an ion-to-electron transducer. The solid contact was examined in terms of morphology and physical properties using XRD, EDS (energy-dispersive X-ray spectroscopy), SEM (scanning electron microscope) and TEM (transmission electron microscopy). Novelization of the electrode structure with hydrophobic Co_3O_4 nanoparticles prevented the formation of an aqueous layer and improved its response in relation to ISE without a Co_3O_4NP layer. This electrode also exhibited a good slope of −56.8 mV/decade, as well as a low limit of detection of 1.04×10^{-8} M. According to the EIS spectra, a reduction of membrane resistance was observed for the electrode modified with oxide nanoparticles. As was confirmed by comparing the results of SC-ISEs and spectrophotometer determinations, this electrode is certainly suitable for the analysis of various types of water samples.

The subsequent two electrodes were based on a glassy carbon electrode using polypyrrole (PPy) doped with NO_3^- ions as the ion carrier. The electrodes differed in their solid contact layer, which, in the case of the electrode presented in the publication [29], was a pericarpium granati-derived biochar (PGCP) activated with phosphoric acid. PGCP was applied together with PPy, forming a bilayer membrane. Based on morphological studies PGCP was found to have a porous structure and, on the basis of the EIS measurements, also a low resistance, which translates into the possibility of using it as a transducer material for the charge transfer. In the second article, a double-layer structure consisting of gold nanoparticles (AuNPs) (solid contact) and a conductive polymer PPy-NO$_3^-$ (ion-selective membrane) was successfully used [30]. AuNPs solid contact layer and nitrate-doped polypyrrole molecularly imprinted polymer membranes were prepared by electrodeposition. Both of these electrodes had a slope that deviated from the ideal Nernstian slope and respectively amounted to -50.86 mV/decade for PGCP-ISE and -50.4 mV/decade for AuNPs-ISE. For the PGCP-modified electrode, the linearity range was in the range of 1.0×10^{-5}–5.0×10^{-1} M and the LOD was 4.64×10^{-6} M, which is very similar to the results obtained for the gold nanoparticle-modified electrode, where the LOD was an order of magnitude higher. A positive aspect of the electrode, with solid contact in the form of AuNPs, is the favorable result of the aqueous layer test because no potential drift towards higher values of potentials was observed for this electrode compared to the unmodified AuNPs electrode. Finally, both the PGCP- and AuNPs-modified electrodes exhibited excellent long- and short-term stability and significantly better electrical parameters than the unmodified electrode. The PGCP-ISE was also successfully used for the determination of nitrates in samples from Shenzhen OCT wetland and laboratory wastewater with 4% RSD.

In publications [31–33] are presented electrodes that were connected by the presence of the same ionophore, which was nitrate ionophore VI. The ISE described by [31] was developed using screen-printing technology by applying inkjet printing gold onto the substrate, thereby forming a screen-printed gold electrode. Afterwards, the entire sensor was secured with Teflon tape and the membrane containing the ionophore was applied. The parameters of this potentiometric sensor were determined, i.e., the slope of the characteristic -54.1 ± 2.1 mV/decade and the linearity range between 1.0×10^{-5} M and 1.0×10^{-1} M. The water layer test was carried out with positive results. Unfortunately, the LOD for the obtained electrode was not included in the publication, and the electrical parameters were not determined.

Another article [33] described an SC-ISE that was produced by applying a membrane mixture containing varying amounts of PTFE (poly(tethrafluoroethylene)) (0%, 2.5%, 5%. 7.5%, 10%) onto a screen-printed electrode. The best results were obtained for an ISE in which PTFE constituted 5% of the membrane cocktail. This electrode was characterized by a slope of -58.0 mV/decade, which unfortunately reduced to -35.0 mV/decade after 20 days of measurements, which means that the lifetime of this electrode is not long, and further studies are required to eliminate this drawback.

The last discussed electrode using nitrate ionophore VI is an SC-ISE based on the gold electrode (AuE) and thiol-functionalized reduced graphene oxide (TRGO) as solid contact material, which was used for the first time in this role [32]. This electrode was characterized by an almost Nernst response as the slope equaled -60.0 ± 0.5 mV/decade. The range of linearity of the described ISE was similar to electrodes possessing the same type of ionophore. An additional advantage of this SC-ISE is the low detection limit reaching micromole values and the wide working range of pH 2.0–10.0. Moreover, the use of TRGO improved the potential reversibility, which indicates that reduced graphene oxide performs well in the solid contact function due to its good conductivity. In addition, according to the EIS tests, the membrane resistance values of the electrode with TRGO are lower than those for the unmodified electrode. The aqueous layer test came off satisfactorily as no potential drift was observed. After improving the stability of the electrode in real samples, it will be possible to use it for the determination of ions in blood samples.

A new ionophore as the active ingredient in an ion-selective membrane used in the construction of nitrate(V) ion-selective electrodes was proposed by our group in the publication [34]. In the construction of SC-ISEs, a cobalt(II) complex with 4,7-diphenyl-1,10-phenanthroline (Co(Bphen)$_2$(NO$_3$)$_2$(H$_2$O)$_2$) was used, which performed very well in this role and cooperated with the Ag | AgCl inner electrode. Trihexyltetradecylphosphonium chloride was selected as the ionic component of the membrane, which provided a constant concentration of chloride ions and represented a reversible redox system with the internal electrode Ag/AgCl/Cl-(silver/silver chloride electrode/chloride). A satisfying electrode response was obtained for this ISE, which was characterized by a slope of

−56.34 mV/decade, a linearity range of 1.0×10^{-5}–1.0×10^{-1} M and an LOD lower than the micromole value. The electrode exhibited a constant potential over a wide pH range of 5.4–10.6. Furthermore, the reversibility of the potential and its drift were at a very satisfactory level. This electrode was successfully applied to the determination of the concentration of NO_3^- ions in tap, mineral and river water samples with reproducibility close to 100%. The same ionophore was used in an ion-selective electrode based on a glassy carbon electrode [35]. A nanocomposite consisting of multi-walled carbon nanotubes and the ionic liquid trihexyltetradecylphosphonium chloride (THTDPCl) was proposed as a solid contact and used as a membrane component. Various composites obtained from MWCNTs differing in structure were tested. The electrode with a nanocomposite containing nanotubes characterized by the highest porosity and homogeneity of the structure showed the best performance. In comparison to the previous electrodes with the same ionophore [34] based on Ag/AgCl/Cl, the electrode with a nanocomposite achieved a better slope closer to Nernst's value, which was −57.1 mV/decade, a lower detection limit of 5.0×10^{-7} M and a linearity range that was an order of magnitude larger, as well as a wider working pH range. Interestingly, the addition of the nanocomposite directly to the membrane did not cause the redox sensitivity of the electrodes. Moreover, this simultaneously increased their hydrophobicity, so that no water film formation was observed between the membrane and the inner electrode in these SC-ISEs.

The ionophore in the form of $TDANO_3$ (tetradecylammonium nitrate) was used both in a nitrate ion-selective electrode with a solid contact [36] and a liquid contact [37]. In the solid-contact electrode, the leading electrode was a screen-printed carbon electrode, and the solid-contact layer was a reduced graphite oxide aerogel (rGOA) [36]. A Nernst response of −59.1 mV/decade, a linearity range of 0.1 M to micromoles and a detection limit of 7.59×10^{-7} M were obtained. With success, this electrode was used for the determination of NO_3^- ions in perilla leaves, where the results achieved by ISE were compared with chromatographic results. In the case of liquid contact, the inner electrode was one of the more commonly selected Ag | AgCl electrodes and the inner electrolyte—solution of 0.01 M KNO_3 or 0.001 M KCl [37]. There is no doubt that the team carrying out the study on SC-ISEs succeeded in improving the parameters obtained for the competing electrode with LC. The slope of the characteristic differed from the theoretical value and was -53.7 ± 0.4 mV/decade, and, additionally, the values for linearity range and LOD were less satisfactory, so in this case the solid-contact ISE worked much more effectively.

In order to improve the performance of nitrate ion-selective electrodes with solid contact, the application of a nitron–nitrate complex (Nit^+/NO_3^-) that acted as an ionophore was proposed [38]. A novel form of an ion-selective membrane that had the shape of a 'sandwich membrane' (bilayer membrane) was presented. This type of membrane was formed by pressing together two previously dried membranes, where one of them contained an ionophore and the other did not. The membrane prepared in this way was then applied onto the surface of the glassy carbon electrode with the non-ionophore side, where the part with the ionophore had direct contact with the testing sample. Multi-walled carbon nanotubes were used as the solid contact. The presented electrode showed a very wide linearity range of 8.0×10^{-8}–1.0×10^{-1} M, and a low LOD of 2.8×10^{-8} M. The response of the GCE/MWCNTs/NO_3^--ISM electrode was equal to -55.1 ± 2.1 mV/dec. The working pH range of such an electrode was between 3.5 and 10.0 pH and the lifetime was 8 weeks. The proposed ISE exhibited a very high tolerance to the presence of interfering ions, as we could observe from the rather low values of the selectivity coefficients. However, the authors did not report a value of this parameter for highly lipophilic ions, i.e., ClO_4^- or SCN^-, which are well-known interferents for nitrate electrodes. The ISE was successfully used to measure the concentration of nitrate ions in wastewater samples.

The article [39] presents liquid-contact nitrate ion-selective electrodes used for the determination of nitrates in hydroponic solutions. A design with two internal solutions and two membranes was used. The internal electrode was a silver chloride electrode placed in a PP tube filled with 0.1 M LiCl, which ended with a Nafion membrane—this constituted part 1. Part 1 was immersed in part 2, which was a second PP tube that ended with a PVC membrane, which, in turn, contained a 0.1 M LiNO$_3$ solution (Figure 5). The ionophore included in the PVC membrane was THANO$_3$ (tetraheptylammonium nitrate), which had already been used in nitrate ion-selective electrodes. On the basis of the calibration curve, a slope of -53.3 ± 0.1 mV/decade was determined, which deviates slightly from the ideal value for monovalent ions. The range of linearity and the detection limit took on average values and did not stand out from other nitrate electrodes. It was concluded that this type of electrode could be used in studies of NO$_3^-$ ions for more than four weeks in order to determine their concentration in hydroponic solutions.

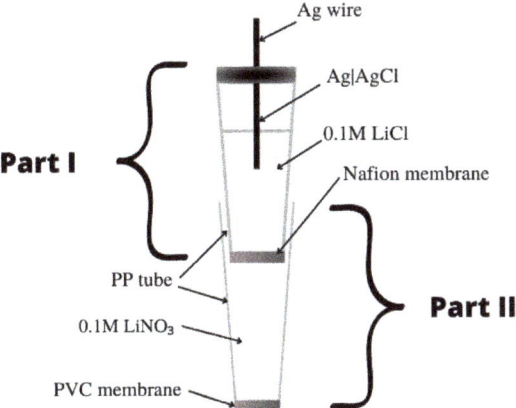

Figure 5. Ion-selective electrode construction based on [39].

The paper [40] presents a new form of solid contact in the form of TTF-TCNQ (tetrahiafulvalene-tetracyanoquinodimethane) illustrated in Figure 6. Here, the inner electrode was a glassy carbon disc onto which an intermediate layer was applied, which was followed by an ion-selective membrane containing an active ingredient, which was a nitrate ionophore V. The electrode showed great potential stability and was not sensitive to changes in the redox potential. The response of the electrode was almost Nernstian and amounted to -58.47 mV/decade, while the LOD was equal to 1.6×10^{-6} M. The presence of the proposed solid contact improved the selectivity of the nitrate electrodes and their electrical parameters, i.e., decreased the membrane resistance and increased the electrical capacitance.

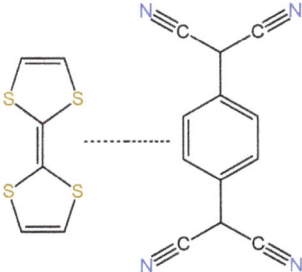

Figure 6. Tetrathiafulvalene 7,7,8,8-tetracyanoquinodimethane TTF-TCNQ salt used as solid contact in nitrate SC-ISE [40].

A comparison of the analytical parameters of various nitrate electrodes is presented in Table 1.

Table 1. Nitrate ion-selective electrodes.

E. No	Type of Contact	Ionophore/Ion Carrier	Intermediate/Transducer Layer	Type of Internal Electrode	Slope [mV/decade]	Range of Linearity [M]	Limit of Detection [M]	pH Range	Application/Samples	References
1	SC	TDMAN	Graphite	GrE	-55.4 ± 0.7	2.9×10^{-4}–1.7×10^{-1}	2.04×10^{-4}	4.0–11.0	Industrial and environmental.	[23]
2	SC	TDMAN	LIG	-	-58.2 ± 4.2	5.0×10^{-4}–1.0×10^{-1}	6.01×10^{-6}	6.0–8.0	Agricul-ture and surface water.	[24]
3	SC	TDMAN	Co_3O_4NPs	SPCE	-56.8	1.0×10^{-7}–1.0×10^{-2}	1.04×10^{-8}	3.0–8.0	Aquaculture, river, domestic and tap water.	[28]
4	SC	TDMAN	PANINFs-Cl	GCE	-56.8	1.0×10^{-6}–1.0×10^{-1}	3.16×10^{-7}	4.0–12.5	Environmental samples.	[26]
5	SC	TDMAN	PANINFs-NO_3		-57.8	1.0×10^{-6}–1.0×10^{-1}	1.12×10^{-6}	4.0–11.5		
6	SC	TDMAN	MCB	Ag wire	-54.8	5.0×10^{-5}–1.0×10^{-1}	2.5×10^{-6}	-	-	[27]
7	SC	TDMAN	POT-MoS_2	AuE	-64.0	7.1×10^{-4}–1.0×10^{-1}	9.2×10^{-5}	-	Soil.	[25]
8	SC	PPy-NO_3^-	PGCP	GCE	-50.9	1.0×10^{-5}–5.0×10^{-1}	4.64×10^{-6}	3.5–9.5	Environmental and clinical laboratories.	[29]
9	SC	PPy-NO_3^-	AuNPs	GCE	-50.4	5.3×10^{-5}–1.0×10^{-1}	5.25×10^{-5}	-	Water samples and aqueous solutions of fertilizers.	[30]
10	SC	Nitrate ionophore VI	-	PAuE	-54.1	5.0×10^{-5}–1.0×10^{-1}	-	-	Field soils.	[31]
11	SC	Nitrate ionophore VI	TRGO	AuE	-60.0 ± 0.5	4.0×10^{-5}–1.0×10^{-1}	4.0×10^{-6}	2.0–10.0	Blood.	[32]
12	SC	Nitrate ionophore VI	PTFE	SPCE	-58.0	-	-	4.0–11.0	Wastewater.	[33]
13	SC	$Co(Bphen)_2(NO_3)_2(H_2O)_2$	MWCNTs-THTDPCl	GCE	-57.1	1.0×10^{-6}–1.0×10^{-1}	5.0×10^{-7}	6.0–8.0	-	[35]
14	SC	$Co(Bphen)_2(NO_3)_2(H_2O)_2$	Ag/AgCl/Cl$^-$	Ag/AgCl	-56.3	1.0×10^{-5}–1.0×10^{-1}	3.98×10^{-6}	5.4–10.6	Mineral, tap and river water.	[34]
15	SC	TDANO_3	rGOA	SPCE	-59.1	1.0×10^{-6}–1.0×10^{-1}	7.59×10^{-7}	-	Plant sap e.g., perilla leaf.	[36]
16	LC	TDANO_3	0.01 M KNO_3 and 0.001 M KCl	Ag/AgCl	-53.7 ± 0.4	1.0×10^{-5}–1.0×10^{-1}	1.3×10^{-6}	-	-	[37]
17	SC	Nit$^+$/NO_3^-	MWCNTs	GCE	-55.1 ± 1.0	8.0×10^{-8}–1.0×10^{-2}	2.8×10^{-8}	3.5–10.0	Environmental samples.	[38]

Table 1. Cont.

E. No	Type of Contact	Ionophore/Ion Carrier	Intermediate/Transducer Layer	Type of Internal Electrode	Slope [mV/decade]	Range of Linearity [M]	Limit of Detection [M]	pH Range	Application/Samples	References
18	LC	THANO$_3$	0.1 M LiCl and 0.1 M LiNO$_3$	Ag|AgCl	−53.3 ± 1.0	1.0×10^{-5}–1.0×10^{-1}	1.0×10^{-6}	-	Hydroponic solutions.	[39]
19	SC	Nitrate ionophore V	TTF-TCNQ	GCD	−58.5	1.0×10^{-5}–1.0×10^{-1}	1.6×10^{-6}	-	Aqueous samples.	[40]

5. Ion-Selective Electrodes Sensitive to Fluoride Ions

Excessive amounts of fluoride ions become a threat not only to plant and animal organisms but also to humans. Increased content of this ion can cause disturbances in the normal sequence of the biological chain, and as a result, the ecological balance will be upset. In order to check how an excess of fluoride ions affects organisms, a number of studies were carried out over several years, which confirmed its neurotoxicity. High quantities of F^- ions damage cell organelles, i.e., mitochondria, and possess a mutagenic function, which in turn can lead to changes in gene expression. Furthermore, due to the ubiquitous presence of fluoride in drinking water and the large-scale use of fluoride, e.g., in toothpaste, it is necessary to monitor its concentration in the surrounding environment [41,42]. Consequently, as potentiometry is a rather cheap, fast and accurate method, it can be used to monitor fluoride ion concentrations in different types of samples. Over the past five years, 13 fluoride ion-selective electrodes have been proposed, and their design has been continuously improved to obtain better results.

The article [43] describes a new ion-selective electrode based on an electrode constituted by a titanium film, where lanthanum fluoride (LaF_3) nanocrystals doped with europium (Eu) were proposed as the ion-conducting layer to improve the conductivity of the membrane. No solid contact intermediate layer was used here. The results obtained for this electrode were average, which was reflected in a typical linearity range of 1.0×10^{-5}–1.0×10^{-1} M and a slope of -56 mV/decade. The deviation from the slope measurements was quite large and as high as -13 mV/decade. It is worth mentioning that the addition of Eu increased the slope of the characteristic by -10 mV/decade. In addition, it was possible to obtain LOD results at the micromole level. The disadvantage of these electrodes is the loss during the measurement of some Eu-doped LaF_3 nanocrystals, which is the reason for the instability and non-repeatability of the potential.

The paper [44] presents a comparison of the performance of fluoride electrodes, each of which had a LaF_3 single crystal membrane as an ion carrier. Two of the tested electrodes had the same silver inner electrode (AgE) to which different solid contact layers were applied. In the first case, PEDOT was used as the SC, which performed well in this role and undoubtedly contributed to stabilizing the electrode potential. It showed a slope of -56.0 ± 0.9 mV/decade. The second SC-ISE had an intermediate layer in the form of Ag paste. A super-normal slope was obtained, but the detection limit was the worst of all the fluoride electrodes described, with a slope of 2.0×10^{-2} M. Both SC-ISEs had the same linearity range. The third fluoride electrode presented in this article was a liquid-contact electrode, where the IE was Ag|AgCl and the internal solution was a mixture of phosphate buffer solution (PBS)—0.01 M Na_2HPO_3 and 0.02 M KH_2PO_3. For this electrode, the response slope of -38.6 ± 9.1 mV/decade was unsatisfactory and far from the book value. The best performing of the three presented electrodes was the PEDOT-modified SC-ISE, which had not only the lowest potential drift of 0.06 mV/min but also the highest electrical capacitance of 937 µF, which is 70 times higher than the Ag paste-modified electrode.

Novel fluoride electrodes with a LaF_3 single crystalline membrane and different types of intermediate layers (solid contact or electrolyte solution) with the addition of Fe_xO_y nanoparticles were presented in [45]. The intermediate layer consisted of Fe_xO_yNPs, which were incorporated into a membrane. For the SC-ISE, the internal electrode was a stainless-steel disc (SSDE), and for the liquid contact electrode, it was a silver chloride electrode, which was immersed in an electrolyte that was a mixture of KCl, HCl, 0.1 M KNO_3. The working pH range was the same for both types of electrodes. The LC-ISE had a lower detection limit of 7.4×10^{-8} M and an order of magnitude wider range of linearity compared to the SC-ISEs, which in turn had a higher slope response. An additional advantage of the LC electrode was the high stability of the sensitivity over a period of two years.

A novel construction of a wearable spandex textile-based solid-contact fluoride sensor is presented in the article [46]. The scientific team proposed an SC-ISE based on a screen-printed carbon electrode (screen-printing spandex electrode) to which an intermediate layer of MWCNTs-COOH (carboxyl-functionalized multi-walled carbon nanotubes) was deposited, followed by a membrane containing the fluoride ionophore bis(fluorodioctylstannyl)methane shown in Figure 7. This electrode exhibited good reproducibility, stability and potential reversibility. The achieved Nernst characteristic response of −59.2 mV/decade and the nanomole value of the detection limit only confirmed that this sensor is suitable for the determination of even trace concentrations of fluoride ions. An additional advantage of this electrode is its high selectivity. An SC-ISE of this type can be used for the determination of either chemicals, biologicals or DFP (diisopropyl fluorophosphate).

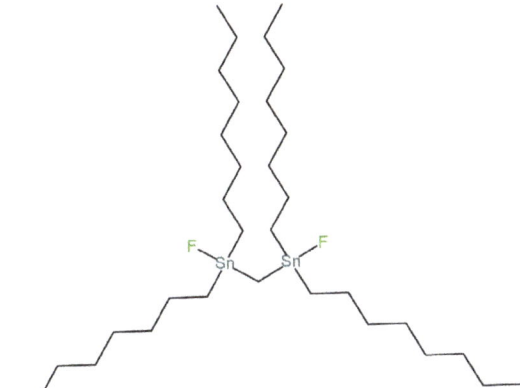

Figure 7. The structure of bis(fluorodioctylstannyl)methane.

Another electrode proposed for the determination of fluoride ions is SC-ISE, which in its design had a crystalline membrane consisting of a single crystal cadmium(II) Schiff base complex (CdLI$_2$—cadmium iodide complex) (Figure 8) [47]. The electrode was obtained by impregnating the carbon paste electrode (CPE) with the Schiff ligand (E)-N$_1$-(2-nitrobenzylidene)-N$_2$-(2-((E)-(2-nitrobenzylidene)amino)ethyl)ethane-1,2-diamine(L) and its complex CdI$_2$ (CdLI$_2$). CdLI$_2$ performed of the function of fixed-charge carriers. The presented electrode exhibited a good membrane resistance determined by the EIS method, a satisfactory slope of −58.9 mV/decade and a rather low detection limit of 1.2×10^{-7} M. In addition, very good selectivity against the ions such as IO$_4^-$, SCN$^-$, SO$_4^{2-}$, CN$^-$, ClO$_3^-$, Br$^-$, Cl$^-$ and I$^-$ was recorded.

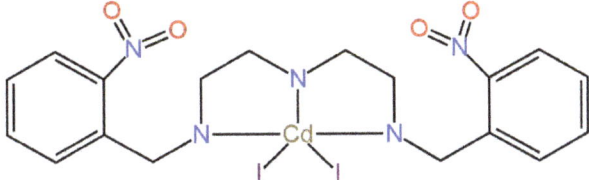

Figure 8. The structure of CdLI$_2$—cadmium iodide complex [47].

Fluoride solid contact electrodes with a new type of ionophore belonging to Lewis acidic organo-antimony(V) compounds were presented in [48]. A silver chloride electrode was used as the internal electrode, and a solution containing 0.2 M Gly/H$_2$PO$_4$ buffer and 0.001 M NaF was used as the internal electrolyte. All four ionophores are shown in Figure 9. The electrode in which tetrakis-(pentafluorophenyl)stibonium (Figure 9b) was used had the most Nernstian electrode response of −59.2 mV/decade and the lowest detection limit of 5.0×10^{-6} M. The electrode in which tetraphenylstibonium fluoride (Ph$_4$SbF) (Figure 9a) acted as the ionophore had a slightly lower slope and higher LOD. The ionophore shown in Figure 9d was a component of the LC-ISE membrane cocktail with a slope of −57.8 mV/decade. All three electrodes had the same linearity range of 1.0×10^{-5}–1×10^{-1} M. The electrode using tetrachloro-substituted organoantimony(V) (Figure 9c) had the weakest performance of all the LC-ISEs described in this publication. This ISE showed the lowest slope, the narrowest linearity range and the highest detection limit. The difficulty in carrying out measurements with these electrodes is that the pH of the samples must be within the range of three, as this is the most optimal value, so the acidity of the solution is recommended. These electrodes can be used to measure F^- ions, for example, in tap water.

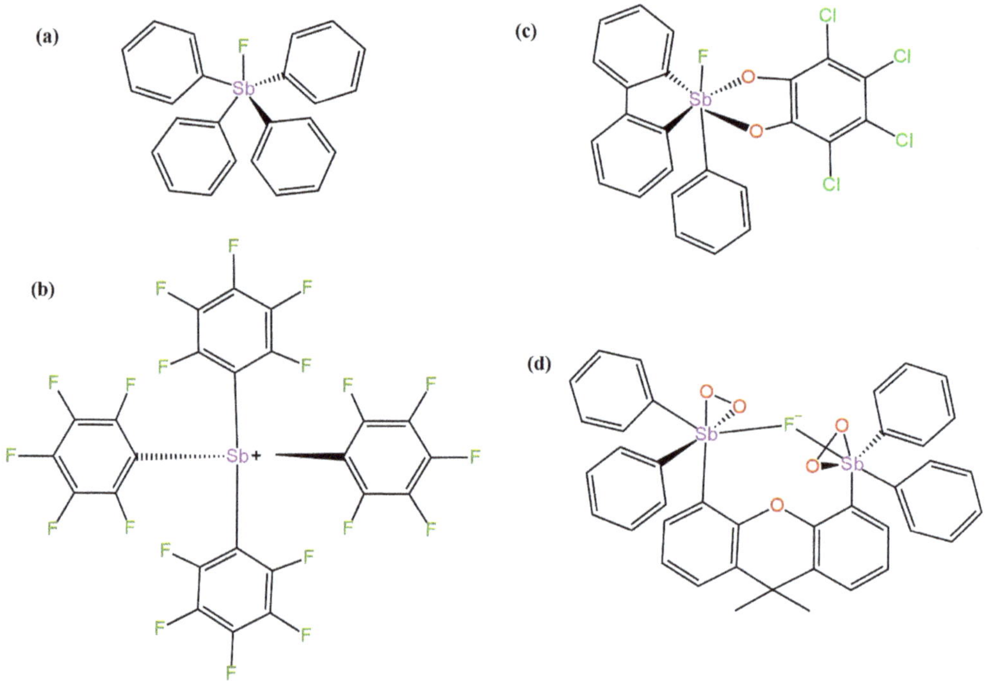

Figure 9. Lewis acidic organo-antimony(V) compounds used as fluoride ionophores: tetraphenyl stibonium fluoride (**a**), tetrakis-(pentafluorophenyl)stibonium (**b**), tetrachloro-substituted organoantimony(V) compound with fluoride (**c**) and bidentate organoantimony(V) compound with fluoride (**d**) [48].

A comparison of the analytical parameters of various fluoride electrodes is presented in Table 2.

Table 2. Fluoride ion-selective electrodes.

E. No	Type of Contact	Ionophore/Ion Carrier	Intermediate/Transducer Layer	Type of Internal Electrode	Slope [mV/decade]	Range of Linearity [M]	Limit of Detection [M]	pH Range	Application/Samples	References
1	SC	Eu-doped LaF_3 nanocrystals	-	TlE	-56 ± 13	1.0×10^{-5}–1.0×10^{-1}	1.0×10^{-6}	-	-	[43]
2	SC	LaF_3 single crystal	PEDOT	AgE	-56.0 ± 0.9	1.0×10^{-5}–1.0×10^{-1}	2.0×10^{-5}	5.0–11.0	-	[44]
3	SC	LaF_3 single crystal	Ag paste	AgE	-62.8 ± 3.8	1.0×10^{-5}–1.0×10^{-1}	1.0×10^{-2}	-	-	[44]
4	LC	LaF_3 single crystal	PBS, 0.01 M Na_2HPO_3 and 0.02 M KH_2PO_3	Ag∣AgCl	-38.6 ± 9.1	1.0×10^{-5}–1.0×10^{-1}	-	-	-	[44]
5	SC	LaF_3 single crystal	Fe_xO_y NPs	SSDE	-52.9–-57.3	6.3×10^{-6}–1.0×10^{-1}	3.6×10^{-7}	4.0–7.0	-	[45]
6	LC	LaF_3 single crystal	KCl + HCl + 0.1 M $AgNO_3$	Ag∣AgCl	-50.8–-52.7	3.9×10^{-7}–1.0×10^{-1}	7.4×10^{-8}	4.0–7.0	-	[45]
7	SC	Bis(fluorodioctylstannyl)methane	MWCNTs-COOH	SPCE	-59.2	-	1.7×10^{-9}	-	-	[46]
8	LC	tetrakis-(pentafluorophenyl)stibonium			-59.3	1.0×10^{-5}–4.0×10^{-2}	5.0×10^{-6}	3.0		
9	LC	$[Ph_4Sb]^+$	0.2 M Gly/H_3PO_4 buffer and 0.001 M NaF	Ag∣AgCl	-58.2	1.0×10^{-5}–4.0×10^{-2}	2.0×10^{-5}	-	Tap water samples.	[48]
10	LC	tetrachloro-substituted organoantimony(V)			-54.6	1.0×10^{-5}–4.0×10^{-2}	2.0×10^{-4}	-		
11	LC	Organo antimony(V) compound			-57.8	1.0×10^{-5}–4.0×10^{-2}	3.0×10^{-5}	-		
12	SC	CdI_2	-	CPE	-58.9	1.5×10^{-6}–5.5×10^{-3}	1.2×10^{-7}	5.0–7.0	River water samples.	[47]

6. Ion-Selective Electrodes Sensitive to Chloride and Perchloride Ions

Chloride anions are quite common ions in drinking, ground and surface waters. Not only they are responsible for their saltiness, but they also affect the physiology of plants. Chlorine-containing compounds are used in the production of food or fertilizers. Chlorine is an important component for the human organism, and it is not toxic in small quantities, but the excess can cause, e.g., hyper-chloremia, which results in dehydration, diarrhea and metabolic problems, which, in turn, increases blood pressure and can lead to the damage of certain organs, e.g., the kidneys [49,50]. Perchlorate anions are very widespread in the world around us. They are present not only in various types of water, including ground and surface water, but also in plants and soil. In addition, they are used on a large scale in the industry for the production of explosives and pyrotechnics, and their increasing growth in the environment is linked to intensified rocket testing as perchlorates are used in rocket engine fuels. One of the main effects of overexposure to perchlorate ions is reported to be the disruption of the thyroid gland, due to the replacement of iodine by perchlorate, caused by a very similar ionic radius [51,52]. This makes it very necessary to monitor these ions. Other techniques are already used for this purpose, i.e., chromatography and spectroscopy, but potentiometry is an equally good and inexpensive way to determine the concentration of chloride and perchlorate anions. Since 2018, 16 Cl^--ISEs have been proposed with the use of novel charge transfer mediating layers and three leading ionophores (TDMACl, chloride ionophore(III), chloride ionophore(I)) and various types of nanocomposites (Figure 10). Whereas for the ClO_4^- anions, three new SC-ISEs with different types of compounds used as an ionophore for the first time are presented.

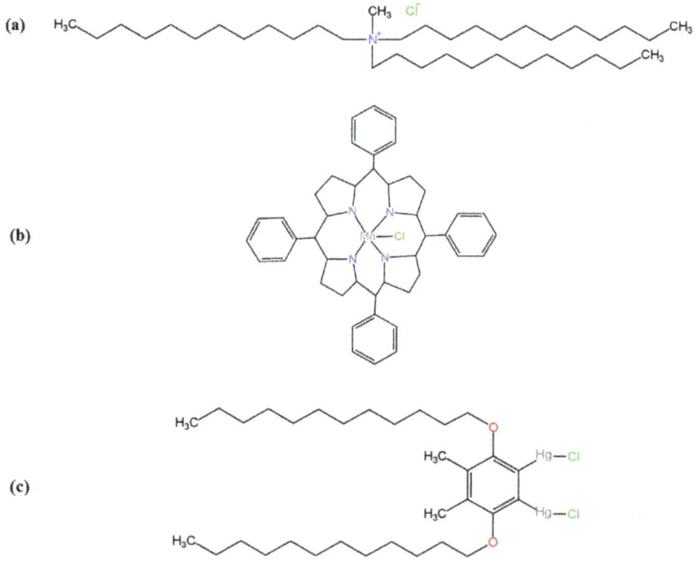

Figure 10. Active substances used in chloride ISEs: TDMACl (a), chloride ionophore(I) (b) and chloride ionophore(III) (c).

In 2022, a paper that introduced novel Cl^--SC-ISEs was published. In the role of solid contact, the polyaniline nanofibers doped with Cl^- ions, multi-walled carbon nanotubes and three types of nanocomposites with different ratios of constituents that were PANINFs-Cl- and MWCNTs (2:1, 1:1, 1:2) were used [53]. The ionophore that was used was chloride ionophore(III) (Figure 10c), and the solid contact was applied by dropping an appropriate volume of the components directly onto the glassy carbon electrode. The proposed intermediate layers were investigated both from a morphological point of view, where their structure was presented on SEM images, and in terms of electrical properties that were determined by chronopotentiometry and EIS measurements. On the basis of these studies, the intermediate layer capacitance and resistance were determined, and the best results were obtained for a nanocomposite with a component ratio of 2:1, where C (electric capacitance) = 7.16 mF and R (resistance) = 0.21 kΩ. The same measurements were also carried out with the SC-ISEs after application of the membrane and again the best electrical performance was represented by the electrode containing this composite as an intermediate layer. All of the modified electrodes showed the same range of linearity and very similar LOD values of the order of 10^{-6} M and slopes close to the Nernstian value. Another advantage of these ISEs is their insensitivity to environmental changes, i.e., the influence of light or the presence of gases (O_2 and CO_2). The electrodes also exhibited very good selectivity, which makes them good devices for determining the concentration of Cl^- ions. This is confirmed by the determination of chloride anions using the proposed ISE, whose results were compared with those obtained using the classic Mohr method.

The next solution implemented in chloride potentiometric sensors is the use of an ion exchanger in the form of an ionophore, which is TDMACl (Figure 10a). This method was used, among others, by Kalayci [54] in a liquid contact electrode and by Pięk et al. [55] in an ISE with solid contact. In the Cl^--LC-ISE, a classical design of this type of electrode was used, where the IE was Ag | AgCl. The performance of this electrode was not outstanding and was average compared to the other electrodes listed in Table 3. On the other hand, in chloride SC-ISEs, novel conductive materials (redox mediators) belonging to the group of molecular organic materials (MOMs), i.e., TTF (tetrathiafulvalene), the chloride salt TTFCl and TTF-TCNQ (tetrathia-fulvalene-tetracyanoquinodimethane), as well as a combination of these materials with carbon black (CB) showing good hydrophobicity, conductivity and high specific surface area, were used as solid contacts. The modification of the electrodes improved their linearity range and slightly increased their slope but had little effect on LOD. Nanocomposites of CBs and MOMs turned out to be the best solid contact material as they exhibited better electrical performance, i.e., lower resistance and higher double-layer capacitance, and had a slightly better electrode response compared to electrodes based on molecular organic materials' SCs. An additional advantage of the presented electrodes is their good selectivity and potential reversibility.

The paper [56] describes the novelization of SC-ISE with a new layer providing sensitivity to chloride ions. For this purpose, the carbon paste electrode (CPE) was modified with an ion-sensitive layer, which was a composite of graphitic carbon nitride (g-C_3N_4) that was anchored to a crystalline AgCl structure. The obtained structure was examined in terms of physical and morphological properties using techniques such as XRD, SEM and FTIR. After measurements, the best option was found to be the modification of the CPE with the addition of a 5% g-C_3N_4/AgCl composite. For this electrode, the parameters that were determined showed very good performance compared to the other Cl^--ISEs. The detection limit was at the lowest level compared to all the chloride electrodes presented in Table 3. A further advantage of this electrode is its very fast time of response and long-term potential stability (more than two months), as well as its good selectivity in the presence of interfering ions, i.e., I^-, Br^- and CN^-.

A comparison of the analytical parameters of various perchlorate electrodes is presented in Table 4.

Table 3. Chloride ion-selective electrodes.

E. No	Type of Contact	Ionophore/Ion Carrier	Intermediate/Transducer Layer	Type of Internal Electrode	Slope [mV/decade]	Range of Linearity [M]	Limit of Detection [M]	pH Range	Application/Samples	References
1		Chloride ionophore(III)	MWCNTs	GCE	−59.6		2.6×10^{-6}		Inspection of the efficiency of water desalination.	[53]
2	SC		PANINFs-Cl		−60.3	5.0×10^{-6}–1.0×10^{-1}	2.8×10^{-6}	4.0–9.0		
3			PANINFs-MWCNTs		−61.2		2.7×10^{-6}			
4	LC	TDMACl	0.5 M NaCl	Ag|AgCl	−55.0 ± 2	1.0×10^{-5}–1.0×10^{-1}	1.0×10^{-6}	2.0–8.0	Wastewater samples.	[54]
5			-		−57.1 ± 0.66	1.0×10^{-4}–1.0×10^{-1}	1.6×10^{-5}			
6			TTF		−58.2 ± 0.27	1.0×10^{-5}–1.0×10^{-1}	5.0×10^{-6}			
7			TTF-TCNQ		−58.3 ± 0.16	1.0×10^{-5}–1.0×10^{-1}	4.0×10^{-6}			
8			TTFCl		−58.4 ± 0.14	1.0×10^{-5}–1.0×10^{-1}	4.0×10^{-6}			
9	SC	TDMACl	CB	GCD	−59.6 ± 0.11	1.0×10^{-5}–1.0×10^{-1}	2.5×10^{-6}	-	Water samples.	[55]
10			CB-TTF		−58.5 ± 0.13	1.0×10^{-5}–1.0×10^{-1}	3.2×10^{-6}			
11			CB-TTF-TCNQ		−58.7 ± 0.10	1.0×10^{-5}–1.0×10^{-1}	2.5×10^{-6}			
12			CB-TTFCl		−59.1 ± 0.09	1.0×10^{-5}–1.0×10^{-1}	2.0×10^{-6}			
13	SC	g-C_3N_4/AgCl	-	CPE	−55.4 ± 0.3	1.0×10^{-6}–1.0×10^{-1}	4.0×10^{-7}	-	Aqueous samples.	[56]
14	SC with ISM	Chloride ionophore(I)	-	Ag|AgCl	−61.7 ± 2.4	1.0×10^{-5}–1.0×10^{-1}	1.1×10^{-5}	-	Sweat.	[57]
15	SC	AgCl:Ag_2S:PTFE	Fe_xO_y NPs	multi-purpose solid state electrode made from stainless steel	−44.4	2.0×10^{-6}–1.0×10^{-1}	1.42×10^{-6}	-	-	[58]
16			ZnO NPs		−40.5	3.6×10^{-6}–1.0×10^{-1}	1.0×10^{-6}	-		

Table 4. Perchlorate ion-selective electrodes.

E. No	Type of Contact	Ionophore/Ion Carrier	Intermediate/Transducer Layer	Type of Internal Electrode	Slope [mV/decade]	Range of Linearity [M]	Limit of Detection [M]	pH Range	Application/Samples	References
1	SC	Dixanthylium dye	-	Pt wire	−57.4	1.0×10^{-6}–6.1×10^{-2}	5.0×10^{-7}	1.5–11.0	Aqueous samples.	[59]
2	SC	Bn$_{12}$BU [6]	PEDOT	GCE	−59.9 ± 1.1	1.0×10^{-6}–1.0×10^{-1}	1.0×10^{-6}	-	Real water samples.	[60]
3	SC	In^{III}-porphyrin	SWCNTs	GCE	−56.0 ± 1.1	1.1×10^{-6}–1.0×10^{-2}	1.8×10^{-7}	-	Fireworks and propellants.	[61]

Two kinds of electrodes differing in the type of layer providing sensitivity to chloride ions were reported in [57]. One of the electrodes had an ion-selective membrane with chloride ionophore(I) (Figure 10b), which was applied to an Ag|AgCl electrode, while the other ISE was a classical silver chloride electrode without an ionophore and ISM. It turned out that the electrode without the ion-selective membrane showed better stability and reproducibility of potential and kept a constant potential in solutions containing interfering ions than the proposed SC-ISE. Such an electrode can be used for the determination of chloride in human sweat.

Another example of chloride electrodes is the new SC-ISEs that have a membrane enriched with an $AgCl:Ag_2S$:PTFE nanocomposite (two copies with different component ratios (1:1:2 and 2:1:2)), which provided selectivity of the electrode [58]. Additionally, nanoparticles of metal oxides, i.e., zinc oxide II (ZnONPs) or iron oxide (Fe_xO_yNPs), were used in the electrode construction as layers supporting charge transfer. These oxides were investigated using FTIR. In Table 3, the ISEs with the best analytical performance of the electrodes depending on the type of used oxide are compared. Iron oxides proved to be the best option for the studied electrodes. Despite the modifications, the slope of the produced electrodes was quite low and deviated from the theoretical value of -59.16 mV/decade, which undoubtedly indicates that the sensitivity of the proposed electrodes was quite weak.

One of three electrodes used for the determination of perchlorate ions was an all-solid-state coated wire electrode (CWE), where a platinum wire was immersed in a membrane cocktail to apply a membrane (opaque membrane—dixanthylium dye) [59]. The structure of the new ionophore is shown in Figure 11. The proposed electrode had a wide range of linearity, a fairly low detection limit of 5.0×10^{-7} M and an excellent working pH range of 1.5–11.0. The advantage of this ISE is a fast response time oscillating around 4 s and good selectivity towards many interfering ions. This electrode has been successfully used in the determination of perchlorate ions in the samples of mineral and tap water.

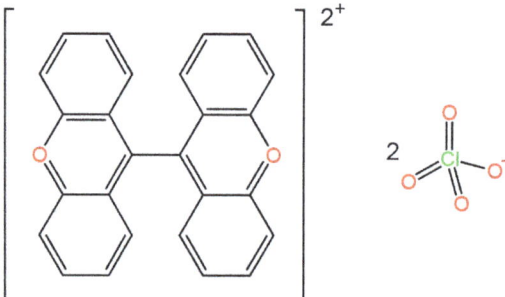

Figure 11. The structure of dixanthylium dye [59].

The next two SC-ISEs had a GCE as the basic electrode. In the first one, PEDOT was used as an intermediate layer and a dodecabenzylbambus[6]uril ($Bn_{12}BU$ [6]) as an ionophore (Figure 12) [60]. The proposed selectophore binds perchlorate ions very well due to an almost perfect match between the ion size and the receptor hole. Potentiometric, cyclic voltammetry, chronopotentiometry and electrochemical impedance spectroscopy measurements were carried out to verify if the proposed membrane component succeeds in its role. On the basis of the ISE response, the slope was determined, which equaled -59.9 mV/decade for the best electrode, which shows that this electrode has excellent sensitivity, a six-order linearity range and an LOD of 1.0×10^{-6} M. SC-ISE showed good stability and selectivity towards inorganic ions such as Br^-, Cl^- and NO_3^-.

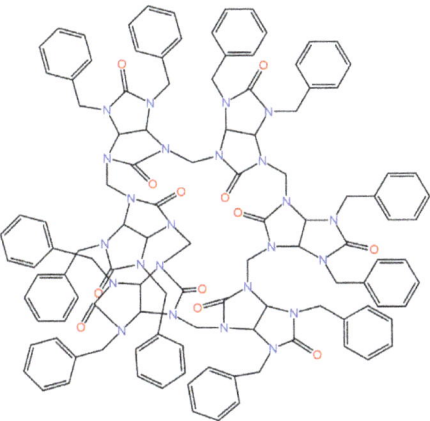

Figure 12. The structure of novel ionophore—Bn$_{12}$BU [6].

The second electrode with a GCE was the ISE with a solid contact in the form of single-walled carbon nanotubes (SWCNTs), where indium(III) 5,10,15,20-(tetraphenyl) porphyrin chloride (In^{III}-porph) (Figure 13) was responsible for the selectivity towards the main ion [61]. The selectivity of the described electrode was at a high level and the analytical parameters of the ISE were very close to the SC-ISE proposed by Babaei et al. The electrode showed very good short-term stability and satisfying electrical parameters, e.g., the electrical capacitance equal to 27.6 ± 0.7 µF. The ClO$_4^-$-ISE response time is less than 10 s, and the lifetime oscillates around 8 weeks, which is not a very long period of time. The electrode was used for the determination of ClO$_4^-$ ions in firework samples, and the results were compared with those obtained by ion chromatography with satisfactory accuracy. In addition, the measurement of perchlorate anions in urea perchlorate, hydrazine, ethylenediamine and ammonium was also successful.

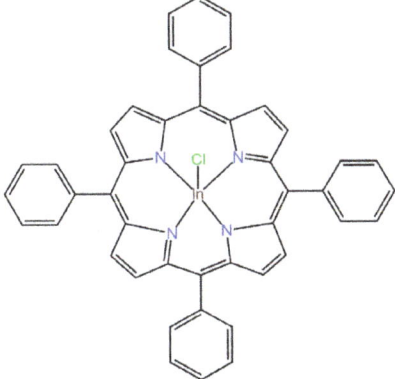

Figure 13. The structure of In^{III}-porph.

7. Bromide Ion-Selective Electrodes

Bromine is a fairly common element that occurs in large quantities in the biosphere. More often than in its free state, it is found in a bound form, e.g., in inorganic salts, as a decomposition product of hydrocarbons saturated with bromine, e.g., methyl bromide, which is a pesticide with toxic effects on humans and the rest of the ecosystem [62]. As bromine belongs to the group of halides, it has similar properties to chlorine or iodine. This is quite dangerous because in human organisms the substitution of iodine by bromide ions can occur, for example, in thyroid cells, which can cause dysfunction of this organ and of other functioning on the basis of thyroid hormone homeostasis. In addition, bromine as a toxin causes nervous system and neuropsychiatric disorders, problems with excessive muscle tremors and induces dermatological diseases, i.e., dermatitis and rashes [63,64]. Due to its toxicity, it is necessary to determine this element, e.g., in water and food, in order to verify whether the potential product can be safely consumed by humans. One of the best solutions involves the use of cheap and easy-to-use potentiometric sensors that selectively and accurately determine the concentration of this ion in samples. Several Br-ISEs have been proposed in the last five years, two of those described below have a SC and three have a LC.

One of the discussed liquid contact electrodes is a classical ISE with a silver chloride electrode, which was filled with 0.5 M NaBr solution [54]. The role of the bromide ion carrier was played by TDMABr (tridodecylmethylammonium bromide). This electrode was different from other Br$^-$-ISEs because of its very wide working pH range of 1.0–11.0. The sensitivity of this electrode was remarkably high and the measurement results in the real sample were almost identical to those performed by ion chromatography. Another ISE with the same Ag|AgCl internal electrode is the electrode where Pt(II) 5,10,15,20-tetra(4-methoxyphenyl)-porphyrin (PtTMeOPP) was the selective ion carrier (Figure 14) [65]. In order to characterize the newly developed ionophore, both ^{1}H-NMR and UV-Vis analyses were performed. A significant advantage of this electrode is its good selectivity and high sensitivity, as the slope of the response curve reached −64.4 mV/decade. It was successfully applied to the determination of Br$^-$ ions in drug samples.

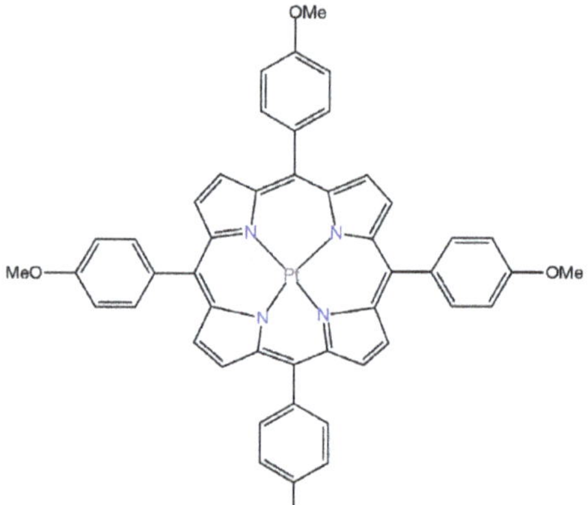

Figure 14. Structure of PtTMeOPP [65].

A two orders of magnitude wider range of linearity and a lower detection limit of 7.1×10^{-8} M was observed for LC-ISE in which the membrane was a core-shell nanocomposite based on boron-doped graphene oxide-aluminium fumarate metal organic framework (BGO/AlFu MOF) (Figure 15) [66]. The nanocomposite was characterized by XRD, FTIR and SEM techniques, and its absorption properties were investigated by UV-Vis spectroscopy technique. The response time of the proposed electrode was about 13 s and the obtained slope equaled −54.5 mV/decade, which is the worst value among all described Br$^-$-LC-ISEs.

In the subsequent article [67] three variants of bromide electrodes with a solid contact, represented by POT sprinkled directly onto the glassy carbon electrode, were presented. The electrodes differed in the composition of the ion-selective membrane. The mesotetraphenylporphyrin manganese(III)-chloride complex (ionophore 1) was used as the ionophore in the first option and 4,5-dimethyl-3,6-diacetyl-o-phenylene-bis(mercuritrifluoroacetate) (ionophore 2) in the second. Membrane III contained in its composition, alongside DOS, ionophore

2 and PVC, the ion exchanger TDMACl. ISE I showed sensitivity to chloride ions; ISE with membrane III was rejected because it had significantly worse selectivity than ISE with membrane II. Electrode II, which exhibited the best performance, had a rather low detection limit of 2.0×10^{-9} M but a narrow linearity range of 1.0×10^{-8}–1.0×10^{-6} M.

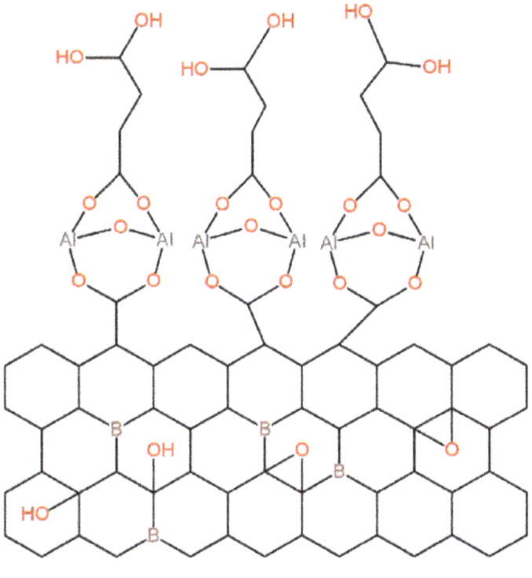

Figure 15. The structure of BGO/AlFu MOF [66].

A comparison of the analytical parameters of various bromide electrodes is presented in Table 5.

Table 5. Bromide ion-selective electrodes.

E. No	Type of Contact	Ionophore/Ion Carrier	Intermediate/Transducer Layer	Type of Internal Electrode	Slope [mV/decade]	Range of Linearity [M]	Limit of Detection [M]	pH Range	Application/Samples	References
1	LC	TDMABr	0.5 M NaBr	Ag|AgCl	-57.4 ± 0.3	$1.0 \times 10^{-6} - 1.0 \times 10^{-1}$	1.2×10^{-6}	1.0–11.0	Water samples e.g., tea samples.	[54]
2	LC	BGO/AlFu-MOF	0.05 M KBr	GE	-54.5 ± 0.2	$1.0 \times 10^{-7} - 1.0 \times 10^{-1}$	7.1×10^{-8}	4.0–9.0	Environmental samples.	[66]
3	—	Mesotetraphenylporphyrin manganese(III)-chloride complex	POT	GCE	-	$1.0 \times 10^{-6} - 1.0 \times 10^{-2}$	2.0×10^{-9}	-	Water samples.	[67]
4	SC	4,5-dimethyl-3,6-dioctyloxy-o-phenylene-bis(mercurytrifluoroacetate)								
5	LC	PtTMeOPP	0.01 M KCl	Ag|AgCl	-64.4 ± 0.4	$1.0 \times 10^{-5} - 1.0 \times 10^{-1}$	8.0×10^{-6}	6.0–12.0	Pharmaceutical samples.	[65]

8. Iodide Ion-Selective Electrodes

Iodine is one of the most crucial elements for the human body. It not only influences the proper functioning of the neurological system but also regulates metabolic processes. It ensures the proper functioning of the thyroid gland and the development of bones or muscles. It is very important to supply the body with adequate amounts of this micronutrient via food and liquids. A deficiency or excess of this element leads to hypothyroidism or hyperthyroidism, which, in turn, causes a number of other health problems, such as circulatory disorders, weakness and mental illnesses [68–70]. Therefore, it is essential to monitor the amount of iodine in the human diet, and, for this purpose, potentiometric sensors characterized by selectivity and good sensitivity to iodine can be used, as well as other techniques. Thirteen ISEs are described below, which differed in the type of ionophore, the intermediate layer, the type of contact and the lead electrode.

For the determination of I^- ions, a new iodide electrode was proposed with an $AgI:Ag_2S:PTFE$ ion-sensitive membrane in a ratio of 1:1:2, which was enriched with nanoparticles of zinc oxide (ZnO) [71]. Several electrodes differing in the amount of ZnO in the range of 1–5 wt.% were fabricated. The best of the proposed membranes turned out to be the one marked as M2, which contained 10 mg ZnO (it was examined by SEM and EDS). Of all the LC-ISEs, this I^--ISE had the slope that was the closest to Nernstian's. The range of linearity for the presented potentiometric sensor in comparison to measurements using voltammetric methods was actually wider by as much as two orders of magnitude. This electrode was successfully used for the determination of penicillin in pharmacological samples.

Classic electrodes with liquid contact and silver chloride internal electrodes are described in the works of [54,65,72]. They differed in the type of ion-sensitive component that provided good sensitivity towards iodide ions. Kalayci used tridodecyl-methylammonium iodide (TDMAI) for I^- ion capture. The performance of this electrode was not significantly different from other I^--LC-ISEs but was better than that for the sensor proposed by [65], where PtTMeOPP was used as the ionophore. This electrode had a lower slope of -52.3 ± 0.4 mV/decade, while the other parameters, i.e., linearity range or LOD, were not significantly different. A final example of LC-ISEs with Ag|AgCl as the internal electrode are electrodes modified with newly proposed active ingredients to provide selectivity to iodide ions. The first such material is the metallic complex platinum(IV) tetra-tertbutylphthalocyanine dichloride (Pc^tPtCl_2) (Figure 16a) [72]. In addition, three other composites consisting of the above-mentioned complex and the ionic liquid cetylpyridinim chloride (CPCl) (Figure 16d) (CPCl + Pc^tPtCl_2) or cetylpyridinuim bromide (CPBr) (Figure 16c) (CPBr + Pc^tPtCl_2) or 1,3-dicetylimidazolium iodide (DCImI) (Figure 16b) (DCImI + Pc^tPtCl_2) were used. These modifications did not have the desirable effect as these electrodes had the lowest detection limit and narrowest linearity range compared to the other cited LC-ISEs, and the slope of the characteristics was less than -50 mV/decade for all electrodes except the one modified with the CPBr + Pc^tPtCl_2 composite. Moreover, in the publication, electrodes with a solid contact that had the same active components in the ion-selective membrane as those mentioned above, i.e., Pc^tPtCl_2, CPCl + Pc^tPtCl_2, CPBr + Pc^tPtCl_2, DCImI + Pc^tPtCl_2, were also described. The parameters of the proposed SC-ISEs were clearly more satisfactory than those for the classic ISEs. The best results were obtained for electrodes modified with a composite of the metal complex and ionic liquid DCImI + Pc^tPtCl_2 and CPBr + Pc^tPtCl_2. The second one was used for the determination of iodide ion content in pharmaceuticals, i.e., Iodomarine 100 and Iodobalanse 100.

Two novel ionophores providing sensitivity to I^- ions were characterized in a publication [73]. They were used as a component of an ion-selective membrane in an SC-ISE, where a combination of SPE with PANI was used as a solid contact. The first of the new active ingredients is XB_1 (tripodal halogen bonding (XB) ionophore) (Figure 17) and the second is HB_1 (H-triazal analogue of XB_1) (Figure 18). Both ionophores were investigated using NMR techniques (the effect of XB interactions between I^- ions and the ionophore was determined). It is also worth mentioning that the determined selectivity was incompatible for the SCN⁻ ion with the Hofmeister series, which indicates that the ionophore binds with the I⁻ ions, as a result of the NMR measurements. Among the two proposed structures, the SC-ISE with HB$_2$ as the ionophore proved to be the better solution with higher sensitivity, but it was not significantly different from the competitive ISE with XB$_1$ in the ISM structure.

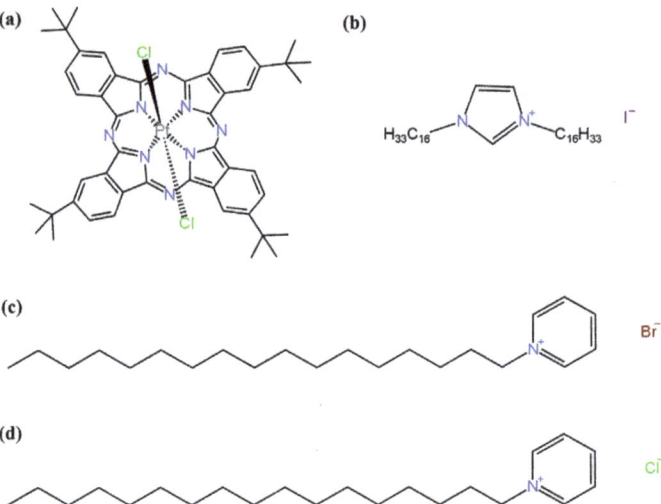

Figure 16. Structure of novel iodic ionophores: platinum (IV) tetra-tertbutylphthalocyanine dichloride (**a**), 1,3-dicetylimidazolium iodide (**b**), cetylpyridinuim bromide (**c**) and cetylpyridinuim chloride (**d**) [72].

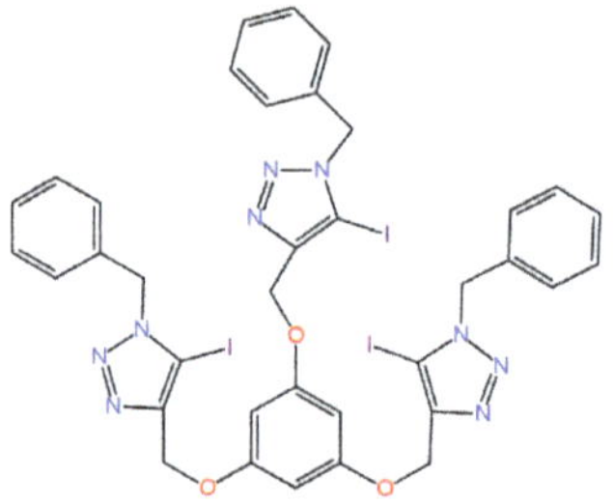

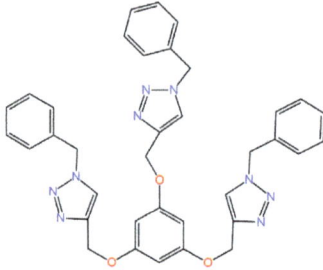

Figure 17. Structure of XB1 [73].

Figure 18. Structure of HB$_1$ [73].

A comparison of the analytical parameters of various iodide electrodes is presented in Table 6.

Table 6. Iodide ion-selective electrodes.

E. No	Type of Contact	Ionophore/Ion Carrier	Intermediate/Transducer Layer	Type of Internal Electrode	Slope [mV/decade]	Range of Linearity [M]	Limit of Detection [M]	pH Range	Application/Samples	References
1	LC	TDMAI	0.1 M KI and 0.1 M KCl	Ag|AgCl	-54 ± 1	1.0×10^{-5}–1.0×10^{-1}	1.3×10^{-6}	2.0–8.0	Wastewater samples.	[54]
2	SC	XBI	PANI	SPE	-54 ± 1	1.0×10^{-5}–1.0×10^{-1}	1.3×10^{-6}	-	-	[73]
3		HB2			-51.9	1.0×10^{-6}–1.0×10^{-1}	1.3×10^{-6}			
4		Pc^{t}PtCl$_2$			-54.9	1.0×10^{-6}–1.0×10^{-1}	1.0×10^{-6}			
5	LC	CPCl + Pc^{t}PtCl$_2$	0.001M KI and 0.1 KCl	Ag|AgCl	-26 ± 3	1.0×10^{-3}–1.0×10^{-1}	1.8×10^{-4}	-	Medicaments such as "Iodomarine 100" and other pharmaceuticals.	[72]
6		CPBr + Pc^{t}PtCl$_2$			-45 ± 1	1.0×10^{-4}–1.0×10^{-1}	2.1×10^{-5}			
7		CPBr			-54 ± 1	1.0×10^{-4}–1.0×10^{-1}	3.5×10^{-5}			
8		CPCl + Pc^{t}PtCl$_2$			-46 ± 2	1.0×10^{-3}–1.0×10^{-1}	3.0×10^{-4}			
9	SC	CPBr + Pc^{t}PtCl$_2$	graphite	SPPE	-51 ± 1	1.0×10^{-4}–1.0×10^{-1}	5.3×10^{-5}			
10		CPBr			-54 ± 1	1.0×10^{-4}–1.0×10^{-1}	1.9×10^{-5}			
11		DCImI + Pc^{t}PtCl$_2$			-50 ± 1	1.0×10^{-3}–1.0×10^{-1}	1.0×10^{-4}			
12	LC	AgCl·Ag$_2$S·PTFE + ZnO NPs	-	-	-57 ± 2	1.0×10^{-4}–1.0×10^{-1}	1.8×10^{-5}	-	Penicillamine in real samples.	[71]
13	LC	PtMeOPP	0.01M KCl	Ag|AgCl	-57.4 ± 0.3	2.5×10^{-6}–1.0×10^{-2}	2.2×10^{-6}	3.0–12.0	Pharmaceutical such as potassium iodide tablets.	[65]

9. Ion-Selective Electrodes Sensitive to S^{2-}, SO_3^{2-} and SO_4^{2-} Ions

Sulfur is one of the basic micronutrients that are essential for the human organism. It is not only a component of amino acids and proteins, but it is also an important element crucial and necessary for metabolic processes that take place in the body [74]. Sulfur is, apart from other things, a component of insulin, and a deficiency of this element can therefore lead to hyper- or hypoglycemia; it has an anti-inflammatory effect and is involved in the synthesis of connective tissue. Sulfur compounds are also important nutrients for the growth and development of plants [75]. Both a deficiency and an excess of this element, its compounds and ions are not beneficial to humans and animals, as well as to the flora and the rest of the environment. Therefore, a variety of techniques are used to determine the amount of these molecules. One of the leading methods is potentiometry, so newer and newer ISEs are being constructed to provide accurate and sensitive measurements. This review focuses on ten electrodes that are designed to monitor the concentration of sulfur-containing ions, where four are sensitive to S^{2-} ions, two to SO_3^{2-} ions and four to SO_4^{2-} ions.

For the detection of S^{2-} ions, four different ISEs were recently constructed, including three with a liquid contact and one with a layer of solid contact. One of the LC-ISEs described in the publication [76] had an ion-selective membrane made of Ag_2S, which was obtained by sedimentation. The membrane in the form of a disk was placed at one end of a PVC case containing a 1×10^{-6} M Na_2S (electrolyte solution), which provided a conductivity between the membrane and the Ag|AgCl inner electrode. The parameters that were obtained for the proposed electrode were quite satisfying, i.e., a low limit of detection of 2.3×10^{-7} M, a six-order linearity range and an almost Nernstian slope of -28.2 mV/decade. A fairly fast electrode response of 5–17 s was also obtained. This electrode can be used for the determination of sulfide ions in various types of solutions and industrial water used in oil refineries.

Another two LC-ISEs were described by Matveichuk et al. [77]. In this study, the electrodes differing in an ion carrier were compared. One of these was the previously used 4-(trifluoroacetyl)heptyl ester of benzoic acid (TFA-BAHE), and the other was the new PVC-modified 4-(trifluoroacetyl)benzoate (TFAB-PVC) shown in Figure 19. The use of TFAB-PVC resulted in an extension of the electrode lifetime in comparison to ISE, where TFABAHE was used. This was probably due to the fact that the covalent bonding between the TFAB and the PVC matrix prevented its elution from the membrane. Depending on the environment, this process has a different course. Both of the presented electrodes had almost identical parameters, i.e., slope, linearity range and LOD. The change of ion carrier had a positive effect on the electrode's lifetime, and an additional advantage is the possibility to perform measurements in alkaline and acidic environments.

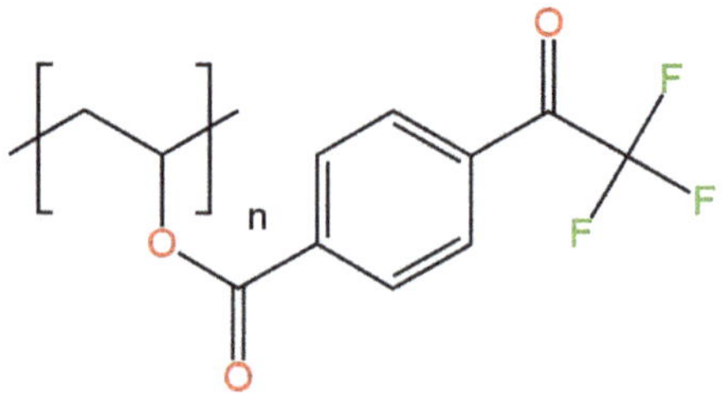

Figure 19. The structure of TFAB-PVC [77].

The last S^{2-}-ISE is an electrode composed of a silver wire (inner electrode) to which a transducer layer in the form of reduced graphene sheets (RGSs) (solid contact) was electrodeposited by reducing graphene oxide [78]. The final step in the preparation of this electrode was the electrodeposition of Ag_2S constituting the ISM. The proposed electrode showed good 7-day potential stability, as well as satisfactory selectivity, but had an order of magnitude more narrow linearity range than the other presented ISEs. The electrode exhibited a super-Nerstian slope of characteristics of about −200 mV/decade. The authors explain this by the phenomenon of ion bonding in the solid contact layer that results in increased transport of ions from the diffusion layer.

Another potentiometrically monitored anion is the SO_3^{2-} ion, and the electrodes sensitive to it were described in the work [79]. Two screen-printed electrodes with a polymeric membrane containing cobalt(II) phthalocyanine (CoPC) as a material for selective detection of SO_3^{2-} ions were proposed. These two electrodes differed in the solid contact layer, which was constituted by organic conductors, i.e., carboxyl functionalized multi-walled carbon nanotubes (MWCNTs-COOH) or polyaniline nanofibers (PANINFs). Both described electrodes had a rather narrow range of linearity. The MWCNTs-COOH-modified electrode had a better electrode response of −29.8 mV/decade and a lower LOD, as well as better electrical parameters of C = 26.1 µF and R = 5.3 kΩ compared to the one in which the intermediate layer was PANINFs. In addition, the ISEs were not sensitive to changes in environmental conditions, i.e., the presence of light and gases (O_2 and CO_2). They also positively passed the water layer test.

Among the four ISEs intended for the determination of SO_4^{2-} ions, the CPE (carbon paste electrode) showed the best predisposition in this direction, where the function of the ionophore was performed by a Schiff base complex with nickel (Figure 20) [80]. To verify whether there are no sulfide ions in the complex and if it is suitable to be an ionophore, UV-Vis spectra were determined. Good parameter values were obtained: a Nernst response of −29.7 mV/decade, a nanomole LOD value of 5.0×10^{-9} M and a satisfactory linearity range of 7.5×10^{-9}–1.5×10^{-3} M. The effect of temperature and the sensitivity of the potential to changes of pH were also tested. The electrode was successfully used for the determination of SO_4^{2-} ions in real samples, i.e., mineral water or blood serum.

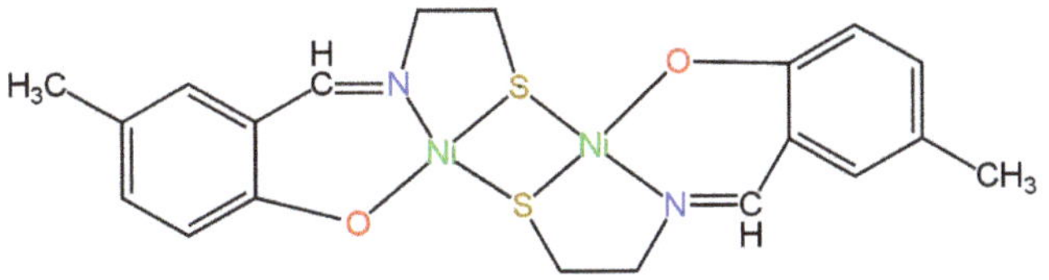

Figure 20. Schiff base complex with nickel used as a sulfate ionophore [80].

Three further electrodes that differed in the type of ion carrier were described in articles [77,81]. In all electrodes, a solution of 0.01 M Na_2SO_4 and 0.001 M KCl was used as the internal electrolyte. The ingredient providing selectivity in the electrode described in the paper [81] was the higher quaternary ammonium salt 3,4,5-tris(dodecyloxy)benzyl(oxyethyl)$_3$ trimethylammonium chloride ((oxyethyl)$_3$TM) which was the ion exchanger (Figure 21). Meanwhile, in the work [77], the anionic ion carrier TFABAHE and the neutral ion carrier TFAB-PVC were used for this purpose. Among the above-mentioned electrodes, the highest slope of the characteristic of −27 mV/decade and the lowest limit of detection showed the ISE in which the ion exchange (oxyethyl)$_3$TM was used as the ionophore. Furthermore, the lifetime of this electrode was approximately 1 month. The second in terms of quality was the TFABAHE-modified electrode, which slightly outperformed the competitive ISE with TFAB-PVC as the ion carrier.

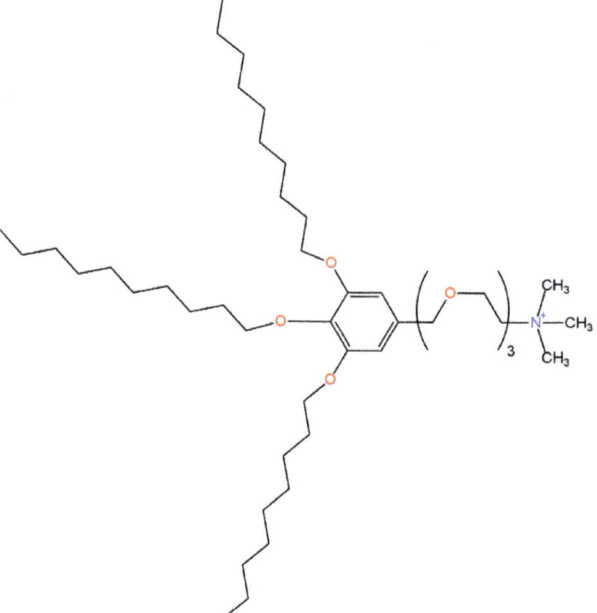

Figure 21. The structure of (oxyethyl)$_3$TM.

The basic analytical parameters of ISEs sensitive to S^{2-}, SO_3^{2-} and SO_4^{2-} ions are presented in Table 7.

Table 7. Ion-selective electrodes sensitive to S^{2-}, SO_3^{2-} and SO_4^{2-} ions.

E. No	Ion	Type of Contact	Ionophore/Ion Carrier	Intermediate/Transducer Layer	Type of Internal Electrode	Slope [mV/decade]	Range of Linearity [M]	Limit of Detection [M]	pH Range	Application/Samples	References
1	S^{2-}	LC	Ag_2S	10^{-6} M Na_2S	Ag∣AgCl	−28.2	1.0×10^{-6}–1.0×10^{-1}	2.3×10^{-7}	6.0–12.0	Industrial water, e.g., petroleum industries.	[76]
2	S^{2-}	SC	Ag_2S	RGSs	Ag wire	-	5.0×10^{-7}–1.0×10^{-3}	1.8×10^{-7}	-	Sea and tap water.	[78]
3	S^{2-}	LC	TFAB-PVC	0.01 M Na_2S and 0.001 M KCl	-	−27.1	1.0×10^{-6}–1.0×10^{-1}	6.0×10^{-7}	-	Water samples.	[77]
4			TFABAHE			−27.1	1.0×10^{-6}–1.0×10^{-1}	3.8×10^{-7}			
5	SO_3^{2-}	SC	CoPC	MWCNTs-COOH	SPCE	−29.8 ± 0.4	2.0×10^{-6}–2.3×10^{-3}	1.1×10^{-6}	5.0–7.2	Various samples.	[79]
6	SO_3^{2-}	SC		PANINFs		−26.5 ± 0.6	5.0×10^{-6}–2.3×10^{-3}	1.5×10^{-6}	4.8–7.7		
7	SO_4^{2-}	SC	Shiff base complex with nickel	-	CPE	−29.7	7.5×10^{-9}–1.5×10^{-3}	5.0×10^{-9}	3.0–9.0	Water and blood samples.	[80]
8	SO_4^{2-}	LC	(oxyethyl)$_3$TM	0.01 M Na_2SO_4 and 0.001 M KCl	-	−27.0	-	6.7×10^{-7}	-	Water samples.	[81]
9	SO_4^{2-}	LC	TFAB-PVC	0.01 M Na_2S and 0.001 M KCl	-	−25.7	1.0×10^{-6}–1.0×10^{-2}	1.0×10^{-6}	-	Water samples.	[77]
10	SO_4^{2-}	LC	TFABAHE		-	−26.5	1.0×10^{-6}–1.0×10^{-2}	7.0×10^{-7}	-		

10. Ion-Selective Electrodes Sensitive to Phosphates

Phosphorus ions are a fairly common constituent contained in food, drinking water and soil. They have a major impact not only on human health but also on the economy, flora, fauna and industry. As a micronutrient, phosphorus is involved in the synthesis of phospholipids and other nucleic proteins [82]. In addition, phosphorus compounds are used quite abundantly in the production of fertilizers. They permeate from the soil into the groundwater and then further, which results in surface eutrophication of water reservoirs [83,84]. This process reduces the oxygen saturation of water and causes the death of many aquatic organisms. Phosphorus ions are also present in processed foods, e.g., meats, beverages and vegetables, where they have a preservative function, but an excess of phosphorus can cause a negative response in the human body, i.e., circulatory problems or problems with the urological system, e.g., kidneys. The maximum number of phosphates in drinking water according to the. WHO is 1 mg/L [85]. For this reason, it is extremely important to improve the methods used to detect phosphate concentrations. One possibility is the use of ion-selective electrodes, which are not only a simple and inexpensive tool, but also give satisfactory, accurate results at a micromole level for phosphates. Over the past few years, four ISEs sensitive to $H_2PO_4^-$ ions, seven selectively detecting HPO_4^{2-} ions and three sensitive to PO_4^{3-} ions have been developed.

One of the four electrodes used to determine the $H_2PO_4^-$ ion was a type of ISE with an internal electrolyte. In the publication [86], a new ionophore 1,3-phenylenebis(methylene)[3-(N,N-diethyl)carbamoylpyridinium] hexafluorophosphate (bis-meta-NICO-PF$_6$) (Figure 22) that provides selectivity for ISE was presented. The same type of ionophores was also presented in bis-, tris- and para-isomerism, but it was the meta-isomer that had the highest affinity for phosphates, which was confirmed by NMR studies. In this paper, two types of the applied membrane were presented: one of them was a classic polymeric membrane, while the other one was a double membrane consisting of a polymeric membrane coated with a silicon rubber (SR) solution layer that was in direct contact with the solution after evaporation of the solvent. The SR layer was introduced to prevent the washing-out of the ionophore from the membrane. Thanks to this, the life of the electrode was extended to approximately 40 days and the electrode response was improved by −15.4 mV/decade compared to an electrode without SR. An additional advantage of this ISE is its good selectivity.

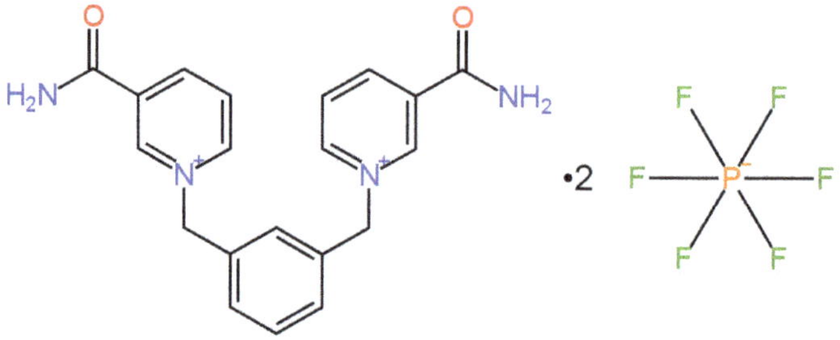

Figure 22. The structure of bis-meta-NICO-PF6 [86].

The next three electrodes are electrodes without IE. They have, as leading electrodes, glassy carbon (GCE) [87], carbon paste electrode (CPE) [88] and Cu wire [89], respectively. The publication [87] presents a novel nanocomposite consisting of polypyrrole (PPy), cobalt and mesoporous ordered carbon (Co-PPY-OMC), which was deposited on GCE in a one-step electrodeposition at constant potential. In order to determine the appropriate component ratios of the selective layer, Response Surface Methodology (RSM) combined with Box Behnken Design (BBD) was used. EIS measurements were carried out and showed that the use of the nanocomposite resulted in a reduction of resistance. Moreover, a satisfactory LOD of 6.8×10^{-6} M and a fairly short response time of 9 s were observed. The ISE had a relatively low potential drift of 1.45 µV/s but unfortunately showed a variable slope depending on the pH value. On the other hand, in the paper [88], the preparation of the hydrogen phosphate ion-imprinted polymer nanoparticles (nano-IIP) in ACN/H2O (acetonitrile/water) with the help of a matrix in the form of H_2PO_4 was presented. Nano-IIP is a novel component providing the carbon paste ISEs selectivity to H_2PO_4 ions. The LOD of the presented electrode reached micromole values. In addition, the electrode is not resistant to pH changes and its potential varies with the pH value. It is associated with the deprotonation that occurred at the interface between the solution and the nano-IIP due to the presence of pyridinium groups. For both Co-PPY-OMC/ISEs and nano-IIP/ISEs the slope is quite weak, deviating fairly strongly from the Nernst value: −31.6 mV/decade and −30.6 mV/decade, respectively. This is quite a disadvantage of such ISEs and may be related, for instance, to the leaching of ionophores from the membranes. The electrode described in the publication [89] had a solid crystalline membrane obtained from a compressed mixture of Ba_3PO_4, Cu_2S and Ag_2S salts connected to a Cu wire. The electrode showed a much higher sensitivity because the slope was similar to Nernstian: −57 mV/decade. In addition, it had the widest linearity range of 1.0×10^{-6}–1.0×10^{-1} M, as well as the lowest detection limit of 2.4×10^{-7} M. The slope value was determined for the electrode conditioned in 0.1 M NaH_2PO_4 buffer solution, while for the electrodes stored in air and water, the slope was −20.6 mV/decade and −27.5 mV/decade, respectively. Unfortunately, in the publication, there is no explicit statement as to which form of phosphate ion the electrode is sensitive to.

Among the HPO_4^{2-} ion-selective electrodes, there are six electrodes with solid contact and two with an internal electrolyte. Two of the SC-type electrodes had a Cu wire as the leading electrode. In the carbon paste ISE construction, a new copolymer (whose structure is shown in Figure 23) and MWCNTs were used [90]. The newly proposed layer was characterized using FTIR, XPS, TG/DTG-DTA and SEM techniques. A satisfying response of the electrode for the divalent main ion of -30.7 ± 0.4 mV/decade was obtained. The ISE exhibited a potential drift of 0.48 mV. In addition, the lifetime of HPO_4^{2-}-ISE was 17 weeks, so it is possible to use it repeatably. The solid-state monohydrogen phosphate sensor was used to determine the content of HPO_4^{2-} ions in many samples with very good reproducibility, i.e., tap water, dam water and river water.

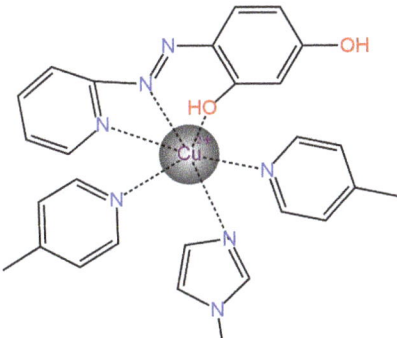

Figure 23. Cross-linked copper(II) doped copolymer [90].

The article [91] presents a new SC-type electrode in which Bi particles were electrolytically deposited onto a platinum wire (Pt wire) from a solution of potassium citrate and bismuth using a Shoddy diode and a function generator. The ion-sensitive function was performed by a $BiPO_4$ membrane, which was deposited by electroplating using chronoamperometry. The surface of the resulting structures was studied by the SEM technique, which confirmed that both materials formed a homogeneous layer. The described electrode was characterized by good analytical performance with a slope of −30.3 mV/decade, a six-order linearity range and an LOD equal to 7.7×10^{-7} M. Moreover, the advantage of the electrode is its response time of 1–2 s and also good reversibility and stability of the potential in different concentrations. The electrode showed a lifetime longer than 90 days. The most convenient working range of the pH value is between 5–9. The electrode was successfully applied to the determination of HPO_4^{2-} ions in natural water. The disadvantage of this electrode is its high sensitivity to chloride ions, and, therefore, it cannot be used in chlorinated solutions, e.g., seawater samples, glacial water samples, etc.

A new electrode based on a molybdenum wire was also proposed for the determination of HPO_4^{2-} ions [92]. The Mo wire was placed in a silicon tube in which one end was covered with a resin in order to prevent it from entering the sample solution. Subsequently, electrochemical deposition of MoO_2 + $PMo_{12}O_{40}^{3-}$ (molybdenum dioxide and molybdophosphate) was carried out until the electrode turned black and the solution turned green. This layer was the ion-sensitive layer towards the hydrogen phosphate ions. The measurements were performed in solutions of the main ion at different pH values, and pH = 9 was found to be the most optimal value. On the basis of the calibration curve, a slope of −27 mV/decade, an LOD equal to 1 µM as well as a linearity range of 1.0×10^{-5}–1.0×10^{-1} M were determined. The disadvantage of this electrode is its rather long response time of approximately 5 min.

In the work [93], two kinds of hydrogen-phosphate electrodes in which two types of ion-selective membranes were used were proposed. Both electrodes contained a mixture of silver salts (Ag_3PO_4 + Ag_2S) but differed in the dopants, the method of membrane preparation and the type of electrode material. In the electrode marked 'type 1 membrane', the inner electrode was a stainless steel disc (SSD) and the third membrane component was polytetrafluoroethylene (PTFE) constituting 50% of the membrane weight. The components of the mixture were pressed together then placed in a Teflon casing and connected to the SSD. In the electrode described as a 'type 2 membrane', the inner electrode was a copper wire while the salt mixture that was used was enriched with multi-walled carbon nanotubes (MWCNTs) (2%). The membrane was prepared by homogenizing the composite mixture with linseed oil, which was then placed in a plastic tube. Subsequently, the Cu wire was immersed in it and then allowed to dry for three days. Both electrodes had the same linearity range and a similar LOD but differed significantly in the slope, which was −21.0 mV/decade for the PTFE-modified electrode and −32.6 mV/decade for the ISE with MWCNTs, which was considerably better. For some parameters, the results were only presented for one electrode, i.e., the response time was only presented for the ISE 'type 2 electrode' and amounted to nearly 60 s. According to the authors' description, the type 1 electrode showed a lifetime of two years, while the type 2 electrode showed a lifetime of only a few days. Definitely, a better comparison of the properties of the conductive materials would be if the same type of internal electrode and method of electrode formation was used because a comparison of the parameters of two completely different electrodes is not adequate.

Only two of the seven electrodes sensitive to HPO_4^{2-} ions had a construction using an internal electrolyte; they are described in [77], where a new material performing the function of a neutral ion carrier was presented, which had already been used for the S^{2-} and SO_4^{2-} ion-sensitive electrodes. In this case, ion-sensitive layers in the form of TFAB-PVC and TFABAHE worked similarly well, as can be seen by the good response and LOD of the electrodes. The TFAB-PVC-modified electrode showed a significantly longer lifetime, while the presence of TFABAHE reduced the lifetime of the electrodes by 60 days compared to the unmodified electrode with this carrier. The presence of chemically modified PVC improved the selectivity and limit of detection.

The last type of electrodes described was PO_4^{3-}-ISEs, which are presented in [94]. All three electrodes were made up of Cu wire immersed in an internal electrolyte of 0.001 M KCl + 0.001 M Na_3PO_4. To ensure selectivity, membranes were prepared using phosphate-imprinted polymers as ionophores: IIP-1 (chitosan-La(III)-PO_4^{3-}), IIP-2 (chitosan-La(III)-AAPTS-PO_4^{3-}) and IIP-3 (AAPTS-La(III)-PO_4^{3-}). The applied materials were investigated using FTIR. Calibration curves were obtained, based on which the slope,

LOD and linearity range were determined. The slope of the characteristics was extremely different from Nernsian's and equaled, respectively, −3.2 mV/decade, −1.9 mV/decade and −3.7 mV/decade for IIP-1/ISE, IIP-2/ISE and IIP-3/ISE, indicating their very weak sensitivity. Despite of the inclusion of information in the publication that each electrode had a linearity range of 1.0×10^{-6}–1.0×10^{-2} M, this value is only true for IIP-1-ISE, while this value was incorrectly reported for the other ISEs. The electrodes showed high sensitivity to changes in the pH value and extremally long response times of 150, 130 and 30 min sequentially for the ISEs modified with IIP-1, IIP-2 and IIP-3.

A comparison of the analytical parameters of various phosphate electrodes is presented in Table 8.

Table 8. Ion-selective electrodes sensitive to phosphate ions.

E. No	Ion	Type of Contact	Ionophore/Ion Carrier	Intermediate/ Transducer Layer	Type of Internal Electrode	Slope [mV/decade]	Range of Linearity [M]	Limit of Detection [M]	pH Range	Application/ Samples	References
1	$H_2PO_4^-$	LC	Bis-meta-NiCO-PF_6	-	-	−53.3	1.0×10^{-6}–1.0×10^{-2}	0.9×10^{-6}	-	Environmental and other real samples.	[86]
2	$H_2PO_4^-$	SC	Co-PPY-OMC	-	GCE	−31.6	1.0×10^{-5}–5.0×10^{-2}	6.8×10^{-6}	3.0–5.0	Water samples for example in human urine or wastewater.	[87]
3	$H_2PO_4^-$	SC	nano-IIP	-	CPE	−30.6 ± 0.5	1.0×10^{-5}–1.0×10^{-1}	4.0×10^{-6}	9.0–12.0	Water samples.	[88]
4	HPO_4^{2-}	SC	Ba_3PO_4 + Cu_2S + Ag_2S pellet	-	Cu wire	−57.0	1.0×10^{-6}–1.0×10^{-1}	2.4×10^{-7}	7.0–9.0	Food samples e.g., meat, vegetables and fruits.	[89]
5	HPO_4^{2-}	SC	Cu(II)-DCP	MWCNTs + graphite	Cu wire	−30.7 ± 0.4	1.0×10^{-6}–1.0×10^{-1}	6.5×10^{-7}	-	Water samples.	[90]
6	HPO_4^{2-}	SC	$BiPO_4$	Bi particles	Pt wire	−30.3	1.0×10^{-6}–1.0×10^{-1}	7.7×10^{-7}	5.0–9.0	Drinking water.	[91]
7	HPO_4^{2-}	SC	MoO_2 + $PMo_{12}O_{40}^{3-}$	-	Mo wire	−27.8 ± 0.5	1.0×10^{-5}–1.0×10^{-1}	1.0×10^{-6}	8.0–9.5	Wastewater, nutrient solution and Coca-Cola.	[92]
8	HPO_4^{2-}	SC	Ag_3PO_4 + Ag_2S	PTFE	SSD	−21.0	1.0×10^{-5}–1.0×10^{-1}	5.3×10^{-6}	3.0–7.0	Solution of pH range 3–7.	[93]
9	HPO_4^{2-}			MWCNTs	Cu wire	−32.6	1.0×10^{-5}–1.0×10^{-1}	5.5×10^{-6}			
10	HPO_4^{2-}		TFAB-PVC	0.01 M Na_2S and 0.001 M KCl	-	−27.5	1.0×10^{-7}–1.0×10^{-2}	7.0×10^{-7}	-	Water samples	[77]
11	HPO_4^{2-}		TFABAHE			−28.7	1.0×10^{-7}–1.0×10^{-2}	5.0×10^{-7}			
12	PO_4^{3-}	LC	IIP-1 (chitosan-La(III)-PO_4^{3-})			−3.2	1.0×10^{-6}–1.0×10^{-2}	7.6×10^{-6}			
13	PO_4^{3-}	LC	IIP-2 (chitosan-La(III)-AAPTS-PO_4^{3-})	0.001 M KCl + 0.001 M Na_3PO_4	Cu wire	−1.9	1.0×10^{-6}–1.0×10^{-2}	5.1×10^{-6}	5.0–7.0	Household wastewater.	[94]
14	PO_4^{3-}	LC	IIP-3 (AAPTS-La(III)-PO_4^{3-})			−3.7	1.0×10^{-6}–1.0×10^{-2}	2.5×10^{-6}			

11. Ion-Selective Electrodes Sensitive to Tiocyanate Ions

The thiocyanate anions have toxic impacts on humans and other living organisms. One of the effects of the excessive amount of SCN^- ions, e.g., in milk, is the inhibition of iodine uptake by the thyroid gland, loss of consciousness and intense dizziness. Newborns and pregnant women are the most susceptible to the influence of SCN^-, as well as organisms living in communities located far from saline water reservoirs, which are the source of iodine [95,96]. Increased thiocyanide concentrations in the human body are associated with excessive smoking (passive and direct smokers) or industrial pollution. The thiocyanates are the compounds that have the greatest environmental impact on iodine metabolism in the body and therefore on the occurrence of diseases associated with thyroid malfunction [97,98]. Consequently, it is necessary to monitor the concentration of these ions in order to protect ourselves from being excessively exposed to them. For this purpose, we can use low-cost and fast-measuring instruments such as ion-selective electrodes. Over the past few years, five new modifications of SCN^--ISEs have been proposed to improve their performance.

In the article [99], liquid contact electrodes in the form of a 0.01 M KSCN solution were used to determine SCN^- ions in the human saliva of smokers and non-smokers. Several ionophores for the detection of thiocyanides were described in the present study, but ultimately 3,4,5-tris(dodecyloxy)benzyltrilauryl ammonium (TL) bromide was used as the membrane component of the presented ISE (Figure 24). The electrode had a wide pH range of 0.5–12.5 and a good detection limit of 5.6×10^{-6} M. The ISE was also characterized by an average slope and linearity range.

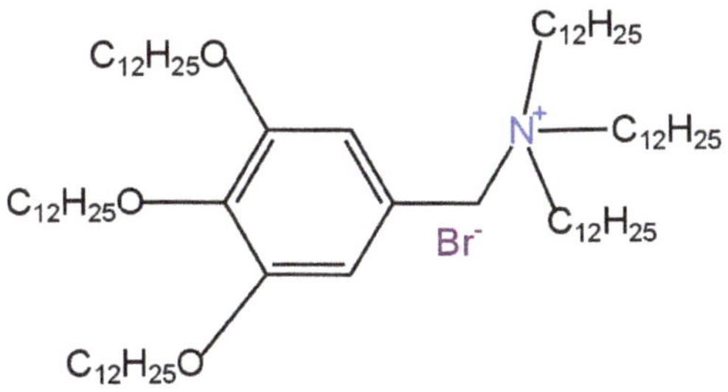

Figure 24. Structure of 3,4,5-tris(dodecyloxy)benzyltrilauryl ammonium (TL) bromide [99].

Urbanowicz et al. presented three types of SCN^--ISEs, where one was a liquid and two were solid contact electrodes [100]. All three electrodes in the membrane had as an ionophore a tetrakis-(4-diphenylmethylphosphonium-butoxy)-tetrakis-p-tert-butylcalix[4]arene tetrathiocyanate (Figure 25). For the liquid-contact electrode, the ionophore membrane was placed in the Ag|AgCl electrode body and then immersed in 0.001 M KCl, which was the internal electrolyte. On the other hand, electrodes without an internal electrolyte were created by dropping an appropriate volume of membrane mixture on the surface contact of GC electrodes or Au rods. In this study, membranes with three plasticizers o-NPOE, BBPA and chloroparaffins were investigated. Among all three electrodes, the best performance was obtained for SCN^-/GCE/ISE, where a Nernstian slope of -59.9 ± 0.3 mV/decade was achieved. Both SC-type electrodes showed an order of magnitude wider range of linearity with respect to LC-ISE. Analogous to the electrode described in the previous paper, the concentration of SCN^- ions in the saliva of non-smokers and smokers was investigated.

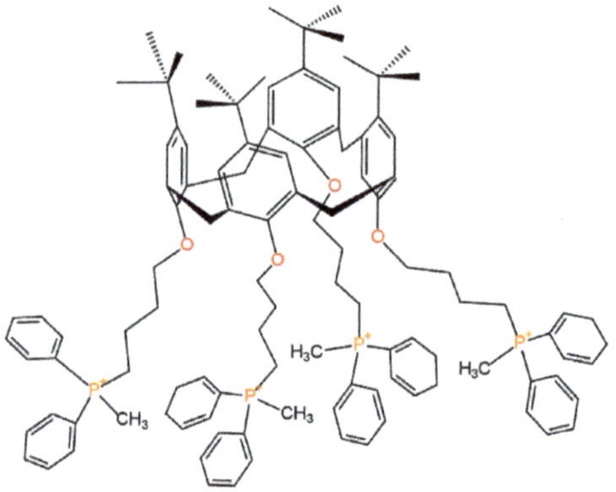

Figure 25. The structure of a new ionophore—tetrakis-(4-diphenylmethylphosphonium-butoxy)-tetrakis-p-tert-butylcalix [4]arene tetrathiocyanate [100].

The last SCN^--ISE described in this review is the LC electrode (Ag | AgCl immersed in an electrolyte constituted by a mixture of 0.01 M NaSCN and 0.1 M NaCl) [101]. The ion-selective membrane was a sol-gel-based matrix in which the active ingredient was tricaprylylmethylammonium thiocyanate (Aliquat336-SCN) (ionophore). The results obtained for this electrode were compared with those for an ISE using a conventional polymeric PVC membrane. Better selectivity towards anions, i.e., ClO_4^- and SiO_4^-, was obtained. Compared to the other liquid contact electrodes, it had the highest slope value (highest sensitivity), while the other parameters were very similar. This electrode was also used for the determination of SCN^- ions in human saliva.

A comparison of the analytical parameters of various thiocyanate electrodes is presented in Table 9.

Table 9. Ion-selective electrodes sensitive to SCN$^-$ ions.

E. No	Type of Contact	Ionophore/Ion Carrier	Intermediate/ Transducer Layer	Type of Internal Electrode	Slope [mV/decade]	Range of Linearity [M]	Limit of Detection [M]	pH Range	Application/ Samples	References
1	LC	TL	0.01 M KSCN	-	−53.9	1.0×10^{-6}–1.0×10^{-1}	5.6×10^{-6}	0.5–12.5	Human saliva.	[99]
2	LC	Tetrakis-(4-diphenylmethylphosphonium-butoxy)-tetrakis-p-tert-butylcalix[4]arene tetrathiocyanate	0.001 M KCl	Ag|AgCl	−55.5 ± 2.1	1.0×10^{-5}–1.0×10^{-1}	6.3×10^{-6}	-	Saliva and other medical measurements.	[100]
3	SC		-	GCE	−59.9 ± 0.3	1.0×10^{-6}–1.0×10^{-1}	1.6×10^{-6}			
4	SC		-	Au rods	−53.3 ± 0.3	1.0×10^{-6}–1.0×10^{-1}	3.2×10^{-6}			
5	LC	Aliquat336-SCN	0.01 M NaSCN and 0.1 M NaCl	Ag|AgCl	−56.3	3.2×10^{-5}–5.0×10^{-1}	6.3×10^{-6}	-	Human saliva.	[101]

12. Ion-Selective Electrodes Sensitive to Other Ions

The development of ion-selective electrodes is still in progress and newer and newer sensors that selectively detect a large range of ions were introduced by researchers. Over the past five years, single solutions have been presented for the determination of ions such as AsO_4^{3-}, BO_3^{3-}, CH_3COO^-, CO_3^{2-} and SiO_3^{2-}.

An electrode sensitive to AsO_4^{3-} ions was presented in 2018 by Khan et al. [102]. A new fibrous poly-methylmethacrylate-ZnO (PMMA-ZnO) ion exchanger (Figure 26) was proposed, which was prepared by a solid-gel method. The newly formed material was then characterized by techniques, i.e., FTIR, SEM, TEM, XRD, TGA and EDX. It showed very good properties and was successfully applied as an active component of an ion-selective membrane providing selectivity to arsenic ions. The presented LC-ISE exhibited a super-Nerstian response of the electrode (slope of −28.6 mV/decade for the trivalent ion), a nine-order linearity range and a nanomole detection limit of 1.0×10^{-9} M. The response time of such an electrode is about 35 s, whereas a great advantage is the possibility of using it for a period of 12 months.

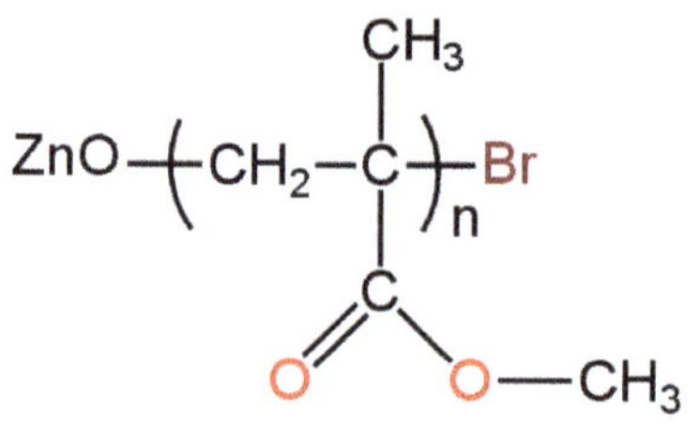

Figure 26. The structure of PMMA-ZnO [102].

In order to monitor glucose and glycate in blood and to determine the boron in real samples, an electrode sensitive to BO_3^{3-} ions was invented [103]. This electrode is based on a composite of multi-walled carbon nanotubes and $Ag_2B_4O_7$, which simultaneously acts as a solid contact layer and a BO_3^{3-} ion trapping material. The MWCNTs provide good ionic conductivity, while the other component of the composite provides ISE selectivity. The composite membrane was placed in a tube and compressed before being placed on a Cu wire. The performance of the electrode was not remarkable because the LOD = 5.6×10^{-5} M and the linearity range was equal to 1.0×10^{-4}–1.0×10^{-1} M. The advantage is that the electrode is not sensitive to the presence of interfering ions and its lifetime extends to 18 weeks. Moreover, good potential reversibility of such an electrode was reported. It was successfully used for the determination of boron in rocks, soil and water because comparable results to those achieved by the ICP-MS technique were obtained.

The determination of CH_3COO^- ions in aqueous solutions was presented in [104], where a solid contact electrode with 1,3-bic(carbosyl)urea derivate acting as the ionophore was described (Figure 27). The function of the solid contact was performed by the conductive polymer PEDOT, which was deposited on the GCE surface by galvanostatic electropolymerization. A polymeric ion-selective membrane containing the proposed active ingredient was then dropped on the prepared surface. The impedance measurements using EIS as well as the parameters determined by potentiometric measurements were carried out for CH_3COO^--ISE, all of which were not impressive. A narrow linearity range and a low LOD were obtained, as well as a slope = −51.3 mV/decade. Measurements cannot be performed in alkaline solutions due to the high probability of OH^- ions interference. In comparison to the electrode that contained the TDMACl ion exchanger, a weaker influence of interfering ions, i.e., SCN^-, I^-, Br^- and NO_3^-, was obtained.

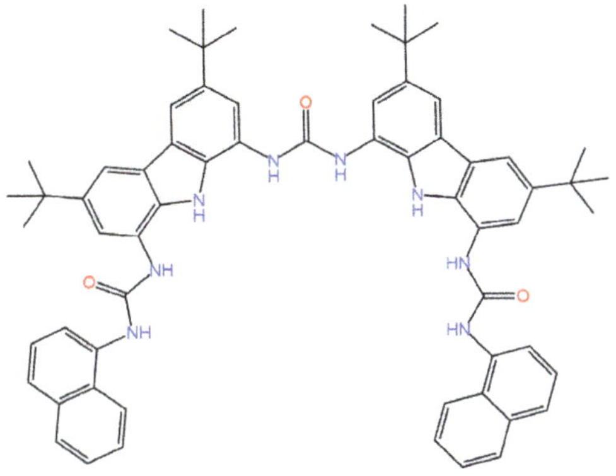

Figure 27. The structure of 1,3-bic(carbozyl)urea derivate [104].

The new carbonate electrode proposed by the team of Zhang et al. can be used for the exploration of deep-sea hydrothermal activity [105]. It has carbonate film as the solid contact and carbonate ionophore VII as the ionophore. In order to make such an electrode selective for CO_3^{2-} ions, a carbon film was first applied electrochemically to the Ni wire and then that electrode was immersed in a solution containing the ion-sensitive component. To characterize the formed film on the electrode morphologically, the SEM technique was used. Subsequently, analytical parameters that were at a very good level were determined, including slope = -30.4 mV/decade and LOD = 2.8×10^{-6} M. The electrode also showed good reproducibility and potential stability, so the carbon film proved to work well as a solid contact.

To ensure the possibility of measuring silicates in aqueous solutions containing small amounts of chlorine, an ISE which was Ag wire coated by the Pb film was proposed [106]. Selectivity for silicates was provided by a $PbSiO_3$ membrane, which was electrochemically applied to the surface of the sensor. The electrode was characterized by good sensitivity, as is evidenced by an over-Nernst slope for this type of ion, as well as a satisfying detection limit of 2.8×10^{-6} M. An advantage of this electrode is its fast response time of 5 s. It showed good selectivity towards SiO_3^{2-} ions and was not affected by the interfering ions NO_3^-, CH_3COO^- and CO_3^{2-}. The electrode needs to be improved in terms of its sensitivity to Cl^- ions so that it can be applied to different types of samples.

The basic analytical parameters of other ISEs described above are presented in Table 10.

Table 10. Ion-selective electrodes sensitive to other ions.

E. No	Ion	Type of Contact	Ionophore/Ion Carrier	Intermediate/Transducer Layer	Type of Internal Electrode	Slope [mV/decade]	Range of Linearity [M]	Limit of Detection [M]	pH Range	Application/Samples	References
1	AsO_4^{3-}	LC	PMMA-ZnO	0.05 M Na_3AsO_4	SCE	−28.6	1.0×10^{-9}–1.0×10^{-1}	1.0×10^{-9}	4.0–7.0	Water solutions.	[102]
2	BO_3^{3-}	SC	$Ag_2B_4O_7$	MWCNTs	Cu wire	−34.0 ± 1.0	1.0×10^{-4}–1.0×10^{-1}	5.6×10^{-5}	5.0–8.0	Rock, soil.	[103]
3	CH_3COO^-	SC	1,3-bis(carbazolyl)urea	PEDOT	GCE	−51.3	3.2×10^{-5}–7.9×10^{-2}	1×10^{-5}	6.0–8.0	Aqueous samples.	[104]
4	CO_3^{2-}	SC	Carbonate ionophore VII	Carbon film	Ni wire	−30.4	1.0×10^{-5}–1.0×10^{-1}	2.8×10^{-6}	-	Exploration of deep-sea hydrothermal activity.	[105]
5	SiO_3^{2-}	SC	$PbSiO_3$	$PbSiO_3$	Ag wire coated by the Pb film	−31.3	1.0×10^{-5}–1.0×10^{-1}	-	-	Aqueous samples with low-chloride content.	[106]

13. Conclusions

This has been a review of the new materials used in the construction of ion-selective electrodes for anion determination. From this overview of more than 100 different anion-selective electrodes, two main research directions can be distinguished, one concerning the development of new ionophores and the other concerning the improvement of the construction, mainly related to the introduction of new materials of solid contact or components of paste electrodes. Depending on the type of ion, one or the other direction is dominant, e.g., in the case of nitrate electrodes, most of the papers concern solid contact materials, while in the co-publications concerning new phosphate electrodes, papers describing new ionophores predominate. Among the active substances of the membrane used to obtain the selectivity of a specific ion, ion exchange substances and sparingly soluble salts still dominate. Conductive materials used as intermediate layers or membrane modifiers in ASS-ISEs are mainly conductive polymers, carbon nanomaterials, metal nanoparticles and composite or hybrid materials.

The use of new materials in the construction of ISEs allowed to obtain electrodes with better performance and more convenient to use. This is a valuable achievement considering the numerous advantages of potentiometry, and it is all the more important that there are far fewer alternative determination methods for anions.

In the area of anion-selective electrodes, the development of effective ionophores for hydrophilic anions (carbonates, phosphates and sulfates) is still a current research topic. Another prospective direction of research includes the use of composite materials, especially nanocomposites, based on carbon nanomaterials and metal oxide nanoparticles. These materials combine the valuable properties of the constituent components, which opens up new fields for their effective applications.

Author Contributions: Conceptualization, C.W. and K.M.; methodology and formal analysis, K.P.; investigation, K.P. and K.M.; data curation K.M. and K.P.; writing—original draft preparation, K.M.; writing—review and editing, C.W.; literature survey, K.P., K.M. and C.W.; supervision, C.W. All authors have read and agreed to the published version of the manuscript.

Funding: This research received no external funding.

Institutional Review Board Statement: Not applicable.

Informed Consent Statement: Not applicable.

Data Availability Statement: Not applicable.

Conflicts of Interest: The authors declare no conflict of interest.

References

1. Morf, W.E. *The Principles of Ion-Selective Electrodes and of Membrane Transport*; Elsevier: New York, NY, USA, 1981.
2. Głąb, S.; Maj-Żurawska, M.; Hulanicki, A. Ion-Selective Electrodes | Glass Electrodes. In *Reference Module in Chemistry, Molecular Sciences and Chemical Engineering*; Elsevier: Amsterdam, The Netherlands, 2013.
3. Sohail, M.; De Marco, R. ELECTRODES | Ion-Selective Electrodes. In *Reference Module in Chemistry, Molecular Sciences and Chemical Engineering*; Elsevier: Amsterdam, The Netherlands, 2013.
4. Lyu, Y.; Gan, S.; Bao, Y.; Zhong, L.; Xu, J.; Wang, W.; Liu, Z.; Ma, Y.; Yang, G.; Niu, L. Solid-Contact Ion-Selective Electrodes: Response Mechanisms, Transducer Materials and Wearable Sensors. *Membranes* **2020**, *10*, 128. [CrossRef] [PubMed]
5. Paczosa-Bator, B.; Cabaj, L.; Piech, R.; Skupień, K. Platinum Nanoparticles Intermediate Layer in Solid-State Selective Electrodes. *Analyst* **2012**, *137*, 5272. [CrossRef]
6. Schwarz, J.; Trommer, K.; Mertig, M. Solid-Contact Ion-Selective Electrodes Based on Graphite Paste for Potentiometric Nitrate and Ammonium Determinations. *Am. J. Anal. Chem.* **2018**, *9*, 591–601. [CrossRef]
7. Wardak, C.; Morawska, K.; Paczosa-Bator, B.; Grabarczyk, M. Improved Lead Sensing Using a Solid-Contact Ion-Selective Electrode with Polymeric Membrane Modified with Carbon Nanofibers and Ionic Liquid Nanocomposite. *Materials* **2023**, *16*, 1003. [CrossRef]
8. Wardak, C.; Pietrzak, K.; Morawska, K.; Grabarczyk, M. Ion-Selective Electrodes with Solid Contact Based on Composite Materials: A Review. *Sensors* **2023**, *23*, 5839. [CrossRef] [PubMed]
9. Wardak, C. Solid Contact Cadmium Ion-Selective Electrode Based on Ionic Liquid and Carbon Nanotubes. *Sens. Actuators B Chem.* **2015**, *209*, 131–137. [CrossRef]

10. Wardak, C.; Pietrzak, K.; Morawska, K. Nanocomposite of Copper Oxide Nanoparticles and Multi-Walled Carbon Nanotubes as a Solid Contact of a Copper-Sensitive Ion-Selective Electrode: Intermediate Layer or Membrane Component–Comparative Studies. *Appl. Nanosci.* **2023**. [CrossRef]
11. Ma, S.; Wang, Y.; Zhang, W.; Wang, Y.; Li, G. Solid-Contact Ion-Selective Electrodes for Histamine Determination. *Sensors* **2021**, *21*, 6658. [CrossRef]
12. Bobacka, J. Conducting Polymer-Based Solid-State Ion-Selective Electrodes. *Electroanalysis* **2006**, *18*, 7–18. [CrossRef]
13. Lenik, J.; Wesoły, M.; Ciosek, P.; Wróblewski, W. Evaluation of Taste Masking Effect of Diclofenac Using Sweeteners and Cyclodextrin by a Potentiometric Electronic Tongue. *J. Electroanal. Chem.* **2016**, *780*, 153–159. [CrossRef]
14. Bobacka, J.; Ivaska, A.; Lewenstam, A. Potentiometric Ion Sensors. *Chem. Rev.* **2008**, *108*, 329–351. [CrossRef]
15. Fibbioli, M.; Morf, W.E.; Badertscher, M.; de Rooij, N.F.; Pretsch, E. Potential Drifts of Solid-Contacted Ion-Selective Electrodes Due to Zero-Current Ion Fluxes Through the Sensor Membrane. *Electroanalysis* **2000**, *12*, 1286–1292. [CrossRef]
16. Suman, S.; Singh, R. Anion Selective Electrodes: A Brief Compilation. *Microchem. J.* **2019**, *149*, 104045. [CrossRef]
17. Bobacka, J.; Lindfors, T.; McCarrick, M.; Ivaska, A.; Lewenstam, A. Single-Piece All-Solid-State Ion-Selective Electrode. *Anal. Chem.* **1995**, *67*, 3819–3823. [CrossRef]
18. Wardak, C.; Grabarczyk, M. Single-Piece All-Solid-State Co(II) Ion-Selective Electrode for Cobalt Monitoring in Real Samples. *Int. Agrophys* **2019**, *1*, 17–24. [CrossRef]
19. Bobacka, J. Potential Stability of All-Solid-State Ion-Selective Electrodes Using Conducting Polymers as Ion-to-Electron Transducers. *Anal. Chem.* **1999**, *71*, 4932–4937. [CrossRef] [PubMed]
20. Lindner, E.; Umezawa, Y. Performance Evaluation Criteria for Preparation and Measurement of Macro- and Microfabricated Ion-Selective Electrodes (IUPAC Technical Report). *Pure Appl. Chem.* **2008**, *80*, 85–104. [CrossRef]
21. Zdrachek, E.; Bakker, E. Potentiometric Sensing. *Anal. Chem.* **2019**, *91*, 2–26. [CrossRef] [PubMed]
22. Bobacka, J.; Lewenstam, A.; Ivaska, A. Equilibrium Potential of Potentiometric Ion Sensors under Steady-State Current by Using Current-Reversal Chronopotentiometry. *J. Electroanal. Chem.* **2001**, *509*, 27–30. [CrossRef]
23. Paré, F.; Visús, A.; Gabriel, G.; Baeza, M. Novel Nitrate Ion-Selective Microsensor Fabricated by Means of Direct Ink Writing. *Chemosensors* **2023**, *11*, 174. [CrossRef]
24. Hjort, R.G.; Soares, R.R.A.; Li, J.; Jing, D.; Hartfiel, L.; Chen, B.; Van Belle, B.; Soupir, M.; Smith, E.; McLamore, E.; et al. Hydrophobic Laser-Induced Graphene Potentiometric Ion-Selective Electrodes for Nitrate Sensing. *Microchim. Acta* **2022**, *189*, 122. [CrossRef]
25. Ali, A.; Wang, X.; Chen, Y.; Jiao, Y.; Mahal, N.K.; Moru, S.; Castellano, M.J.; Schnable, J.C.; Schnable, P.S.; Dong, L. Continuous Monitoring of Soil Nitrate Using a Miniature Sensor with Poly(3-Octyl-Thiophene) and Molybdenum Disulfide Nanocomposite. *ACS Appl. Mater. Interfaces* **2019**, *11*, 29195–29206. [CrossRef] [PubMed]
26. Pietrzak, K.; Wardak, C.; Malinowski, S. Application of Polyaniline Nanofibers for the Construction of Nitrate All-Solid-State Ion-Selective Electrodes. *Appl. Nanosci.* **2021**, *11*, 2823–2835. [CrossRef]
27. Zhang, Z.; Papautsky, I. Miniature Ion-selective Electrodes with Mesoporous Carbon Black as Solid Contact. *Electroanalysis* **2021**, *33*, 2143–2151. [CrossRef]
28. Thuy, N.T.D.; Wang, X.; Zhao, G.; Liang, T.; Zou, Z. A Co_3O_4 Nanoparticle-Modified Screen-Printed Electrode Sensor for the Detection of Nitrate Ions in Aquaponic Systems. *Sensors* **2022**, *22*, 9730. [CrossRef] [PubMed]
29. Fozia; Zhao, G.; Nie, Y.; Jiang, J.; Chen, Q.; Wang, C.; Xu, X.; Ying, M.; Hu, Z.; Xu, H. Preparation of Nitrate Bilayer Membrane Ion-Selective Electrode Modified by Pericarpium Granati-Derived Biochar and Its Application in Practical Samples. *Electrocatalysis* **2023**, *14*, 534–545. [CrossRef]
30. Zhang, L.; Wei, Z.; Liu, P. An All-Solid-State NO_3^- Ion-Selective Electrode with Gold Nanoparticles Solid Contact Layer and Molecularly Imprinted Polymer Membrane. *PLoS ONE* **2020**, *15*, e0240173. [CrossRef]
31. Baumbauer, C.L.; Goodrich, P.J.; Payne, M.E.; Anthony, T.; Beckstoffer, C.; Toor, A.; Silver, W.; Arias, A.C. Printed Potentiometric Nitrate Sensors for Use in Soil. *Sensors* **2022**, *22*, 4095. [CrossRef]
32. Liu, Y.; Liu, Y.; Meng, Z.; Qin, Y.; Jiang, D.; Xi, K.; Wang, P. Thiol-Functionalized Reduced Graphene Oxide as Self-Assembled Ion-to-Electron Transducer for Durable Solid-Contact Ion-Selective Electrodes. *Talanta* **2020**, *208*, 120374. [CrossRef]
33. Fan, Y.; Huang, Y.; Linthicum, W.; Liu, F.; Beringhs, A.O.; Dang, Y.; Xu, Z.; Chang, S.-Y.; Ling, J.; Huey, B.D.; et al. Toward Long-Term Accurate and Continuous Monitoring of Nitrate in Wastewater Using Poly(Tetrafluoroethylene) (PTFE)–Solid-State Ion-Selective Electrodes (S-ISEs). *ACS Sens.* **2020**, *5*, 3182–3193. [CrossRef]
34. Pietrzak, K.; Wardak, C.; Łyszczek, R. Solid Contact Nitrate Ion-selective Electrode Based on Cobalt(II) Complex with 4,7-Diphenyl-1,10-phenanthroline. *Electroanalysis* **2020**, *32*, 724–731. [CrossRef]
35. Pietrzak, K.; Wardak, C. Comparative Study of Nitrate All Solid State Ion-Selective Electrode Based on Multiwalled Carbon Nanotubes-Ionic Liquid Nanocomposite. *Sens. Actuators B Chem.* **2021**, *348*, 130720. [CrossRef]
36. Kim, M.-Y.; Lee, J.-W.; Park, D.J.; Lee, J.-Y.; Myung, N.V.; Kwon, S.H.; Lee, K.H. Highly Stable Potentiometric Sensor with Reduced Graphene Oxide Aerogel as a Solid Contact for Detection of Nitrate and Calcium Ions. *J. Electroanal. Chem.* **2021**, *897*, 115553. [CrossRef]
37. Ivanova, A.; Mikhelson, K. Electrochemical Properties of Nitrate-Selective Electrodes: The Dependence of Resistance on the Solution Concentration. *Sensors* **2018**, *18*, 2062. [CrossRef] [PubMed]

38. Hassan, S.S.M.; Eldin, A.G.; Amr, A.E.-G.E.; Al-Omar, M.A.; Kamel, A.H.; Khalifa, N.M. Improved Solid-Contact Nitrate Ion Selective Electrodes Based on Multi-Walled Carbon Nanotubes (MWCNTs) as an Ion-to-Electron Transducer. *Sensors* **2019**, *19*, 3891. [CrossRef]
39. Fukao, Y.; Kitazumi, Y.; Kano, K.; Shirai, O. Construction of Nitrate-Selective Electrodes and Monitoring of Nitrates in Hydroponic Solutions. *Anal. Sci.* **2018**, *34*, 1373–1377. [CrossRef]
40. Pięk, M.; Piech, R.; Paczosa-Bator, B. TTF-TCNQ Solid Contact Layer in All-Solid-State Ion-Selective Electrodes for Potassium or Nitrate Determination. *J. Electrochem. Soc.* **2018**, *165*, B60–B65. [CrossRef]
41. Zuo, H.; Chen, L.; Kong, M.; Qiu, L.; Lü, P.; Wu, P.; Yang, Y.; Chen, K. Toxic Effects of Fluoride on Organisms. *Life Sci.* **2018**, *198*, 18–24. [CrossRef]
42. Guth, S.; Hüser, S.; Roth, A.; Degen, G.; Diel, P.; Edlund, K.; Eisenbrand, G.; Engel, K.-H.; Epe, B.; Grune, T.; et al. Toxicity of Fluoride: Critical Evaluation of Evidence for Human Developmental Neurotoxicity in Epidemiological Studies, Animal Experiments and in Vitro Analyses. *Arch. Toxicol.* **2020**, *94*, 1375–1415. [CrossRef]
43. Yamada, T.; Kanda, K.; Yanagida, Y.; Mayanagi, G.; Washio, J.; Takahashi, N. Fluoride Ion Sensor Based on LaF$_3$ Nanocrystals Prepared by Low-Temperature Process. *J. Ceram. Soc. Jpn.* **2023**, *131*, 22127. [CrossRef]
44. Yamada, T.; Kanda, K.; Yanagida, Y.; Mayanagi, G.; Washio, J.; Takahashi, N. All-solid-state Fluoride Ion-selective Electrode Using LaF$_3$ Single Crystal with Poly(3,4-ethylenedioxythiophene) as Solid Contact Layer. *Electroanalysis* **2023**, *35*, e202200103. [CrossRef]
45. Radić, J.; Bralić, M.; Kolar, M.; Genorio, B.; Prkić, A.; Mitar, I. Development of the New Fluoride Ion-Selective Electrode Modified with FexOy Nanoparticles. *Molecules* **2020**, *25*, 5213. [CrossRef] [PubMed]
46. Goud, K.Y.; Sandhu, S.S.; Teymourian, H.; Yin, L.; Tostado, N.; Raushel, F.M.; Harvey, S.P.; Moores, L.C.; Wang, J. Textile-Based Wearable Solid-Contact Flexible Fluoride Sensor: Toward Biodetection of G-Type Nerve Agents. *Biosens. Bioelectron.* **2021**, *182*, 113172. [CrossRef]
47. Biyareh, M.N.; Rezvani, A.R.; Dashtian, K.; Montazerozohori, M.; Ghaedi, M.; Masoudi Asl, A.; White, J. Potentiometric Ion-Selective Electrode Based on a New Single Crystal Cadmium(II) Schiff Base Complex for Detection of Fluoride Ion: Central Composite Design Optimization. *IEEE Sens. J.* **2019**, *19*, 413–425. [CrossRef]
48. Li, L.; Zhang, Y.; Li, Y.; Duan, Y.; Qian, Y.; Zhang, P.; Guo, Q.; Ding, J. Polymeric Membrane Fluoride-Selective Electrodes Using Lewis Acidic Organo-Antimony(V) Compounds as Ionophores. *ACS Sens.* **2020**, *5*, 3465–3473. [CrossRef] [PubMed]
49. Ke, X. Micro-Fabricated Electrochemical Chloride Ion Sensors: From the Present to the Future. *Talanta* **2020**, *211*, 120734. [CrossRef]
50. de Graaf, D.B.; Abbas, Y.; Gerrit Bomer, J.; Olthuis, W.; van den Berg, A. Sensor–Actuator System for Dynamic Chloride Ion Determination. *Anal. Chim. Acta* **2015**, *888*, 44–51. [CrossRef]
51. Kosaka, K.; Asami, M.; Kunikane, S. Perchlorate: Origin and Occurrence in Drinking Water. In *Reference Module in Earth Systems and Environmental Sciences*; Elsevier: Amsterdam, The Netherlands, 2013.
52. Srinivasan, A.; Viraraghavan, T. Perchlorate: Health Effects and Technologies for Its Removal from Water Resources. *Int. J. Environ. Res. Public Health* **2009**, *6*, 1418–1442. [CrossRef]
53. Pietrzak, K.; Morawska, K.; Malinowski, S.; Wardak, C. Chloride Ion-Selective Electrode with Solid-Contact Based on Polyaniline Nanofibers and Multiwalled Carbon Nanotubes Nanocomposite. *Membranes* **2022**, *12*, 1150. [CrossRef]
54. Kalayci, S. Analysis of Halogens in Wastewater with a New Prepared Ion Selective Electrode. *Monatshefte Chem.-Chem. Mon.* **2022**, *153*, 1137–1141. [CrossRef]
55. Pięk, M.; Paczosa-Bator, B.; Smajdor, J.; Piech, R. Molecular Organic Materials Intermediate Layers Modified with Carbon Black in Potentiometric Sensors for Chloride Determination. *Electrochim. Acta* **2018**, *283*, 1753–1762. [CrossRef]
56. Alizadeh, T.; Rafiei, F.; Akhoundian, M. A Novel Chloride Selective Potentiometric Sensor Based on Graphitic Carbon Nitride/Silver Chloride (g-C$_3$N$_4$/AgCl) Composite as the Sensing Element. *Talanta* **2022**, *237*, 122895. [CrossRef]
57. Liao, C.; Zhong, L.; Tang, Y.; Sun, Z.; Lin, K.; Xu, L.; Lyu, Y.; He, D.; He, Y.; Ma, Y.; et al. Solid-Contact Potentiometric Anion Sensing Based on Classic Silver/Silver Insoluble Salts Electrodes without Ion-Selective Membrane. *Membranes* **2021**, *11*, 959. [CrossRef]
58. Prkić, A. New Sensor Based on AgCl Containing Iron Oxide or Zinc Oxide Nanoparticles for Chloride Determination. *Int. J. Electrochem. Sci.* **2019**, *14*, 861–874. [CrossRef]
59. Babaei, M.; Alizadech, N. Highly Selective Perchlorate Coated-Wire Electrode (CWE) Based on an Electrosynthesized Dixanthylinum Dye and Its Application in Water Samples. *Chem. Rev. Lett.* **2022**, *5*, 241–249. [CrossRef]
60. Itterheimová, P.; Bobacka, J.; Šindelář, V.; Lubal, P. Perchlorate Solid-Contact Ion-Selective Electrode Based on Dodecabenzylbambus[6]Uril. *Chemosensors* **2022**, *10*, 115. [CrossRef]
61. Hassan, S.S.M.; Galal Eldin, A.; Amr, A.E.-G.E.; Al-Omar, M.A.; Kamel, A.H. Single-Walled Carbon Nanotubes (SWCNTs) as Solid-Contact in All-Solid-State Perchlorate ISEs: Applications to Fireworks and Propellants Analysis. *Sensors* **2019**, *19*, 2697. [CrossRef]
62. Pavelka, S. Metabolism of Bromide and Its Interference with the Metabolism of Iodine. *Physiol. Res.* **2004**, *53* (Suppl. S1), S81–S90. [CrossRef] [PubMed]
63. de Souza, A.; Narvencar, K.P.S.; Sindhoora, K.V. The Neurological Effects of Methyl Bromide Intoxication. *J. Neurol. Sci.* **2013**, *335*, 36–41. [CrossRef] [PubMed]

64. Rayamajhi, S.; Sharma, S.; Iftikhar, H. Unexplained Bromide Toxicity Presenting as Hyperchloremia and a Negative Anion Gap. *Cureus* **2023**, *15*, e36218. [CrossRef] [PubMed]
65. Vlascici, D.; Plesu, N.; Fagadar-Cosma, G.; Lascu, A.; Petric, M.; Crisan, M.; Belean, A.; Fagadar-Cosma, E. Potentiometric Sensors for Iodide and Bromide Based on Pt(II)-Porphyrin. *Sensors* **2018**, *18*, 2297. [CrossRef] [PubMed]
66. Kaur, N.; Kaur, J.; Badru, R.; Kaushal, S.; Singh, P.P. BGO/AlFu MOF Core Shell Nano-Composite Based Bromide Ion-Selective Electrode. *J. Environ. Chem. Eng.* **2020**, *8*, 104375. [CrossRef]
67. Kou, L. Detection of Bromide Ions in Water Samples with a Nanomolar Detection Limit Using a Potentiometric Ion-Selective Electrode. *Int. J. Electrochem. Sci.* **2019**, *14*, 1601–1609. [CrossRef] [PubMed]
68. Maruthupandi, M.; Chandhru, M.; Rani, S.K.; Vasimalai, N. Highly Selective Detection of Iodide in Biological, Food, and Environmental Samples Using Polymer-Capped Silver Nanoparticles: Preparation of a Paper-Based Testing Kit for On-Site Monitoring. *ACS Omega* **2019**, *4*, 11372–11379. [CrossRef] [PubMed]
69. Laurberg, P.; Cerqueira, C.; Ovesen, L.; Rasmussen, L.B.; Perrild, H.; Andersen, S.; Pedersen, I.B.; Carlé, A. Iodine Intake as a Determinant of Thyroid Disorders in Populations. *Best. Pract. Res. Clin. Endocrinol. Metab.* **2010**, *24*, 13–27. [CrossRef]
70. Zimmermann, M.B. Iodine Deficiency. *Endocr. Rev.* **2009**, *30*, 376–408. [CrossRef]
71. Prkić, A. Development of a New Potentiometric Sensor Based on Home Made Iodide ISE Enriched with ZnO Nanoparticles and Its Application for Determination of Penicillamine. *Int. J. Electrochem. Sci.* **2018**, *13*, 10894–10903. [CrossRef]
72. Shvedene, N.V.; Abashev, M.N.; Arakelyan, S.A.; Otkidach, K.N.; Tomilova, L.G.; Pletnev, I.V. Highly Selective Solid-State Sensor for Iodide Based on the Combined Use of Platinum (IV) Phthalocyanine and Solidified Pyridinium Ionic Liquid. *J. Solid. State Electrochem.* **2019**, *23*, 543–552. [CrossRef]
73. Seah, G.E.K.K.; Tan, A.Y.X.; Neo, Z.H.; Lim, J.Y.C.; Goh, S.S. Halogen Bonding Ionophore for Potentiometric Iodide Sensing. *Anal. Chem.* **2021**, *93*, 15543–15549. [CrossRef]
74. Dordevic, D.; Capikova, J.; Dordevic, S.; Tremlová, B.; Gajdács, M.; Kushkevych, I. Sulfur Content in Foods and Beverages and Its Role in Human and Animal Metabolism: A Scoping Review of Recent Studies. *Heliyon* **2023**, *9*, e15452. [CrossRef]
75. Fatima, U.; Okla, M.K.; Mohsin, M.; Naz, R.; Soufan, W.; Al-Ghamdi, A.A.; Ahmad, A. A Non-Invasive Tool for Real-Time Measurement of Sulfate in Living Cells. *Int. J. Mol. Sci.* **2020**, *21*, 2572. [CrossRef] [PubMed]
76. Abdullad, Z.K.; Al-Samarrai, S.Y. Modified Solid Ion-Selective Electrode for Potentiometric Determination of Sulfide in Oil Refineries Water. *Sci. Rev. Eng. Environ. Stud. SREES* **2021**, *30*, 98–105. [CrossRef]
77. Matveichuk, Y.; Rakhman'ko, E.; Akayeu, Y. Chemically Modified (Poly)Vinyl Chloride with Built-in Neutral Carrier Function as a New Material for Ion Selective Electrodes. *Chem. Pap.* **2018**, *72*, 1315–1323. [CrossRef]
78. Ye, X.; Qi, P.; Sun, Y.; Zhang, D.; Zeng, Y. A High Flexibility All-Solid Contact Sulfide Selective Electrode Using a Graphene Transducer. *Anal. Methods* **2020**, *12*, 3151–3155. [CrossRef]
79. Abd-Rabboh, H.S.M.; Amr, A.E.-G.E.; Kamel, A.H.; Al-Omar, M.A.; Sayed, A.Y.A. Integrated All-Solid-State Sulfite Sensors Modified with Two Different Ion-to-Electron Transducers: Rapid Assessment of Sulfite in Beverages. *RSC Adv.* **2021**, *11*, 3783–3791. [CrossRef] [PubMed]
80. Ghassab, N.; Soleymanpour, A.; Shafaatian, B. Development of an Ultrasensitive Chemically Modified Carbon Paste Electrode for Selective Determination Trace Amount of Sulfate Ion. *Measurement* **2022**, *205*, 112231. [CrossRef]
81. Matveichuk, Y.V.; Rakhman'ko, E.M.; Okaev, E.B. Effect of the Nature of a Quaternary Ammonium Salt and the Addition of a Neutral Carrier on Analytical Characteristics of Sulfate-Selective Electrodes. *J. Anal. Chem.* **2018**, *73*, 374–382. [CrossRef]
82. Forano, C.; Farhat, H.; Mousty, C. Recent Trends in Electrochemical Detection of Phosphate in Actual Waters. *Curr. Opin. Electrochem.* **2018**, *11*, 55–61. [CrossRef]
83. Barus, C.; Romanytsia, I.; Striebig, N.; Garçon, V. Toward an in Situ Phosphate Sensor in Seawater Using Square Wave Voltammetry. *Talanta* **2016**, *160*, 417–424. [CrossRef]
84. Zeitoun, R.; Biswas, A. Review—Potentiometric Determination of Phosphate Using Cobalt: A Review. *J. Electrochem. Soc.* **2020**, *167*, 127507. [CrossRef]
85. Ding, X.; Behbahani, M.; Gruden, C.; Seo, Y. Characterization and Evaluation of Phosphate Microsensors to Monitor Internal Phosphorus Loading in Lake Erie Sediments. *J. Environ. Manag.* **2015**, *160*, 193–200. [CrossRef] [PubMed]
86. Jeong, B.; Oh, J.S.; Kim, D.Y.; Kim, D.G.; Kim, Y.I.; Heo, J.; Lee, H.-K. Ion-Selective Electrode Based on a Novel Biomimetic Nicotinamide Compound for Phosphate Ion Sensor. *Polymers* **2022**, *14*, 3392. [CrossRef] [PubMed]
87. Zhao, G.; Fozia; Wen, H.; Dai, Z.; Nie, Y.; Jiang, J.; Xu, X.; Ying, M.; Hu, Z.; Xu, H. Preparation of a Phosphate Ion-Selective Electrode Using One-Step Process Optimized with Response Surface Method and Its Application in Real Sample Detections. *Electrocatalysis* **2022**, *13*, 641–652. [CrossRef]
88. Alizadeh, T.; Atayi, K. Synthesis of Nano-Sized Hydrogen Phosphate-Imprinted Polymer in Acetonitrile/Water Mixture and Its Use as a Recognition Element of Hydrogen Phosphate Selective All-Solid State Potentiometric Electrode. *J. Mol. Recognit.* **2018**, *31*, e2678. [CrossRef] [PubMed]
89. Kalayci, S. A New Phosphate Selective Electrode and Its Application in Some Foods. *Int. J. Electrochem. Sci.* **2021**, *16*, 210949. [CrossRef]
90. Topcu, C.; Coldur, F.; Caglar, B.; Ozdokur, K.V.; Cubuk, O. Solid-state Electrochemical Sensor Based on a Cross-linked Copper(II)-doped Copolymer and Carbon Nanotube Material for Selective and Sensitive Detection of Monohydrogen Phosphate. *Electroanalysis* **2022**, *34*, 474–484. [CrossRef]

91. Wu, J. An All-Solid-State Phosphate Ion-Selective Electrode Using BiPO$_4$ as a Sensitive Membrane. *Int. J. Electrochem. Sci.* **2021**, *16*, 210641. [CrossRef]
92. Xu, K.; Kitazumi, Y.; Kano, K.; Sasaki, T.; Shirai, O. Fabrication of a Phosphate Ion Selective Electrode Based on Modified Molybdenum Metal. *Anal. Sci.* **2020**, *36*, 201–205. [CrossRef]
93. Bralić, M. Preparation of Phosphate Ion-Selective Membrane Based on Silver Salts Mixed with PTFE or Carbon Nanotubes. *Int. J. Electrochem. Sci.* **2018**, *13*, 1390–1399. [CrossRef]
94. Birlik Özkütük, E.; Yıldız, B.; Gündüz, M.; Hür, E. Phosphate-imprinted Polymer as an Efficient Modifier for the Design of Ion-selective Electrode. *J. Chem. Technol. Biotechnol.* **2021**, *96*, 2604–2609. [CrossRef]
95. King, L.; Wang, Q.; Xia, L.; Wang, P.; Jiang, G.; Li, W.; Huang, Y.; Liang, X.; Peng, X.; Li, Y.; et al. Environmental Exposure to Perchlorate, Nitrate and Thiocyanate, and Thyroid Function in Chinese Adults: A Community-Based Cross-Sectional Study. *Environ. Int.* **2023**, *171*, 107713. [CrossRef] [PubMed]
96. Song, J.; Huang, P.-C.; Wan, Y.-Q.; Wu, F.-Y. Colorimetric Detection of Thiocyanate Based on Anti-Aggregation of Gold Nanoparticles in the Presence of Cetyltrimethyl Ammonium Bromide. *Sens. Actuators B Chem.* **2016**, *222*, 790–796. [CrossRef]
97. Calderón, R.; Jara, C.; Albornoz, F.; Palma, P.; Arancibia-Miranda, N.; Karthikraj, R.; Manquian-Cerda, K.; Mejias, P. Exploring the Destiny and Distribution of Thiocyanate in the Water-Soil-Plant System and the Potential Impacts on Human Health. *Sci. Total Environ.* **2022**, *835*, 155502. [CrossRef] [PubMed]
98. Laurberg, P.; Pedersen, I.B.; Carlé, A.; Andersen, S.; Knudsen, N.; Karmisholt, J. The Relationship between Thiocyanate and Iodine. In *Comprehensive Handbook of Iodine*; Elsevier: Amsterdam, The Netherlands, 2009; pp. 275–281.
99. Matveychuk, Y.V. A Thiocyanate-Selective Electrode and Its Analytical Applications. *J. Anal. Chem.* **2020**, *75*, 662–668. [CrossRef]
100. Urbanowicz, M.; Sadowska, K.; Pijanowska, D.G.; Pomećko, R.; Bocheńska, M. Potentiometric Solid-Contact Ion-Selective Electrode for Determination of Thiocyanate in Human Saliva. *Sensors* **2020**, *20*, 2817. [CrossRef] [PubMed]
101. Gyeom Kwon, N. Development of Thiocyanate-Selective Membrane Electrodes by the Sol–Gel Method. *Int. J. Electrochem. Sci.* **2018**, *13*, 9481–9492. [CrossRef]
102. Khan, A.A.; Jamil, B.; Shaheen, S. Electrochemical Sensing Studies of AsO$_4$−3 Selective Poly(Methyl Methacrylate)-Zinc Oxide Fibrous Anion Exchanger. *Adv. Polym. Technol.* **2018**, *37*, 566–574. [CrossRef]
103. Uner Bahar, D. A Novel Borate Ion Selective Electrode Based on Carbon Nanotube-Silver Borate. *Int. J. Electrochem. Sci.* **2020**, *15*, 899–914. [CrossRef]
104. Martin, K.; Kadam, S.A.; Mattinen, U.; Bobacka, J.; Leito, I. Solid-contact Acetate-selective Electrode Based on a 1,3-bis(Carbazolyl)Urea-ionophore. *Electroanalysis* **2019**, *31*, 1061–1066. [CrossRef]
105. Zhang, C.; He, Y.; Wu, J.; Ai, M.; Cai, W.; Ye, Y.; Tao, C.; Zhang, P.; Jin, Q. Fabrication of an All-Solid-State Carbonate Ion-Selective Electrode with Carbon Film as Transducer and Its Preliminary Application in Deep-Sea Hydrothermal Field Exploration. *Chemosensors* **2021**, *9*, 236. [CrossRef]
106. Wu, R.; Chen, X.-G.; Tao, C.; Huang, Y.; Ye, Y.; Wang, Q.; Zhou, Y.; Jin, Q.; Cai, W. An All-Solid-State Silicate Ion-Selective Electrode Using PbSiO$_3$ as a Sensitive Membrane. *Sensors* **2019**, *19*, 525. [CrossRef] [PubMed]

Disclaimer/Publisher's Note: The statements, opinions and data contained in all publications are solely those of the individual author(s) and contributor(s) and not of MDPI and/or the editor(s). MDPI and/or the editor(s) disclaim responsibility for any injury to people or property resulting from any ideas, methods, instructions or products referred to in the content.

www.ingramcontent.com/pod-product-compliance
Lightning Source LLC
LaVergne TN
LVHW070712100526
838202LV00013B/1075

MDPI
St. Alban-Anlage 66
4052 Basel
Switzerland
www.mdpi.com

Materials Editorial Office
E-mail: materials@mdpi.com
www.mdpi.com/journal/materials

Disclaimer/Publisher's Note: The statements, opinions and data contained in all publications are solely those of the individual author(s) and contributor(s) and not of MDPI and/or the editor(s). MDPI and/or the editor(s) disclaim responsibility for any injury to people or property resulting from any ideas, methods, instructions or products referred to in the content.